FINANCIAL ENGINEERING
ADVANCED BACKGROUND SERIES

FE PRESS

New York

Financial Engineering Advanced Background Series

1. A Primer for the Mathematics of Financial Engineering, Second Edition, by Dan Stefanica. FE Press, 2011

2. Solutions Manual – A Primer for the Mathematics of Financial Engineering, Second Edition, by Dan Stefanica. FE Press, 2011

3. A Linear Algebra Primer for Financial Engineering, by Dan Stefanica. FE Press, 2014

4. Solutions Manual – A Linear Algebra Primer for Financial Engineering, by Dan Stefanica. FE Press, 2016

5. A Probability Primer for Mathematical Finance, by Elena Kosygina. FE Press, 2017

Other Titles from FE Press

1. 150 Most Frequently Asked Questions on Quant Interviews, by Dan Stefanica, Radoš Radoičić, and Tai-Ho Wang. FE Press, 2013

2. Elements of Stochastic Processes: A Computational Approach, By C. Douglas Howard. FE Press, 2016

3. 50 Challenging Brainteasers from Quant Interviews, by Radoš Radoičić, Ivan Matič, and Dan Stefanica. FE Press, 2017

4. Stochastic Calculus & Probability Quant Interview Questions, by Ivan Matič, Radoš Radoičić, and Dan Stefanica. FE Press, 2017

SOLUTIONS MANUAL
A LINEAR ALGEBRA PRIMER
for
FINANCIAL ENGINEERING

DAN STEFANICA

Baruch College
City University of New York

FE Press

New York

FE PRESS
New York

www.fepress.org

This edition first published 2016

Printed in the United States of America

ISBN-10 0979757665
ISBN-13 9780979757662

To Eric, Rianna and Miriam,

my amazing family

Contents

Preface

The addition of this Solutions Manual to "A Linear Algebra Primer for Financial Engineering: Covariance Matrices, Eigenvectors, OLS, and more" offers the reader the opportunity to undertake a rigorous self-study of the linear algebra topics presented in the NLA Primer, with the goal of achieving a deeper understanding of the financial applications therein.

Every exercise from the NLA Primer is solved in detail in the Solutions Manual.

Studying the material from the NLA Primer and using the Solutions Manual as a companion is a time effective way to ensure that the reader can solve every exercise, therefore gaining an appropriate depth of understanding of the material in the book and making sure that the pseudocodes from the book were correctly implemented.

As Director of the elite Financial Engineering Masters Program at Baruch College,[1] City University of New York, since its inception in 2002, the author has had the privilege of interacting with students whose knowledge and ability are exceptional. The community that evolved around the alumni, students, and faculty of the program embodies the friendliness and mutual support of everyone involved, in a highly competitive and ultimately very rewarding environment.

This is the fourth book in the Financial Engineering Advanced Background Series, following "A Primer for the Mathematics of Financial Engineering" and its Solutions Manual, and "A Linear Algebra Primer for Financial Engineering".

Books on probability and differential equations for financial engineering applications are forthcoming.

Dan Stefanica

New York, 2016

---

[1]Baruch MFE Program web page: http://mfe.baruch.cuny.edu

Acknowledgments

I have been the Director of the Financial Engineering Masters Program at Baruch College for many wonderful years, and seeing the strong community of alumni, faculty, and students that developed around the Baruch MFE Program is particularly rewarding. It is a privilege to educate so many talented students and to contribute to launching and furthering their quantitative finance careers.

The book "A Linear Algebra Primer for Financial Engineering" and its Solutions Manual were developed as textbooks for the Numerical Linear Algebra for Financial Applications refresher seminars that I have been teaching since 2004, at first, exclusively o students of the Baruch MFE Program, and, since 2011, to a much wider audience in Pre-MFE seminars. Many of the exercises from this Solutions Manual first arose in these refresher seminars. I am grateful to everyone who took these seminars for the creative energy that goes together with teaching them.

The art for the book cover is once again due to the professional help of my friend Max Rumyantsev.

Special thanks go to the students who supported the proofreading effort of the Solutions Manual: Sandeep Bangar, Allen Long Chen, Yueyi Danielle Chen, Xiangyu Will Gu, Xiaobo He, Lanchun Thomas Liu, Yuxi Asana Liu, Jiabei Ma, Xinliannan Peng, Luca Quaglia, and Eldar Urmanov.

This book is dedicated to my amazing family, whose support is making everything possible.

Dan Stefanica

New York, 2016

Chapter 1

Vectors and matrices.

1.1 Exercises

1. Let
$$A = \begin{pmatrix} 1 & -1 & 2 & 5 & 4 \\ 3 & -2 & 1 & 4 & 2 \\ 0 & 1 & 2 & -1 & 3 \\ -5 & 4 & 2 & -4 & 3 \end{pmatrix}.$$
Show that the column rank and the row rank of A are both equal to 3.

2. Let x and y be column vectors of size n, and let I be the identity matrix of size n.

(i) If $y^t x \neq -1$, show that
$$(I + xy^t)^{-1} = I - \frac{1}{1 + y^t x} xy^t.$$
In other words, show that
$$\left(I - \frac{1}{1 + y^t x} xy^t\right)(I + xy^t) = I.$$

(ii) Show that the matrix $I + xy^t$ is nonsingular if and only if $y^t x \neq -1$.

3. (i) Use induction to show that
$$\left(\prod_{i=1}^{n} A_i\right)^t = \prod_{i=1}^{n} A_{n+1-i}^t$$
for any $m_i \times n_i$ matrices A_i, $i = 1 : n$, with $n_i = m_{i+1}$ for $i = 1 : (n-1)$.

(ii) Show that
$$\left(\prod_{i=1}^{n} A_i\right)^{-1} = \prod_{i=1}^{n} A_{n+1-i}^{-1}$$
for any nonsingular square matrices A_i of the same size.

4. Let $D = \text{diag}(d_i)_{i=1:n}$ be a diagonal matrix of size n with distinct diagonal entries, i.e., such that $d_j \neq d_k$, for any $1 \leq j \neq k \leq n$. If A is a square matrix of size n, show that $AD = DA$ if and only if the matrix A is diagonal.

5. Use the fact that $D_1 D_2 = D_2 D_1$ for any two diagonal matrices D_1 and D_2 of the same size to show that

$$\prod_{i=1}^{n} D_i = \prod_{i=1}^{n} D_{p(i)},$$

for any one-to-one function $p : \{1, 2, \ldots n\} \rightarrow \{1, 2, \ldots n\}$, where D_i, $i = 1 : n$, are diagonal matrices of the same size.

6. (i) Let A be an $n \times n$ matrix and let L be an $n \times n$ nonsingular lower triangular matrix. Show that, if LA is a lower triangular matrix, then A is lower triangular. Show that, if AL is a lower triangular matrix, then A is lower triangular.

(ii) Let A be an $n \times n$ matrix and let U be an $n \times n$ nonsingular upper triangular matrix. Show that, if UA is an upper triangular matrix, then A is upper triangular. Show that, if AU is an upper triangular matrix, then A is upper triangular.

7. Let A be a nonsingular matrix, and let k be a positive integer. Define A^{-k} as the k–th power of the inverse matrix of A, i.e., let $A^{-k} = \left(A^{-1}\right)^k$. Show that this definition is consistent, i.e., show that

$$A^k \cdot A^{-k} = A^{-k} \cdot A^k = I.$$

8. (i) Let

$$M = \begin{pmatrix} 0 & 0 & 0 & 0 \\ 3 & 0 & 0 & 0 \\ 1 & -1 & 0 & 0 \\ -1 & 2 & 1 & 0 \end{pmatrix}.$$

Compute M^2, M^3, M^4.

(ii) Let

$$C = I + M = \begin{pmatrix} 1 & 0 & 0 & 0 \\ 3 & 1 & 0 & 0 \\ 1 & -1 & 1 & 0 \\ -1 & 2 & 1 & 1 \end{pmatrix}.$$

Compute C^m, where $m \geq 2$ is a positive integer.

Hint: Recall that, if A and B are square matrices of the same size such that $AB = BA$, then the following version of the binomial formula holds true:

$$(A + B)^m = \sum_{j=0}^{m} \binom{m}{j} A^j B^{m-j}, \tag{1.1}$$

where m is a positive integer and the binomial coefficient $\begin{pmatrix} m \\ j \end{pmatrix}$ is given by

$$\begin{pmatrix} m \\ j \end{pmatrix} = \frac{m!}{j!\,(m-j)!},$$

where $k! = 1 \cdot 2 \cdot \ldots \cdot k$. Also, note that $A^0 = B^0 = I$.

9. Let L be an $n \times n$ lower triangular matrix with entries equal to 0 on the main diagonal, i.e., with $L(i, i) = 0$ for $i = 1 : n$.

(i) Show that $L^n = 0$;

(ii) Compute $(I + L)^m$ in terms of $L, L^2, \ldots, L^{n-1}$, where $m \geq n$ is a positive integer.

Hint: Use the binomial formula (1.1).

10. Let A and B be square matrices of the same size with nonnegative entries and such that the sum of the entries in each row is equal to 1. Show that the matrix AB has the same properties, i.e., show that all the entries of the matrix AB are nonnegative and the sum of the entries in each row of AB is equal to 1.

Note: A matrix with nonnegative entries such that the sum of the entries in each row is equal to 1 is called a probability matrix.

11. The covariance matrix of five random variables is

$$\Sigma = \begin{pmatrix} 1 & -0.525 & 1.375 & -0.075 & -0.75 \\ -0.525 & 2.25 & 0.1875 & 0.1875 & -0.675 \\ 1.375 & 0.1875 & 6.25 & 0.4375 & -1.875 \\ -0.075 & 0.1875 & 0.4375 & 0.25 & 0.3 \\ -0.75 & -0.675 & -1.875 & 0.3 & 9 \end{pmatrix}.$$

Find the correlation matrix of these random variables.

12. The correlation matrix of five random variables is

$$\Omega = \begin{pmatrix} 1 & -0.25 & 0.15 & -0.05 & -0.30 \\ -0.25 & 1 & -0.10 & -0.25 & 0.10 \\ 0.15 & -0.10 & 1 & 0.20 & 0.05 \\ -0.05 & -0.25 & 0.20 & 1 & 0.10 \\ -0.30 & 0.10 & 0.05 & 0.10 & 1 \end{pmatrix}$$

(i) Compute the covariance matrix of these random variables if their standard deviations are 0.25, 0.5, 1, 2, and 4, in this order.

(ii) Compute the covariance matrix of these random variables if their standard deviations are 4, 2, 1, 0.5, and 0.25, in this order.

13. The file *indeces-jul26-aug9-2012.xlsx* from fepress.org/nla-primer contains the July 26, 2012 – August 9, 2012 end of day values of Dow Jones, Nasdaq, and S&P 500.

    (i) Compute the daily percentage returns of the three indices over the given time period.

    (ii) Compute the covariance matrix of the daily percentage returns of the three indices.

    (iii) Compute the daily log returns of the three indices over the given time period.

    (iv) Compute the covariance matrix of the daily log returns of the three indices.

    Note: The percentage return and the log return between times t_1 and t_2 of an asset with price $S(t)$ at time t are given by

    $$\frac{S(t_2) - S(t_1)}{S(t_1)} \quad \text{and} \quad \ln\left(\frac{S(t_2)}{S(t_1)}\right),$$

    respectively.

14. The file *indices-july2011.xlsx* from fepress.org/nla-primer contains the January 2011 – July 2011 end of day values of nine major US indices.

    (i) Compute the sample covariance matrix of the daily percentage returns of the indices, and the corresponding sample correlation matrix.

    Compute the sample covariance and correlation matrices for daily log returns, and compare them with the corresponding matrices for daily percentage returns.

    (ii) Compute the sample covariance matrix of the weekly percentage returns of the indices, and the corresponding sample correlation matrix.

    Compute the sample covariance and correlation matrices for weekly log returns, and compare them with the corresponding matrices for weekly percentage returns.

    (iii) Compute the sample covariance matrix of the monthly percentage returns of the indices, and the corresponding sample correlation matrix.

    Compute the sample covariance and correlation matrices for monthly log returns, and compare them with the corresponding matrices for monthly percentage returns.

    (iv) Comment on the differences between the sample covariance and correlation matrices for daily, weekly, and monthly returns.

15. In three months, the value of an asset with spot price $50 will be either $60 or $45. The continuously compounded risk–free rate is 6%. Consider the one period market model with two securities, i.e., cash and the asset, and two states, i.e., asset value equal to $60 and asset value equal to $45, in three months.

(i) Find the payoff matrix of this model.

(ii) Is this one period market complete, i.e., is the payoff matrix nonsingular?

(iii) How do you replicate a three months at–the–money put option on this asset, using the cash and the underlying asset?

16. In six months, the price of an asset with spot price $40 will be either $30, $35, $40, $42, $45, or $50. Consider a one period market model with six states in six months corresponding to the six possible values of the asset in six months, and with the following four securities:

- cash;

- asset;

- six months at-the-money call option with strike $40 on the asset;

- six months at-the-money put option with strike $40 on the asset.

The continuously compounded risk–free interest rate is constant and equal to 6%.

(i) Find the payoff matrix of this model.

(ii) Is this one period market model complete?

(iii) Are the four securities non–redundant?

1.2   Solutions to Chapter 1 Exercises

Problem 1: Let

$$A = \begin{pmatrix} 1 & -1 & 2 & 5 & 4 \\ 3 & -2 & 1 & 4 & 2 \\ 0 & 1 & 2 & -1 & 3 \\ -5 & 4 & 2 & -4 & 3 \end{pmatrix}.$$

Show that the column rank and the row rank of A are both equal to 3.

Solution: Let

$$(c_1 \mid c_2 \mid c_3 \mid c_4 \mid c_5) = \begin{pmatrix} 1 & -1 & 2 & 5 & 4 \\ 3 & -2 & 1 & 4 & 2 \\ 0 & 1 & 2 & -1 & 3 \\ -5 & 4 & 2 & -4 & 3 \end{pmatrix}$$

be the column form of the matrix A. By doing column reduction for A, we obtain that

$$(c_1 \mid c_1 + c_2 \mid c_3 - 2c_1 \mid c_4 - 5c_1 \mid c_5 - 4c_1)$$
$$= \begin{pmatrix} 1 & 0 & 0 & 0 & 0 \\ 3 & 1 & -5 & -11 & -10 \\ 0 & 1 & 2 & -1 & 3 \\ -5 & -1 & 12 & 21 & 23 \end{pmatrix};$$

$$(c_1 \mid c_1 + c_2 \mid (c_3 - 2c_1) + 5(c_1 + c_2) \mid (c_4 - 5c_1) + 11(c_1 + c_2) \mid$$
$$(c_5 - 4c_1) + 10(c_1 + c_2))$$
$$= (c_1 \mid c_1 + c_2 \mid 3c_1 + 5c_2 + c_3 \mid 6c_1 + 11c_2 + c_4 \mid 6c_1 + 10c_2 + c_5) \quad (1.2)$$
$$= \begin{pmatrix} 1 & 0 & 0 & 0 & 0 \\ 3 & 1 & 0 & 0 & 0 \\ 0 & 1 & 7 & 10 & 13 \\ -5 & -1 & 7 & 10 & 13 \end{pmatrix}. \quad (1.3)$$

From (1.2) and (1.3), it follows that the first three columns c_1, c_2, and c_3 of the matrix A are linearly independent, while

$$\frac{1}{7}(3c_1 + 5c_2 + c_3) = \frac{1}{10}(6c_1 + 11c_2 + c_4) = \frac{1}{13}(6c_1 + 10c_2 + c_5) = \begin{pmatrix} 0 \\ 0 \\ 1 \\ 1 \end{pmatrix}.$$

Then,

$$\frac{1}{10}(6c_1 + 11c_2 + c_4) = \frac{1}{7}(3c_1 + 5c_2 + c_3) \iff c_4 = -\frac{12}{7}c_1 - \frac{27}{7}c_2 + \frac{10}{7}c_3;$$

$$\frac{1}{13}(6c_1 + 10c_2 + c_5) = \frac{1}{7}(3c_1 + 5c_2 + c_3) \iff c_5 = -\frac{3}{7}c_1 - \frac{5}{7}c_2 + \frac{13}{7}c_3,$$

and therefore the columns c_4 and c_5 are linearly dependent on c_1, c_2, c_3. We conclude that the column rank of the matrix A is 3.

Let

$$\begin{pmatrix} r_1 \\ r_2 \\ r_3 \\ r_4 \end{pmatrix} = \begin{pmatrix} 1 & -1 & 2 & 5 & 4 \\ 3 & -2 & 1 & 4 & 2 \\ 0 & 1 & 2 & -1 & 3 \\ -5 & 4 & 2 & -4 & 3 \end{pmatrix}$$

be the row form of the matrix A. By doing row reduction for A, we obtain that

$$\begin{pmatrix} r_1 \\ r_2 - 3r_1 \\ r_3 \\ r_4 + 5r_1 \end{pmatrix} = \begin{pmatrix} 1 & -1 & 2 & 5 & 4 \\ 0 & 1 & -5 & -11 & -10 \\ 0 & 1 & 2 & -1 & 3 \\ 0 & -1 & 12 & 21 & 23 \end{pmatrix} ;$$

$$\begin{pmatrix} r_1 \\ r_2 - 3r_1 \\ r_3 - (r_2 - 3r_1) \\ (r_4 + 5r_1) + (r_2 - 3r_1) \end{pmatrix} = \begin{pmatrix} r_1 \\ r_2 - 3r_1 \\ 3r_1 - r_2 + r_3 \\ 2r_1 + r_2 + r_4 \end{pmatrix} \qquad (1.4)$$

$$= \begin{pmatrix} 1 & -1 & 2 & 5 & 4 \\ 0 & 1 & -5 & -11 & -10 \\ 0 & 0 & 7 & 10 & 13 \\ 0 & 0 & 7 & 10 & 13 \end{pmatrix} . \qquad (1.5)$$

From (1.4) and (1.5), it follows that the first three rows r_1, r_2, and r_3 of the matrix A are linearly independent, while

$$3r_1 - r_2 + r_3 = 2r_1 + r_2 + r_4 = (0 \ 0 \ 7 \ 10 \ 13).$$

Then,

$$r_4 = r_1 - 2r_2 + r_3,$$

and therefore the row r_4 is linearly dependent on r_1, r_2, r_3.
We conclude that the row rank of the matrix A is 3. $\square$

Problem 2: Let x and y be column vectors of size n, and let I be the identity matrix of size n.

(i) If $y^t x \neq -1$, show that

$$(I + xy^t)^{-1} = I - \frac{1}{1 + y^t x} \, xy^t. \qquad (1.6)$$

In other words, show that

$$\left(I - \frac{1}{1 + y^t x} \, xy^t \right) (I + xy^t) = I.$$

(ii) Show that the matrix $I + xy^t$ is nonsingular if and only if $y^t x \neq -1$.

Solution: (i) Note that the right hand side of (1.6) in indeed an $n \times n$ matrix: xy^t is the result of a column vector–row vector multiplication, and therefore it is an $n \times n$ matrix, while $y^t x$ is the result of a row vector–column vector multiplication, and therefore it is a number, which means $\frac{1}{1+y^t x}$ is a number.

Then,

$$\left(I - \frac{1}{1 + y^t x} \, xy^t\right)(I + xy^t) \;=\; I + xy^t - \frac{1}{1 + y^t x} \, xy^t(I + xy^t) \qquad (1.7)$$

$$= \; I + xy^t - \frac{1}{1 + y^t x} \, xy^t - \frac{1}{1 + y^t x} \, (xy^t)(xy^t)$$

$$= \; I + xy^t - \frac{1}{1 + y^t x} \, xy^t - \frac{1}{1 + y^t x} \, x(y^t x)y^t$$

$$= \; I + xy^t - \frac{1}{1 + y^t x} \, xy^t - \frac{y^t x}{1 + y^t x} \, xy^t \qquad (1.8)$$

$$= \; I + xy^t - \frac{1 + y^t x}{1 + y^t x} \, xy^t$$

$$= \; I + xy^t - xy^t$$

$$= \; I, \qquad (1.9)$$

where, for (1.8), we used the fact that $y^t x$ is a number (since it is the result of a row vector–column vector multiplication), and therefore

$$\frac{1}{1 + y^t x} \, x(y^t x)y^t \;=\; \frac{1}{1 + y^t x}(y^t x) \, xy^t \;=\; \frac{y^t x}{1 + y^t x} \, xy^t.$$

From (1.7) and (1.9), it follows that

$$\left(I - \frac{1}{1 + y^t x} \, xy^t\right)(I + xy^t) \;=\; I,$$

and therefore

$$(I + xy^t)^{-1} \;=\; I - \frac{1}{1 + y^t x} \, xy^t.$$

(ii) To show that the matrix $I + xy^t$ is nonsingular if and only if $y^t x \neq -1$, note that we showed in part (i) that, if $y^t x \neq -1$, then the matrix $I + xy^t$ has an inverse and it is therefore nonsingular.

Thus, we only need to show that, if $y^t x = -1$, then the matrix $I + xy^t$ is singular. Assume that $y^t x = -1$. Then,

$$(I + xy^t)x \;=\; x + xy^t x \;=\; x + x(y^t x) \;=\; x + x(-1) \;=\; x - x \;=\; 0.$$

Thus,

$$(I + xy^t)x \;=\; 0. \qquad (1.10)$$

If the matrix $I + xy^t$ were nonsingular, then we multiply (1.10) to the left by the inverse of the matrix $I + xy^t$ and obtain that

$$(I + xy^t)x = 0 \quad \Longleftrightarrow \quad (I + xy^t)^{-1}(I + xy^t)x = 0$$

$$\Longleftrightarrow \quad x = 0,$$

since $(I + xy^t)^{-1}(I + xy^t) = I$.

However, if $x = 0$, then $y^t x = 0$, which is not possible since $y^t x = -1$. This is a contradiction which comes from the assumption that the matrix $I + xy^t$ is nonsingular.

We conclude that, if $y^t x = -1$, then the matrix $I + xy^t$ is singular, which is what we wanted to show. $\quad\square$

Problem 3: (i) Use induction to show that

$$\left(\prod_{i=1}^{n} A_i\right)^t = \prod_{i=1}^{n} A_{n+1-i}^t \tag{1.11}$$

for any $m_i \times n_i$ matrices A_i, $i = 1 : n$, with $n_i = m_{i+1}$ for $i = 1 : (n-1)$.

(ii) Show that

$$\left(\prod_{i=1}^{n} A_i\right)^{-1} = \prod_{i=1}^{n} A_{n+1-i}^{-1} \tag{1.12}$$

for any nonsingular square matrices A_i of the same size.

Solution: (i) Recall that, if B is an $m \times n$ matrix and C is an $n \times p$ matrix, than

$$(BC)^t = C^t B^t. \tag{1.13}$$

Note that (1.11) can be written as

$$(A_1 A_2 \ldots A_{n-1} A_n)^t = A_n^t A_{n-1}^t \ldots A_2^t A_1^t. \tag{1.14}$$

We will prove (1.14) by induction.

For $n = 2$, (1.14) becomes $(A_1 A_2)^t = A_2^t A_1^t$, which holds true; see (1.13).

Assume that (1.14) holds true for n, i.e., assume that

$$(A_1 A_2 \ldots A_{n-1} A_n)^t = A_n^t A_{n-1}^t \ldots A_2^t A_1^t. \tag{1.15}$$

We will show that (1.14) holds true for $n + 1$, i.e.,

$$(A_1 A_2 \ldots A_n A_{n+1})^t = A_{n+1}^t A_n^t \ldots A_2^t A_1^t. \tag{1.16}$$

From (1.13) for $B = A_1 A_2 \ldots A_n$ and $C = A_{n+1}$, and using the induction hypothesis (1.15), we find that

$$\begin{aligned}
(A_1 A_2 \ldots A_n A_{n+1})^t &= ((A_1 A_2 \ldots A_n) \cdot A_{n+1})^t \\
&= A_{n+1}^t (A_1 A_2 \ldots A_n)^t \\
&= A_{n+1}^t A_n^t \ldots A_2^t A_1^t,
\end{aligned}$$

which is the same as (1.16).

We conclude that (1.14), which is the same as (1.11), is proved by induction.

(ii) The proof of (1.12) follows the exact same steps as the proof of (1.11) from (i) and is based on the fact that, if B and C are nonsingular matrices of the same size, then $(BC)^{-1} = C^{-1} B^{-1}$, which mirrors the property (1.13) for matrix transposes. $\square$

Problem 4: Let $D = \text{diag}(d_i)_{i=1:n}$ be a diagonal matrix of size n with distinct diagonal entries, i.e., such that $d_j \neq d_k$, for any $1 \leq j \neq k \leq n$. If A is a square matrix of size n, show that $AD = DA$ if and only if the matrix A is diagonal.

Solution: Let $A = \text{col}\,(a_k)_{k=1:n}$, with $a_k = (A(j,k))_{j=1:n}$, be the column form of the matrix A, and let $A = \text{row}\,(r_j)_{j=1:n}$, with $r_j = (A(j,k))_{k=1:n}$, be the row form of A. Recall that

$$AD = \text{col}\,(d_k a_k)_{k=1:n} = (d_1 a_1 \mid d_2 a_2 \mid \ldots \mid d_n a_n); \tag{1.17}$$

$$DA = \text{row}\,(d_j r_j)_{j=1:n} = \begin{pmatrix} d_1 r_1 \\ d_2 r_2 \\ \vdots \\ d_n r_n \end{pmatrix}. \tag{1.18}$$

From (1.17) and (1.18), we find

$$(AD)(j,k) = d_k A(j,k), \quad \forall\, 1 \leq j,k \leq n; \tag{1.19}$$
$$(DA)(j,k) = d_j A(j,k), \quad \forall\, 1 \leq j,k \leq n. \tag{1.20}$$

From (1.19) and (1.20), we conclude that

$$\begin{aligned}
AD = DA \iff & (AD)(j,k) = (DA)(j,k), \quad \forall\, 1 \leq j,k \leq n \\
\iff & d_k A(j,k) = d_j A(j,k), \quad \forall\, 1 \leq j,k \leq n \\
\iff & (d_k - d_j) A(j,k) = 0, \quad \forall\, 1 \leq j,k \leq n \\
\iff & (d_k - d_j) A(j,k) = 0, \quad \forall\, 1 \leq j \neq k \leq n \\
\iff & A(j,k) = 0, \quad \forall\, 1 \leq j \neq k \leq n, \tag{1.21}
\end{aligned}$$

where (1.21) comes from the fact that $d_j \neq d_k$ for all $1 \leq j \neq k \leq n$, and therefore

$$d_k - d_j \neq 0, \quad \forall\, 1 \leq j \neq k \leq n.$$

In other words, $AD = DA$ if and only if $A(j,k) = 0$ for all $1 \leq j \neq k \leq n$, i.e., if and only if the matrix A is diagonal. $\square$

Problem 5: Use the fact that $D_1 D_2 = D_2 D_1$ for any two diagonal matrices D_1 and D_2 of the same size to show that

$$\prod_{i=1}^{n} D_i = \prod_{i=1}^{n} D_{p(i)}, \tag{1.22}$$

for any one-to-one function $p : \{1, 2, \ldots n\} \to \{1, 2, \ldots n\}$, where D_i, $i = 1 : n$, are diagonal matrices of the same size.

Solution: The intuition behind the proof of this result is that, given a product of n diagonal matrices of the same size, we can move the matrix D_n to the last position in the product without changing the product, by using the fact that the product of two diagonal matrices is commutative. Then, we use the same commutativity property to move the matrix D_{n-1} to the $n-1$ position, followed by moving the matrix D_{n-2} to the $n-2$ position, and so on, until moving matrix D_2 to position 2, which leaves matrix D_1 in the first position, thus concluding that the product of n diagonal matrices, regardless of the initial order, is $D_1 D_2 \ldots D_n$.

A formal proof of (1.22) based on the idea above can be given by induction. Assume that

$$\prod_{i=1}^{n} D_i = \prod_{i=1}^{n} D_{p_n(i)}, \tag{1.23}$$

for any one-to-one function $p_n : \{1, 2, \ldots n\} \rightarrow \{1, 2, \ldots n\}$, where D_i, $i = 1 : n$, are diagonal matrices of the same size.

Let $p_{n+1} : \{1, 2, \ldots n, n+1\} \rightarrow \{1, 2, \ldots n, n+1\}$ be a one-to-one function. We will show that

$$\prod_{i=1}^{n+1} D_i = \prod_{i=1}^{n+1} D_{p_{n+1}(i)}, \tag{1.24}$$

where D_i, $i = 1 : (n+1)$, are diagonal matrices of the same size.

Since p_{n+1} is a one-to-one function from a set of $n+1$ elements to a set of $n+1$ elements, it follows that p_{n+1} is also an onto function, and therefore there exists l with $1 \le l \le n+1$ such that $p_{n+1}(l) = n+1$. Then,

$$\prod_{i=1}^{n+1} D_{p_{n+1}(i)} = \prod_{i=1}^{l-1} D_{p_{n+1}(i)} \cdot D_{p_{n+1}(l)} \cdot \prod_{i=l+1}^{n+1} D_{p_{n+1}(i)}$$

$$= \prod_{i=1}^{l-1} D_{p_{n+1}(i)} \cdot D_{n+1} \cdot \prod_{i=l+1}^{n+1} D_{p_{n+1}(i)}, \tag{1.25}$$

since $p_{n+1}(l) = n+1$.

Note that $\prod_{i=l+1}^{n+1} D_{p_{n+1}(i)}$ is the product of $n-l+1$ diagonal matrices and therefore it is a diagonal matrix. Since the product of any two diagonal matrices is a commutative, it follows that

$$D_{n+1} \cdot \prod_{i=l+1}^{n+1} D_{p_{n+1}(i)} = \prod_{i=l+1}^{n+1} D_{p_{n+1}(i)} \cdot D_{n+1}. \tag{1.26}$$

From (1.25) and (1.26), we find that

$$\prod_{i=1}^{n+1} D_{p_{n+1}(i)} = \prod_{i=1}^{l-1} D_{p_{n+1}(i)} \cdot \prod_{i=l+1}^{n+1} D_{p_{n+1}(i)} \cdot D_{n+1}. \tag{1.27}$$

Since $p_{n+1} : \{1, 2, \ldots n, n+1\} \rightarrow \{1, 2, \ldots n, n+1\}$ is a one-to-one function and since $p_{n+1}(l) = n+1$, we obtain that

$$1 \le p_{n+1}(i) \le n, \quad \forall\ 1 < i \ne l < n+1.$$

Thus, the function $p_n : \{1, 2, \ldots n\} \rightarrow \{1, 2, \ldots n\}$ given by

$$p_n(i) = \begin{cases} p_{n+1}(i) & \text{if } 1 \le i \le l-1; \\ p_{n+1}(i+1) & \text{if } l+1 \le i \le n+1, \end{cases} \tag{1.28}$$

is well defined and one-to-one.

From the definition (1.28), it follows that

$$
\prod_{i=1}^{l-1} D_{p_{n+1}(i)} \cdot \prod_{i=l+1}^{n+1} D_{p_{n+1}(i)} = \prod_{i=1}^{l-1} D_{p_{n+1}(i)} \cdot \prod_{i=l}^{n} D_{p_{n+1}(i+1)}
$$

$$
= \prod_{i=1}^{l-1} D_{p_n(i)} \cdot \prod_{i=l}^{n} D_{p_n(i)}
$$

$$
= \prod_{i=1}^{n} D_{p_n(i)}, \tag{1.29}
$$

and therefore, from (1.27) and (1.29), we find that

$$
\prod_{i=1}^{n+1} D_{p_{n+1}(i)} = \prod_{i=1}^{n} D_{p_n(i)} \cdot D_{n+1}. \tag{1.30}
$$

Recall from the induction hypothesis (1.23) that

$$
\prod_{i=1}^{n} D_{p_n(i)} = \prod_{i=1}^{n} D_i. \tag{1.31}
$$

Then, from (1.30) and (1.31), we obtain that

$$
\prod_{i=1}^{n+1} D_{p_{n+1}(i)} = \prod_{i=1}^{n} D_i \cdot D_{n+1} = \prod_{i=1}^{n+1} D_i,
$$

which is what we wanted to show; see (1.24).
We conclude that (1.22) is proved by induction. $\square$

Problem 6: (i) Let A be an $n \times n$ matrix and let L be an $n \times n$ nonsingular lower triangular matrix. Show that, if LA is a lower triangular matrix, then A is lower triangular. Show that, if AL is a lower triangular matrix, then A is lower triangular.

(ii) Let A be an $n \times n$ matrix and let U be an $n \times n$ nonsingular upper triangular matrix. Show that, if UA is an upper triangular matrix, then A is upper triangular. Show that, if AU is an upper triangular matrix, then A is upper triangular.

Solution: (i) Recall that the inverse of a nonsingular lower triangular matrix is a lower triangular matrix, and that the product of two lower triangular matrices is a lower triangular matrix.

Let $L_1 = LA$ be a lower triangular matrix. Since L is nonsingular, we obtain that

$$
L^{-1}L_1 = L^{-1}(LA) = (L^{-1}L)A = I \cdot A = A,
$$

since $LL^{-1} = I$.

Note that L^{-1} is a lower triangular matrix since it is the inverse of the lower triangular matrix L. Thus, $A = L^{-1}L_1$ is a lower triangular matrix since it is the product of the lower triangular matrices L^{-1} and L_1.

Similarly, let $L_2 = AL$ be a lower triangular matrix. Since L is nonsingular, we obtain that

$$L_2 L^{-1} \;=\; (AL)L^{-1} \;=\; A(LL^{-1}) \;=\; A \cdot I \;=\; A.$$

Thus, $A = L_2 L^{-1}$ is a lower triangular matrix since it is the product of the lower triangular matrices L_2 and L^{-1}.

(ii) Proofs along the lines of the proofs given at (i) for lower triangular matrices can be given to the corresponding results for upper triangular matrices; for completeness, we include them below.

However, if the results from (i) for lower triangular matrices are known, then the corresponding results for upper triangular matrices can be obtained by using matrix transposition as follows:

Recall that the inverse of a nonsingular upper triangular matrix is an upper triangular matrix, and that the product of two upper triangular matrices is an upper triangular matrix.

Let U be a nonsingular upper triangular matrix. Then, the transpose U^t of the matrix U is lower triangular and, since $\det(U^t) = \prod_{k=1}^{n} U(k,k) = \det(U) \neq 0$, the matrix U^t is nonsingular. Assume that the matrix UA is an upper triangular matrix. Then, $(UA)^t = A^t U^t$ is a lower triangular matrix.

Recall from (i) that, if L is a nonsingular lower triangular matrix and if AL is a lower triangular matrix, then A is lower triangular. In our case, U^t is a nonsingular lower triangular matrix and $A^t U^t$ is lower triangular. We conclude that A^t is a lower triangular matrix, and therefore that A is an upper triangular matrix, which is what we wanted to show.

Similarly, let U be a nonsingular upper triangular matrix, and assume that the matrix AU is an upper triangular matrix. Then, $(AU)^t = U^t A^t$ is a lower triangular matrix.

Recall from (i) that, if L is a nonsingular lower triangular matrix and if LA is a lower triangular matrix, then A is lower triangular. In our case, U^t is a nonsingular lower triangular matrix and $U^t A^t$ is lower triangular. We conclude that A^t is a lower triangular matrix, and therefore that A is an upper triangular matrix.

Proofs not using the results for lower triangular matrices are included below:

Let $U_1 = UA$ be an upper triangular matrix, where U is a nonsingular upper triangular matrix. Then,

$$U^{-1} U_1 \;=\; U^{-1}(UA) \;=\; (U^{-1}U)A \;=\; A,$$

since $U^{-1}U = I$.

Note that U^{-1} is an upper triangular matrix since it is the inverse of the upper triangular matrix U. Thus, $A = U^{-1} U_1$ is an upper triangular matrix since it is the product of the upper triangular matrices U^{-1} and U_1.

Similarly, let $U_2 = AU$ be an upper triangular matrix. Since U is nonsingular, we obtain that

$$U_2 U^{-1} \;=\; (AU)U^{-1} \;=\; A(UU^{-1}) \;=\; A.$$

Thus, $A = U_2 U^{-1}$ is an upper triangular matrix since it is the product of the upper triangular matrices U_2 and U^{-1}. $\quad\square$

Problem 7: Let A be a nonsingular matrix, and let k be a positive integer. Define A^{-k} as the k–th power of the inverse matrix of A, i.e., let $A^{-k} = (A^{-1})^k$. Show that this definition is consistent, i.e., show that $A^k \cdot A^{-k} = A^{-k} \cdot A^k = I$.

Solution: Since $A^{-k} = (A^{-1})^k$, we obtain that

$$
\begin{aligned}
A^k \cdot A^{-k} &= A^k \cdot (A^{-1})^k \\
&= A^{k-1} \cdot A \cdot A^{-1} \cdot (A^{-1})^{k-1} \\
&= A^{k-1} \cdot (A^{-1})^{k-1} \\
&= A^{k-2} \cdot A \cdot A^{-1} \cdot (A^{-1})^{k-2} \\
&= A^{k-2} \cdot (A^{-1})^{k-2} \\
&= \cdots \\
&= A^2 \cdot (A^{-1})^2 \\
&= A \cdot A \cdot A^{-1} \cdot A^{-1} \\
&= A \cdot A^{-1} \\
&= I;
\end{aligned}
$$

here we used repeatedly the fact that $A \cdot A^{-1} = I$.

Thus, $A^k \cdot A^{-k} = I$. We conclude that A^{-k} is the inverse matrix of A^k, and therefore $A^{-k} \cdot A^k = I$. $\square$

Problem 8: (i) Let

$$
M = \begin{pmatrix} 0 & 0 & 0 & 0 \\ 3 & 0 & 0 & 0 \\ 1 & -1 & 0 & 0 \\ -1 & 2 & 1 & 0 \end{pmatrix}.
$$

Compute M^2, M^3, M^4.

(ii) Let

$$
C = I + M = \begin{pmatrix} 1 & 0 & 0 & 0 \\ 3 & 1 & 0 & 0 \\ 1 & -1 & 1 & 0 \\ -1 & 2 & 1 & 1 \end{pmatrix}.
$$

Compute C^m, where $m \geq 2$ is a positive integer.

Solution: (i) By direct computation, we find that

$$
M^2 = \begin{pmatrix} 0 & 0 & 0 & 0 \\ 0 & 0 & 0 & 0 \\ -3 & 0 & 0 & 0 \\ 7 & -1 & 0 & 0 \end{pmatrix}; \tag{1.32}
$$

$$
M^3 = \begin{pmatrix} 0 & 0 & 0 & 0 \\ 0 & 0 & 0 & 0 \\ 0 & 0 & 0 & 0 \\ -3 & 0 & 0 & 0 \end{pmatrix}; \tag{1.33}
$$

$$M^4 = \ = \ \begin{pmatrix} 0 & 0 & 0 & 0 \\ 0 & 0 & 0 & 0 \\ 0 & 0 & 0 & 0 \\ 0 & 0 & 0 & 0 \end{pmatrix}. \tag{1.34}$$

(ii) Recall that, if A and B are square matrices of the same size such that $AB = BA$, then the following version of the binomial formula holds true:

$$(A + B)^m = \sum_{j=0}^{m} \binom{m}{j} A^j B^{m-j}, \tag{1.35}$$

where m is a positive integer and the binomial coefficient $\binom{m}{j}$ is given by

$$\binom{m}{j} = \frac{m!}{j!\,(m-j)!}, \tag{1.36}$$

where $k! = 1 \cdot 2 \cdot \ldots \cdot k$. Also, by definition, $A^0 = B^0 = I$.

Let $A = M$ and $B = I$ in (1.35), and note from (1.34) that $M^j = 0$ for all $j \geq 4$. Then,

$$C^m = (M + I)^m = \sum_{j=0}^{m} \binom{m}{j} M^j = \sum_{j=0}^{3} \binom{m}{j} M^j$$

$$= \binom{m}{0} I + \binom{m}{1} M + \binom{m}{2} M^2 + \binom{m}{3} M^3. \tag{1.37}$$

From (1.36), we obtain that

$$\binom{m}{0} = 1; \quad \binom{m}{1} = m; \quad \binom{m}{2} = \frac{m(m-1)}{2}; \quad \binom{m}{3} = \frac{m(m-1)(m-2)}{6}.$$

Then, using the values of M^2 and M^3 from (1.32) and (1.33), respectively, we conclude from (1.37) that

$$C^m = \binom{m}{0} I + \binom{m}{1} M + \binom{m}{2} M^2 + \binom{m}{3} M^3$$

$$= I + mM + \frac{m(m-1)}{2} M^2 + \frac{m(m-1)(m-2)}{6} M^3$$

$$= I + \begin{pmatrix} 0 & 0 & 0 & 0 \\ 3m & 0 & 0 & 0 \\ m & -m & 0 & 0 \\ -m & 2m & m & 0 \end{pmatrix} + \begin{pmatrix} 0 & 0 & 0 & 0 \\ 0 & 0 & 0 & 0 \\ -\frac{3m(m-1)}{2} & 0 & 0 & 0 \\ \frac{7m(m-1)}{2} & -\frac{m(m-1)}{2} & 0 & 0 \end{pmatrix}$$

$$+ \begin{pmatrix} 0 & 0 & 0 & 0 \\ 0 & 0 & 0 & 0 \\ 0 & 0 & 0 & 0 \\ -\frac{3m(m-1)(m-2)}{6} & 0 & 0 & 0 \end{pmatrix}$$

$$= \begin{pmatrix} 1 & 0 & 0 & 0 \\ 3m & 1 & 0 & 0 \\ \frac{5m-3m^2}{2} & -m & 1 & 0 \\ \frac{-33m+30m^2-3m^3}{6} & \frac{5m-m^2}{2} & m & 1 \end{pmatrix} \quad \square$$

Problem 9: Let L be an $n \times n$ lower triangular matrix with entries equal to 0 on the main diagonal, i.e., with $L(i, i) = 0$ for $i = 1 : n$.

(i) Show that $L^n = 0$;

(ii) Compute $(I + L)^m$ in terms of L, L^2, ..., L^{n-1}, where $m \geq n$ is a positive integer.

Solution: (i) To show that $L^n = 0$, we will prove by induction the following result:

"The first i rows of the matrix L^i are zero vectors, for any $1 \leq i \leq n$."

For $i = 1$, the first row of the matrix L is a zero vector, since L is lower triangular and therefore $L(1, k) = 0$ for $2 \leq k \leq n$, and $L(1, 1) = 0$ since the main diagonal entries of L are 0.

Let $2 \leq i \leq n - 1$. Assume that the first i rows of the matrix L^i are zero vectors and let

$$L^i = \begin{pmatrix} 0 \\ \vdots \\ 0 \\ r_{i+1}^{(i)} \\ \vdots \\ r_n^{(i)} \end{pmatrix} \tag{1.38}$$

be the row form of L^i. We will show that the first $(i + 1)$ rows of the matrix L^{i+1} are zero vectors.

Let

$$L^{i+1} = \begin{pmatrix} r_1^{(i+1)} \\ r_2^{(i+1)} \\ \vdots \\ r_n^{(i+1)} \end{pmatrix}$$

be the row form of the matrix L^{i+1}. Note that $L^{i+1} = L \cdot L^i$ and therefore the j-th row of L^{i+1} is a linear combination of the rows of L^i with coefficients the entries of the j-th row of L, i.e.,

$$r_j^{(i+1)} = \sum_{k=1}^{n} L(j, k) \, r_k^{(i)}, \quad \forall \, j = 1 : n. \tag{1.39}$$

Note that $L(j, j) = 0$, since all the main diagonal entries of L are 0, and $L(j, k) = 0$ for $j + 1 \leq k \leq n$, since L is a lower triangular matrix. Then, (1.39) can be written as

$$r_j^{(i+1)} = \sum_{k=1}^{j-1} L(j, k) \, r_k^{(i)}, \quad \forall \, j = 1 : n. \tag{1.40}$$

Since $r_1^{(i)} = r_2^{(i)} = \ldots = r_i^{(i)}$, see (1.38), it follows from (1.40) that

$$r_1^{(i+1)} = 0;$$

$$r_2^{(i+1)} = L(2, 1) \, r_1^{(i)} = 0;$$

$$r_3^{(i+1)} = \sum_{k=1}^{2} L(3, k) \, r_k^{(i)} = L(3, 1) \, r_1^{(i)} + L(3, 2) \, r_2^{(i)} = 0;$$

$$\vdots$$

$$
\begin{aligned}
r_{i+1}^{(i+1)} &= \sum_{k=1}^{i} L(i+1,k)\, r_k^{(i)} \\
&= L(i+1,1)\, r_1^{(i)} + L(i+1,2)\, r_2^{(i)} + \ldots + L(i+1,i)\, r_i^{(i)} \\
&= 0.
\end{aligned}
$$

In other words, $r_1^{(i+1)} = r_2^{(i+1)} = \ldots = r_{i+1}^{(i+1)} = 0$, i.e., the first $(i+1)$ rows of the matrix L^{i+1} are zero vectors.

This completes the proof by induction of the fact that the first i rows of the matrix L^i are zero vectors, for all $1 \le i \le n$.

Thus, the first n rows of the matrix L^n are zero vectors, and, since L^n is an $n \times n$ matrix, we conclude that $L^n = 0$.

(ii) Let $m \ge n$. Let $A = L$ and $B = I$ in the binomial formula (1.35). Since $L^j = 0$ for all $j \ge n$, we obtain that

$$
\begin{aligned}
(I+L)^m &= (L+I)^m = \sum_{j=0}^{m} \binom{m}{j} L^j = \sum_{j=0}^{n-1} \binom{m}{j} L^j \\
&= I + mL + \binom{m}{2} L^2 + \ldots + + \binom{m}{n-1} L^{n-1}. \quad \square
\end{aligned}
$$

Problem 10: Let A and B be square matrices of the same size with nonnegative entries and such that the sum of the entries in each row is equal to 1. Show that the matrix AB has the same properties, i.e., show that all the entries of the matrix AB are nonnegative and the sum of the entries in each row of AB is equal to 1.

Solution: A matrix with nonnegative entries such that the sum of the entries in each row is equal to 1 is called a probability matrix. Thus, the problem can be restated as follows:

Let A and B be probability matrices of the same size. Show that AB is a probability matrix.

We first establish the following equivalent definition for a probability matrix:

The $n \times n$ matrix M is a probability matrix if and only if all the entries of M are nonnegative and

$$M\mathbf{1} = \mathbf{1}, \tag{1.41}$$

where $\mathbf{1}$ is the $n \times 1$ column vector with all entries equal to 1.

To see this, let $M = \text{row}\,(r_j)_{j=1:n}$ be the row form of the matrix M, where r_j is an $1 \times n$ row vector, for $j = 1:n$. The sum of all the entries in the j-th row r_j of M can be written as follows:[1]

$$\sum_{k=1}^{n} r_j(k) = r_j\mathbf{1}. \tag{1.42}$$

---

[1]Note that r_j is an $1 \times n$ vector and $\mathbf{1}$ is an $n \times 1$ vector, and therefore the expression $r_j\mathbf{1}$ from (1.42) is consistent.

Thus, the definition of a probability matrix as a matrix with the sum of the entries in each row equal to 1 can be written as

$$\sum_{k=1}^{n} r_j(k) = 1, \quad \forall\, j = 1:n$$
$$\Longleftrightarrow \quad r_j 1 = 1, \quad \forall\, j = 1:n$$
$$\Longleftrightarrow \quad (r_j 1)_{j=1:n} = 1$$
$$\Longleftrightarrow \quad M1 = 1,$$

since $M1 = (r_j 1)_{j=1:n}$ if $M = \text{row}\,(r_j)_{j=1:n}$ is the row form of M.

In other words, we established that (1.41) is an equivalent condition for M to be a probability matrix.

Let A and B be probability matrices. Then all the entries of A and B are non-negative, and therefore all the entries of AB are also nonnegative.[2] From (1.41), it follows that

$$A1 = 1 \quad \text{and} \quad B1 = 1,$$

and therefore

$$(AB)1 = A(B1) = A1 = 1. \tag{1.43}$$

Then, from (1.41) and (1.43), we conclude that AB is a probability matrix, which is what we wanted to show. ☐

Problem 11: The covariance matrix of five random variables is

$$\Sigma = \begin{pmatrix} 1 & -0.525 & 1.375 & -0.075 & -0.75 \\ -0.525 & 2.25 & 0.1875 & 0.1875 & -0.675 \\ 1.375 & 0.1875 & 6.25 & 0.4375 & -1.875 \\ -0.075 & 0.1875 & 0.4375 & 0.25 & 0.3 \\ -0.75 & -0.675 & -1.875 & 0.3 & 9 \end{pmatrix}. \tag{1.44}$$

Find the correlation matrix of these random variables.

Solution: Recall that the correlation matrix Ω and the covariance matrix Σ of n nonconstant random variables $X_1, X_2, \ldots X_n$ satisfy the following relationship:

$$\Omega = (D_\sigma)^{-1}\Sigma(D_\sigma)^{-1}, \tag{1.45}$$

where D_σ is the diagonal matrix given by $D_\sigma = \text{diag}(\sigma_i)_{i=1:n}$; here, σ_i denotes the standard deviation of the random variable X_i, for $i = 1:n$.

From (1.44), we find that the standard deviations of the five random variables are

$$\sigma_1 = \sqrt{\Sigma(1,1)} = 1; \ \sigma_2 = \sqrt{\Sigma(2,2)} = 1.5; \ \sigma_3 = \sqrt{\Sigma(3,3)} = 2.5;$$

---

[2]Formally, the fact that all the entries of the matrix AB are nonnegative if all the entries of A and B are nonnegative can be proved as follows:

$$(AB)(j,k) = \sum_{i=1}^{n} A(j,i)B(i,k) \geq 0, \quad \forall\, 1 \leq j,k \leq n,$$

if $A(j,i) \geq 0$ for all $1 \leq i,j \leq n$ and $B(i,k) \geq 0$ for all $1 \leq i,k \leq n$.

$$\sigma_4 = \sqrt{\Sigma(4,4)} = 0.5; \quad \sigma_5 = \sqrt{\Sigma(5,5)} = 3,$$

and therefore

$$D_\sigma = \text{diag}(\sigma_i)_{i=1:5} = \begin{pmatrix} 1 & 0 & 0 & 0 & 0 \\ 0 & 1.5 & 0 & 0 & 0 \\ 0 & 0 & 2.5 & 0 & 0 \\ 0 & 0 & 0 & 0.5 & 0 \\ 0 & 0 & 0 & 0 & 3 \end{pmatrix}. \tag{1.46}$$

Then, from (1.45) and (1.46), we obtain that the correlation matrix of the five random variables is

$$\Omega = \begin{pmatrix} 1 & -0.35 & 0.55 & -0.15 & -0.25 \\ -0.35 & 1 & 0.05 & 0.25 & -0.15 \\ 0.55 & 0.05 & 1 & 0.35 & -0.25 \\ -0.15 & 0.25 & 0.35 & 1 & 0.20 \\ -0.25 & -0.15 & -0.25 & 0.20 & 1 \end{pmatrix}. \quad \square$$

Problem 12: The correlation matrix of five random variables is

$$\Omega = \begin{pmatrix} 1 & -0.25 & 0.15 & -0.05 & -0.30 \\ -0.25 & 1 & -0.10 & -0.25 & 0.10 \\ 0.15 & -0.10 & 1 & 0.20 & 0.05 \\ -0.05 & -0.25 & 0.20 & 1 & 0.10 \\ -0.30 & 0.10 & 0.05 & 0.10 & 1 \end{pmatrix} \tag{1.47}$$

(i) Compute the covariance matrix of these random variables if their standard deviations are 0.25, 0.5, 1, 2, and 4, in this order.

(ii) Compute the covariance matrix of these random variables if their standard deviations are 4, 2, 1, 0.5, and 0.25, in this order.

Solution: The covariance matrix Σ and the correlation matrix Ω of n random variables $X_1, X_2, \ldots X_n$ satisfy the following relationship:

$$\Sigma = D_\sigma \Omega D_\sigma, \tag{1.48}$$

where D_σ is the diagonal matrix given by $D_\sigma = \text{diag}(\sigma_i)_{i=1:n}$; here, σ_i denotes the standard deviation of the random variable X_i, for $i = 1 : n$.

(i) If the standard deviations of the random variables are 0.25, 0.5, 1, 2, and 4, then

$$D_\sigma = \begin{pmatrix} 0.25 & 0 & 0 & 0 & 0 \\ 0 & 0.5 & 0 & 0 & 0 \\ 0 & 0 & 1 & 0 & 0 \\ 0 & 0 & 0 & 2 & 0 \\ 0 & 0 & 0 & 0 & 4 \end{pmatrix}, \tag{1.49}$$

and, from (1.48) and using (1.47) and (1.49), we obtain that the covariance matrix of the random variables is

$$\Sigma = \begin{pmatrix} 0.0625 & -0.0312 & 0.0375 & -0.025 & -0.3 \\ -0.0312 & 0.25 & -0.05 & -0.25 & 0.2 \\ 0.0375 & -0.05 & 1 & 0.4 & 0.2 \\ -0.025 & -0.25 & 0.4 & 4 & 0.8 \\ -0.3 & 0.20 & 0.2 & 0.8 & 16 \end{pmatrix}.$$

(ii) If the standard deviations of the random variables are 4, 2, 1, 0.5, and 0.25, then

$$D_\sigma = \begin{pmatrix} 4 & 0 & 0 & 0 & 0 \\ 0 & 2 & 0 & 0 & 0 \\ 0 & 0 & 1 & 0 & 0 \\ 0 & 0 & 0 & 0.5 & 0 \\ 0 & 0 & 0 & 0 & 0.25 \end{pmatrix}, \tag{1.50}$$

and, from (1.48) and using (1.47) and (1.50), we obtain that the covariance matrix of the random variables is

$$\Sigma = \begin{pmatrix} 16 & -2 & 0.6 & -0.1 & -0.3 \\ -2 & 4 & -0.2 & -0.25 & 0.05 \\ 0.6 & -0.2 & 1 & 0.1 & 0.0125 \\ -0.1 & -0.25 & 0.1 & 0.25 & 0.0125 \\ -0.3 & 0.05 & 0.0125 & 0.0125 & 0.0625 \end{pmatrix}. \quad \square$$

Problem 13: The file *indeces-jul26-aug9-2012.xlsx* from fepress.org/nla-primer contains the July 26, 2012 – August 9, 2012 end of day values of Dow Jones, Nasdaq, and S&P 500.

(i) Compute the daily percentage returns of the three indices over the given time period.

(ii) Compute the covariance matrix of the daily percentage returns of the three indices.

(iii) Compute the daily log returns of the three indices over the given time period.

(iv) Compute the covariance matrix of the daily log returns of the three indices.

Solution: The July 26, 2012 – August 9, 2012 end of day values of Dow Jones, Nasdaq, and S&P 500 were:

Date	Dow Jones	NASDAQ	S&P 500
07/26/2012	12887.93	2893.25	1360.02
07/27/2012	13075.66	2958.09	1385.97
07/30/2012	13073.01	2945.84	1385.30
07/31/2012	13008.68	2939.52	1379.32
08/01/2012	12976.13	2920.21	1375.32
08/02/2012	12878.88	2909.77	1365.00
08/03/2012	13096.17	2967.90	1390.99
08/06/2012	13117.51	2989.91	1394.23
08/07/2012	13168.60	3015.86	1401.35
08/08/2012	13175.64	3011.25	1402.22
08/09/2012	13165.19	3018.64	1402.80

(i) The percentage return between times t_1 and t_2 of an asset with price $S(t)$ at time t is given by $\frac{S(t_2)-S(t_1)}{S(t_1)}$. Then, the time series matrix of the daily percentage returns of the three indices between 07/26/2012 and 08/09/2012 is

$$
T_{\mathbf{x}}^{(p)} = \begin{pmatrix}
0.014566 & 0.022411 & 0.019081 \\
-0.000203 & -0.004141 & -0.000483 \\
-0.004921 & -0.002145 & -0.004317 \\
-0.002502 & -0.006569 & -0.002900 \\
-0.007495 & -0.003575 & -0.007504 \\
0.016872 & 0.019978 & 0.019040 \\
0.001629 & 0.007416 & 0.002329 \\
0.003895 & 0.008679 & 0.005107 \\
0.000535 & -0.001529 & 0.000621 \\
-0.000793 & 0.002454 & 0.000414
\end{pmatrix}, \tag{1.51}
$$

where, e.g., the percentage return on 08/08/2012 of NASDAQ is

$$
\frac{3011.25 - 3015.86}{3015.86} = -0.001529 = T_{\mathbf{x}}^{(p)}(9,2).
$$

(ii) The sample means of the daily returns of the three indices are 0.002158 (Dow Jones), 0.004298 (NASDAQ), and 0.003139 (S&P 500). By subtracting the sample mean of each column from $T_{\mathbf{x}}^{(p)}$, see (1.51), we obtain that the time series matrix of mean normalized daily percentage returns of the three indices is

$$
\overline{T}_{\mathbf{x}}^{(p)} = \begin{pmatrix}
0.012408 & 0.018113 & 0.015942 \\
-0.002361 & -0.008439 & -0.003622 \\
-0.007079 & -0.006443 & -0.007456 \\
-0.004661 & -0.010867 & -0.006039 \\
-0.009653 & -0.007873 & -0.010642 \\
0.014713 & 0.015680 & 0.015902 \\
-0.000529 & 0.003118 & -0.000809 \\
0.001736 & 0.004381 & 0.001968 \\
-0.001624 & -0.005826 & -0.002518 \\
-0.002952 & -0.001844 & -0.002725
\end{pmatrix}.
$$

Note that $\overline{T}_{\mathbf{x}}^{(p)}$ is a 10×3 matrix, corresponding to $N = 10$ daily returns computed from the 11 daily closes.

The covariance matrix of the daily percentage returns of the three indices is

$$
\begin{aligned}
\widehat{\Sigma}_{\mathbf{x}}^{(p)} &= \frac{1}{N-1} \left(\overline{T}_{\mathbf{x}}^{(p)}\right)^t \overline{T}_{\mathbf{x}}^{(p)} \\
&= \begin{pmatrix}
0.000061742 & 0.000074276 & 0.000071106 \\
0.000074276 & 0.000103667 & 0.000087988 \\
0.000071106 & 0.000087988 & 0.000082637
\end{pmatrix}.
\end{aligned}
$$

(iii) The log return between times t_1 and t_2 of an asset with price $S(t)$ at time t is given by $\ln\left(\frac{S(t_2)}{S(t_1)}\right)$. Then, the time series matrix of the daily log returns of the three indices between 07/26/2012 and 08/09/2012 is

$$
T_{\mathbf{x}}^{(log)} = \begin{pmatrix}
0.014461 & 0.022163 & 0.018901 \\
-0.000203 & -0.004150 & -0.000484 \\
-0.004933 & -0.002148 & -0.004326 \\
-0.002505 & -0.006591 & -0.002904 \\
-0.007523 & -0.003581 & -0.007532 \\
0.016731 & 0.019781 & 0.018861 \\
0.001628 & 0.007389 & 0.002327 \\
0.003887 & 0.008642 & 0.005094 \\
0.000534 & -0.001530 & 0.000621 \\
-0.000793 & 0.002451 & 0.000414
\end{pmatrix}, \qquad (1.52)
$$

where, e.g., the log return on 08/01/2012 of S&P 500 is

$$
\ln\left(\frac{1375.32}{1379.32}\right) = -0.002904 = T_{\mathbf{x}}^{(log)}(4,3).
$$

(iv) The sample means of the log returns of the three indices are 0.002129 (Dow Jones), 0.004243 (NASDAQ), and 0.003097 (S&P 500). By subtracting the sample mean of each column from $T_{\mathbf{x}}^{(log)}$, see (1.52), we obtain that the time series matrix of mean normalized log percentage returns of the three indices is

$$
\overline{T}_{\mathbf{x}}^{(log)} = \begin{pmatrix}
0.012333 & 0.017921 & 0.015804 \\
-0.002331 & -0.008392 & -0.003581 \\
-0.007061 & -0.006390 & -0.007423 \\
-0.004634 & -0.010833 & -0.006001 \\
-0.009651 & -0.007824 & -0.010629 \\
0.014603 & 0.015538 & 0.015764 \\
-0.000500 & 0.003146 & -0.000771 \\
0.001759 & 0.004399 & 0.001997 \\
-0.001594 & -0.005772 & -0.002476 \\
-0.002922 & -0.001791 & -0.002684
\end{pmatrix}.
$$

Note that $\overline{T}_{\mathbf{x}}^{(log)}$ is a 10×3 matrix, corresponding to $N = 10$ daily returns computed from the 11 daily closes.

The covariance matrix of the daily log returns of the three indices is

$$
\begin{aligned}
\widehat{\Sigma}_{\mathbf{x}}^{(log)} &= \frac{1}{N-1} (\overline{T}_{\mathbf{x}}^{(log)})^t \overline{T}_{\mathbf{x}}^{(log)} \\
&= \begin{pmatrix}
0.000061075 & 0.000073212 & 0.000070216 \\
0.000073212 & 0.000102023 & 0.000086587 \\
0.000070216 & 0.000086587 & 0.000081456
\end{pmatrix}. \quad \square
\end{aligned}
$$

Problem 14: The file *indices-july2011.xlsx* from fepress.org/nla-primer contains the January 2011 – July 2011 end of day values of nine major US indices.

(i) Compute the sample covariance matrix of the daily percentage returns of the indices, and the corresponding sample correlation matrix.

Compute the sample covariance and correlation matrices for daily log returns, and compare them with the corresponding matrices for daily percentage returns.

(ii) Compute the sample covariance matrix of the weekly percentage returns of the indices, and the corresponding sample correlation matrix.

Compute the sample covariance and correlation matrices for weekly log returns, and compare them with the corresponding matrices for weekly percentage returns.

(iii) Compute the sample covariance matrix of the monthly percentage returns of the indices, and the corresponding sample correlation matrix.

Compute the sample covariance and correlation matrices for monthly log returns, and compare them with the corresponding matrices for monthly percentage returns.

(iv) Comment on the differences between the sample covariance and correlation matrices for daily, weekly, and monthly returns.

Solution: (i) The sample covariance matrix of the daily percentage returns of the indices, and the corresponding sample correlation matrix are

$$
10^{-6} \cdot
\begin{pmatrix}
100.04 & 67.04 & 96.67 & 40.04 & 82.77 & 78.90 & 74.35 & 71.11 & 54.56 \\
67.04 & 58.16 & 71.29 & 38.01 & 65.29 & 61.38 & 59.63 & 52.87 & 41.99 \\
96.67 & 71.29 & 135.88 & 45.36 & 86.15 & 83.05 & 77.46 & 60.15 & 57.46 \\
40.04 & 38.01 & 45.36 & 44.84 & 43.20 & 40.10 & 39.37 & 34.60 & 30.43 \\
82.77 & 65.29 & 86.15 & 43.20 & 82.37 & 74.46 & 71.37 & 73.05 & 54.01 \\
78.90 & 61.38 & 83.05 & 40.10 & 74.46 & 69.77 & 66.94 & 61.80 & 46.55 \\
74.35 & 59.63 & 77.46 & 39.36 & 71.37 & 66.94 & 65.17 & 58.66 & 43.66 \\
71.11 & 52.87 & 60.15 & 34.60 & 73.05 & 61.80 & 58.66 & 103.96 & 50.01 \\
54.56 & 41.99 & 57.46 & 30.43 & 54.01 & 46.55 & 43.66 & 50.01 & 110.84
\end{pmatrix}
$$

$$
\begin{pmatrix}
1 & 0.8790 & 0.8292 & 0.5979 & 0.9118 & 0.9444 & 0.9208 & 0.6972 & 0.5181 \\
0.8790 & 1 & 0.8019 & 0.7444 & 0.9433 & 0.9634 & 0.9686 & 0.6799 & 0.5229 \\
0.8292 & 0.8019 & 1 & 0.5812 & 0.8143 & 0.8529 & 0.8232 & 0.5061 & 0.4682 \\
0.5979 & 0.7444 & 0.5812 & 1 & 0.7109 & 0.7169 & 0.7282 & 0.5067 & 0.4316 \\
0.9118 & 0.9433 & 0.8143 & 0.7109 & 1 & 0.9821 & 0.9742 & 0.7894 & 0.5652 \\
0.9444 & 0.9634 & 0.8529 & 0.7169 & 0.9821 & 1 & 0.9927 & 0.7256 & 0.5293 \\
0.9208 & 0.9686 & 0.8232 & 0.7282 & 0.9742 & 0.9927 & 1 & 0.7126 & 0.5137 \\
0.6972 & 0.6799 & 0.5061 & 0.5067 & 0.7894 & 0.7256 & 0.7126 & 1 & 0.4659 \\
0.5181 & 0.5229 & 0.4682 & 0.4316 & 0.5652 & 0.5293 & 0.5137 & 0.4659 & 1
\end{pmatrix}
$$

The sample covariance matrix of the daily log returns of the indices, and the corresponding sample correlation matrix are

$$
10^{-6} \cdot
\begin{pmatrix}
100.40 & 67.27 & 97.25 & 40.15 & 83.13 & 79.21 & 74.63 & 71.38 & 54.60 \\
67.27 & 58.32 & 71.57 & 38.09 & 65.53 & 61.58 & 59.82 & 53.06 & 41.99 \\
97.25 & 71.57 & 136.67 & 45.46 & 86.59 & 83.46 & 77.81 & 60.39 & 57.57 \\
40.15 & 38.09 & 45.46 & 44.90 & 43.34 & 40.20 & 39.47 & 34.76 & 30.56 \\
83.13 & 65.53 & 86.59 & 43.34 & 82.72 & 74.76 & 71.65 & 73.33 & 54.04 \\
79.21 & 61.58 & 83.46 & 40.20 & 74.76 & 70.03 & 67.18 & 62.04 & 46.58 \\
74.63 & 59.82 & 77.81 & 39.47 & 71.65 & 67.18 & 65.39 & 58.88 & 43.66 \\
71.38 & 53.06 & 60.39 & 34.76 & 73.33 & 62.04 & 58.88 & 104.16 & 50.06 \\
54.60 & 41.99 & 57.57 & 30.56 & 54.04 & 46.58 & 43.66 & 50.06 & 111.05
\end{pmatrix}
$$

$$\begin{pmatrix}
1 & 0.8791 & 0.8302 & 0.5980 & 0.9121 & 0.9447 & 0.9210 & 0.6981 & 0.5171 \\
0.8791 & 1 & 0.8017 & 0.7444 & 0.9435 & 0.9635 & 0.9687 & 0.6808 & 0.5218 \\
0.8302 & 0.8017 & 1 & 0.5803 & 0.8144 & 0.8531 & 0.8232 & 0.5062 & 0.4673 \\
0.5980 & 0.7444 & 0.5803 & 1 & 0.7111 & 0.7169 & 0.7283 & 0.5083 & 0.4327 \\
0.9121 & 0.9435 & 0.8144 & 0.7111 & 1 & 0.9822 & 0.9742 & 0.7900 & 0.5638 \\
0.9447 & 0.9635 & 0.8531 & 0.7169 & 0.9822 & 1 & 0.9927 & 0.7264 & 0.5281 \\
0.9210 & 0.9687 & 0.8232 & 0.7283 & 0.9742 & 0.9927 & 1 & 0.7134 & 0.5124 \\
0.6981 & 0.6808 & 0.5062 & 0.5083 & 0.7900 & 0.7264 & 0.7134 & 1 & 0.4654 \\
0.5171 & 0.5218 & 0.4673 & 0.4327 & 0.5638 & 0.5281 & 0.5124 & 0.4654 & 1
\end{pmatrix}$$

Since log returns are excellent approximations for percentage returns in the case of small returns such as daily returns, the sample covariance and correlation matrices for daily percentage returns and for daily log returns are very close to each other.

(ii) The sample covariance matrix of the weekly percentage returns[3] of the indices, and the corresponding sample correlation matrix are

$$10^{-6} \cdot \begin{pmatrix}
499.83 & 344.14 & 453.90 & 208.79 & 398.01 & 391.49 & 378.60 & 411.23 & 317.36 \\
344.14 & 289.74 & 355.05 & 181.11 & 306.84 & 302.12 & 297.39 & 276.41 & 244.01 \\
453.90 & 355.05 & 678.75 & 253.15 & 383.24 & 392.69 & 371.69 & 276.45 & 306.13 \\
208.79 & 181.11 & 253.15 & 222.43 & 179.49 & 192.78 & 190.84 & 104.43 & 188.94 \\
398.01 & 306.84 & 383.24 & 179.49 & 372.05 & 344.68 & 336.28 & 391.27 & 271.59 \\
391.49 & 302.12 & 392.69 & 192.78 & 344.68 & 335.21 & 326.99 & 327.86 & 263.23 \\
378.60 & 297.39 & 371.69 & 190.84 & 336.28 & 326.99 & 323.48 & 316.18 & 256.98 \\
411.23 & 276.41 & 276.45 & 104.43 & 391.27 & 327.86 & 316.18 & 656.36 & 258.95 \\
317.36 & 244.01 & 306.13 & 188.94 & 271.59 & 263.23 & 256.98 & 258.95 & 410.64
\end{pmatrix}$$

$$\begin{pmatrix}
1 & 0.9043 & 0.7793 & 0.6262 & 0.9230 & 0.9564 & 0.9416 & 0.7180 & 0.7005 \\
0.9043 & 1 & 0.8006 & 0.7134 & 0.9346 & 0.9694 & 0.9714 & 0.6338 & 0.7074 \\
0.7793 & 0.8006 & 1 & 0.6515 & 0.7626 & 0.8233 & 0.7932 & 0.4142 & 0.5799 \\
0.6262 & 0.7134 & 0.6515 & 1 & 0.6240 & 0.7060 & 0.7115 & 0.2733 & 0.6252 \\
0.9230 & 0.9346 & 0.7626 & 0.6240 & 1 & 0.9760 & 0.9694 & 0.7918 & 0.6948 \\
0.9564 & 0.9694 & 0.8233 & 0.7060 & 0.9760 & 1 & 0.9930 & 0.6990 & 0.7095 \\
0.9416 & 0.9714 & 0.7932 & 0.7115 & 0.9694 & 0.9930 & 1 & 0.6862 & 0.7051 \\
0.7180 & 0.6338 & 0.4142 & 0.2733 & 0.7918 & 0.6990 & 0.6862 & 1 & 0.4988 \\
0.7005 & 0.7074 & 0.5799 & 0.6252 & 0.6948 & 0.7095 & 0.7051 & 0.4988 & 1
\end{pmatrix}$$

The sample covariance matrix of the weekly log returns of the indices, and the corresponding sample correlation matrix are

$$10^{-6} \cdot \begin{pmatrix}
493.00 & 337.88 & 448.96 & 206.47 & 391.99 & 385.37 & 372.19 & 407.91 & 310.63 \\
337.88 & 284.42 & 349.79 & 178.75 & 301.59 & 296.86 & 291.97 & 272.65 & 237.68 \\
448.96 & 349.79 & 676.39 & 249.89 & 378.35 & 388.15 & 366.51 & 272.85 & 298.99 \\
206.47 & 178.75 & 249.89 & 223.35 & 177.08 & 190.72 & 188.71 & 102.76 & 185.58 \\
391.99 & 301.59 & 378.35 & 177.08 & 366.86 & 339.41 & 330.83 & 388.38 & 265.44 \\
385.37 & 296.86 & 388.15 & 190.72 & 339.41 & 329.97 & 321.52 & 324.24 & 257.08 \\
372.19 & 291.97 & 366.51 & 188.71 & 330.83 & 321.52 & 317.81 & 312.58 & 250.61 \\
407.91 & 272.65 & 272.85 & 102.76 & 388.38 & 324.24 & 312.58 & 655.52 & 256.49 \\
310.63 & 237.68 & 298.99 & 185.58 & 265.44 & 257.08 & 250.61 & 256.49 & 406.67
\end{pmatrix}$$

---

[3] The weekly returns are computed using the end of day price from the last trading day of the prior week and the end of day price from the last trading day of the current week, i.e., the "Friday to Friday" convention.

$$\begin{pmatrix}
1 & 0.9023 & 0.7775 & 0.6222 & 0.9217 & 0.9555 & 0.9403 & 0.7175 & 0.6937 \\
0.9023 & 1 & 0.7975 & 0.7092 & 0.9337 & 0.9690 & 0.9711 & 0.6314 & 0.6989 \\
0.7775 & 0.7975 & 1 & 0.6429 & 0.7595 & 0.8216 & 0.7905 & 0.4098 & 0.5701 \\
0.6222 & 0.7092 & 0.6429 & 1 & 0.6186 & 0.7025 & 0.7083 & 0.2686 & 0.6158 \\
0.9217 & 0.9337 & 0.7595 & 0.6186 & 1 & 0.9755 & 0.9689 & 0.7920 & 0.6872 \\
0.9555 & 0.9690 & 0.8216 & 0.7025 & 0.9755 & 1 & 0.9929 & 0.6972 & 0.7018 \\
0.9403 & 0.9711 & 0.7905 & 0.7083 & 0.9689 & 0.9929 & 1 & 0.6848 & 0.6971 \\
0.7175 & 0.6314 & 0.4098 & 0.2686 & 0.7920 & 0.6972 & 0.6848 & 1 & 0.4968 \\
0.6937 & 0.6989 & 0.5701 & 0.6158 & 0.6872 & 0.7018 & 0.6971 & 0.4968 & 1
\end{pmatrix}$$

Weekly log returns are very good approximations for weekly percentage returns, and therefore the sample covariance and correlation matrices for weekly percentage returns and for weekly log returns are close to each other.

(iii) The monthly[4] percentage returns of the nine indices are

$$\begin{matrix}
0.0178 & 0.0272 & -0.0160 & 0.0108 & 0.0220 & 0.0226 & 0.0236 & -0.0165 & -0.0252 \\
0.0304 & 0.0281 & 0.0119 & 0.0153 & 0.0368 & 0.0320 & 0.0289 & 0.0963 & -0.0162 \\
-0.0004 & 0.0076 & 0.0423 & -0.0061 & -0.0040 & -0.0010 & -0.0054 & -0.0058 & 0.0055 \\
0.0332 & 0.0398 & 0.0406 & 0.0387 & 0.0317 & 0.0285 & 0.0263 & 0.0488 & 0.0246 \\
-0.0133 & -0.0188 & -0.0082 & 0.0170 & -0.0224 & -0.0135 & -0.0181 & -0.0185 & -0.0099 \\
-0.0218 & -0.0124 & -0.0084 & -0.0066 & -0.0187 & -0.0183 & -0.0168 & -0.0383 & -0.0314
\end{matrix}$$

The sample covariance matrix of the monthly percentage returns of the indices and the corresponding sample correlation matrix are

$$10^{-6} \cdot \begin{pmatrix}
529.78 & 526.36 & 244.69 & 267.24 & 584.63 & 503.19 & 497.53 & 967.95 & 218.71 \\
526.36 & 565.13 & 271.06 & 231.25 & 594.65 & 504.06 & 509.56 & 839.03 & 216.96 \\
244.69 & 271.06 & 664.65 & 104.25 & 214.54 & 174.44 & 139.12 & 586.15 & 465.90 \\
267.24 & 231.25 & 104.25 & 285.36 & 254.10 & 230.93 & 221.68 & 480.23 & 209.20 \\
584.63 & 594.65 & 214.54 & 254.10 & 672.48 & 568.58 & 572.66 & 1076.07 & 162.90 \\
503.19 & 504.06 & 174.44 & 230.93 & 568.58 & 488.24 & 488.07 & 899.60 & 151.06 \\
497.53 & 509.56 & 139.12 & 221.68 & 572.66 & 488.07 & 494.77 & 852.96 & 119.97 \\
967.95 & 839.03 & 586.15 & 480.23 & 1076.07 & 899.60 & 852.96 & 2609.89 & 397.89 \\
218.71 & 216.96 & 465.90 & 209.20 & 162.90 & 151.06 & 119.97 & 397.89 & 431.37
\end{pmatrix}$$

$$\begin{pmatrix}
1 & 0.9620 & 0.4124 & 0.6873 & 0.9795 & 0.9894 & 0.9718 & 0.8232 & 0.4575 \\
0.9620 & 1 & 0.4423 & 0.5758 & 0.9646 & 0.9596 & 0.9637 & 0.6909 & 0.4394 \\
0.4124 & 0.4423 & 1 & 0.2394 & 0.3209 & 0.3062 & 0.2426 & 0.4450 & 0.8701 \\
0.6873 & 0.5758 & 0.2394 & 1 & 0.5801 & 0.6187 & 0.5900 & 0.5565 & 0.5963 \\
0.9795 & 0.9646 & 0.3209 & 0.5801 & 1 & 0.9923 & 0.9928 & 0.8123 & 0.3025 \\
0.9894 & 0.9596 & 0.3062 & 0.6187 & 0.9923 & 1 & 0.9930 & 0.7969 & 0.3292 \\
0.9718 & 0.9637 & 0.2426 & 0.5900 & 0.9928 & 0.9930 & 1 & 0.7506 & 0.2597 \\
0.8232 & 0.6909 & 0.4450 & 0.5565 & 0.8123 & 0.7969 & 0.7506 & 1 & 0.3750 \\
0.4575 & 0.4394 & 0.8701 & 0.5963 & 0.3025 & 0.3292 & 0.2597 & 0.3750 & 1
\end{pmatrix}$$

The sample covariance matrix of the monthly log returns of the indices, and the corresponding sample correlation matrix are

$$10^{-6} \cdot \begin{pmatrix}
523.10 & 517.96 & 240.42 & 261.17 & 576.86 & 496.69 & 491.77 & 939.22 & 217.82 \\
517.96 & 554.85 & 265.01 & 223.42 & 586.02 & 496.14 & 502.49 & 814.35 & 213.56 \\
240.42 & 265.01 & 645.86 & 99.41 & 211.91 & 172.23 & 137.85 & 578.67 & 460.57 \\
261.17 & 223.42 & 99.41 & 277.29 & 248.02 & 226.08 & 217.18 & 467.24 & 204.01 \\
576.86 & 586.02 & 211.91 & 248.02 & 663.64 & 560.78 & 565.98 & 1039.55 & 162.04 \\
496.69 & 496.14 & 172.23 & 226.08 & 560.78 & 481.52 & 482.06 & 871.09 & 150.94 \\
491.77 & 502.49 & 137.85 & 217.18 & 565.98 & 482.06 & 489.54 & 827.20 & 119.56 \\
939.22 & 814.35 & 578.67 & 467.24 & 1039.55 & 871.09 & 827.20 & 2464.76 & 403.04 \\
217.82 & 213.56 & 460.57 & 204.01 & 162.04 & 150.94 & 119.56 & 403.04 & 434.42
\end{pmatrix}$$

[4]The monthly returns are computed using the end of day price from the last trading day of the prior month and the end of day price from the last trading day of the current month.

$$\begin{pmatrix}
1 & 0.9614 & 0.4136 & 0.6857 & 0.9791 & 0.9897 & 0.9718 & 0.8272 & 0.4569 \\
0.9614 & 1 & 0.4427 & 0.5696 & 0.9657 & 0.9599 & 0.9642 & 0.6964 & 0.4350 \\
0.4136 & 0.4427 & 1 & 0.2349 & 0.3237 & 0.3088 & 0.2452 & 0.4586 & 0.8695 \\
0.6857 & 0.5696 & 0.2349 & 1 & 0.5782 & 0.6187 & 0.5895 & 0.5652 & 0.5904 \\
0.9791 & 0.9657 & 0.3237 & 0.5782 & 1 & 0.9920 & 0.9930 & 0.8128 & 0.3018 \\
0.9897 & 0.9599 & 0.3088 & 0.6187 & 0.9920 & 1 & 0.9929 & 0.7996 & 0.3300 \\
0.9718 & 0.9642 & 0.2452 & 0.5895 & 0.9930 & 0.9929 & 1 & 0.7531 & 0.2593 \\
0.8272 & 0.6964 & 0.4586 & 0.5652 & 0.8128 & 0.7996 & 0.7531 & 1 & 0.3895 \\
0.4569 & 0.4350 & 0.8695 & 0.5904 & 0.3018 & 0.3300 & 0.2593 & 0.3895 & 1
\end{pmatrix}$$

It is not a priori clear how well do monthly log returns approximate monthly percentage returns, although they should be reasonably close to each other. However, for this particular set of prices, the sample covariance and correlation matrices for monthly percentage returns and for monthly log returns are very close to each other.

(iv) Monthly returns are, generally speaking, likely to be larger than weekly returns, which are likely to be larger than daily returns. We note that the entries of the sample covariance matrices for daily returns are clearly smaller than the entries of the sample covariance matrices for weekly and monthly returns.

The entries of the sample covariance matrices for monthly returns are typically greater than the entries of the sample covariance matrices for weekly returns. However, some of the entries of the sample covariance matrices for weekly returns are larger than the corresponding entries of the sample covariance matrices for monthly returns. These results hold for both percentage and log returns.

The connections between the sample correlation matrices for daily, weekly, and monthly returns are less clear–cut.   □

Problem 15: In three months, the value of an asset with spot price $50 will be either $60 or $45. The continuously compounded risk–free rate is 6%. Consider the one period market model with two securities, i.e., cash and the asset, and two states, i.e., asset value equal to $60 and asset value equal to $45, in three months.

(i) Find the payoff matrix of this model.

(ii) Is this one period market complete, i.e., is the payoff matrix nonsingular?

(iii) How do you replicate a three months at–the–money put option on this asset, using the cash and the underlying asset?

Solution: The three months one period market model considered here has the following two securities and two states:

Securities:
• cash;
• asset;

Market states:
• asset at $60 (state ω^1);
• asset at $45 (state ω^2).

(i) Note that, for a continuously compounded risk–free rate of 6%, the future value of $1 in three months is $e^{0.06 \cdot \frac{3}{12}} = 1.015113$.

Then, the payoff matrix of this one period market model corresponding to a \$1 cash position and to an asset position equal to one unit is

$$M_\tau = \begin{pmatrix} 1.015113 & 1.015113 \\ 60 & 45 \end{pmatrix}.$$

(ii) The payoff matrix M_τ is nonsingular since

$$\det(M_\tau) = 1.015113 \cdot 45 - 1.015113 \cdot 60 = -15.23 \neq 0.$$

We conclude that the one period market model is complete.

(iii) The payoff at maturity of a put option is

$$P(T) = \max(K - S(T), 0) = \begin{cases} 0, & \text{if } S(T) \geq K; \\ K - S(T), & \text{if } S(T) < K. \end{cases} \tag{1.53}$$

From (1.53), we obtain that the values at maturity (i.e., in three months) of at–the–money (ATM) put with strike \$50 on the asset are given by

$$P(1/4) = \max(50 - S(1/4), 0),$$

where $S(1/4)$ denotes the value of the asset in three months, and are as follows:

state w^1 : $S(1/4) = 60$ and $P(1/4) = \max(50 - 60, 0) = 0$;
state w^2 : $S(1/4) = 45$ and $P(1/4) = \max(50 - 45, 0) = 5$.

Thus, the vector value $P_{1/4}$ of the put option in three months is

$$P_{1/4} = (0 \ 5). \tag{1.54}$$

Denote by Θ_1 and Θ_2 the cash and asset positions, respectively, in a portfolio replicating the three months ATM put option on the asset, and let $\Theta = \begin{pmatrix} \Theta_1 \\ \Theta_2 \end{pmatrix}$ be the positions vector. Then, the value of the portfolio in three months is

$$V_{1/4} = \Theta^t M_\tau = (\Theta_1 \ \Theta_2) \begin{pmatrix} 1.015113 & 1.015113 \\ 60 & 45 \end{pmatrix}. \tag{1.55}$$

For the portfolio to replicate the three months ATM put, $V_{1/4}$ and $P_{1/4}$ must be equal, i.e., $V_{1/4} = P_{1/4}$, which can be written as

$$(\Theta_1 \ \Theta_2) \begin{pmatrix} 1.015113 & 1.015113 \\ 60 & 45 \end{pmatrix} = (0 \ 5); \tag{1.56}$$

see (1.54) and (1.55). By taking the transpose on both sides of (1.56), we obtain that

$$\begin{pmatrix} 1.015113 & 60 \\ 1.015113 & 45 \end{pmatrix} \begin{pmatrix} \Theta_1 \\ \Theta_2 \end{pmatrix} = \begin{pmatrix} 0 \\ 5 \end{pmatrix}. \tag{1.57}$$

To solve (1.57),[5] recall that the inverse of a 2×2 matrix is given by

$$\begin{pmatrix} a & b \\ c & d \end{pmatrix}^{-1} = \frac{1}{ad - bc} \begin{pmatrix} d & -b \\ -c & a \end{pmatrix}.$$

---

[5] Alternatively, note that (1.57) is equivalent to the linear system

$$\begin{cases} 1.015113 \, \Theta_1 + 60 \, \Theta_2 &= 0 \\ 1.015113 \, \Theta_1 + 45 \, \Theta_2 &= 5 \end{cases}$$

and can be solved as such.

Then, the solution to (1.57) is

$$
\begin{pmatrix} \Theta_1 \\ \Theta_2 \end{pmatrix} = \begin{pmatrix} 1.015113 & 60 \\ 1.015113 & 45 \end{pmatrix}^{-1} \begin{pmatrix} 0 \\ 5 \end{pmatrix}
$$

$$
= \frac{1}{-15.2267} \begin{pmatrix} 45 & -60 \\ -1.015113 & 1.015113 \end{pmatrix} \begin{pmatrix} 0 \\ 5 \end{pmatrix}
$$

$$
= \begin{pmatrix} 19.7022 \\ -0.3333 \end{pmatrix}.
$$

We conclude that the portfolio replicating the three months ATM put option on the asset is made of a long cash position of $19.7022 and a short $0.3333 = \frac{1}{3}$ position in the asset. □

Problem 16: In six months, the price of an asset with spot price $40 will be either $30, $35, $40, $42, $45, or $50. Consider a one period market model with six states in six months corresponding to the six possible values of the asset in six months, and with the following four securities:

• cash;
• asset;
• six months at-the-money call option with strike $40 on the asset;
• six months at-the-money put option with strike $40 on the asset.

The continuously compounded risk–free interest rate is constant and equal to 6%.

(i) Find the payoff matrix of this model.

(ii) Is this one period market model complete?

(iii) Are the four securities non–redundant?

Solution: (i) This one period market model has the following four assets and six states in six months:

Securities:
• cash;
• asset;
• six months at-the-money call option with strike $40 on the asset;
• six months at-the-money put option with strike $40 on the asset.

States of the market in six months:
• asset price $30 (state ω^1);
• asset price $34 (state ω^2);
• asset price $40 (state ω^3);
• asset price $42 (state ω^4);
• asset price $45 (state ω^5);
• asset price $50 (state ω^6).

The future value of $1 in six months corresponding to a continuously compounded 6% constant interest rate is $e^{0.06 \cdot \frac{1}{2}} = e^{0.03}$.

For $j = 1 : 6$, let $S_{j,1/2}$ be the vector of the six possible prices of asset j in six months. The price vectors $S_{1,1/2}$ of cash and $S_{2,1/2}$ of the asset are

$$S_{1,1/2} = \left(e^{0.03} \ e^{0.03} \ e^{0.03} \ e^{0.03} \ e^{0.03} \ e^{0.03} \right); \qquad (1.58)$$

$$S_{2,1/2} = (30 \ 35 \ 40 \ 42 \ 45 \ 50). \qquad (1.59)$$

Recall that the payoffs at maturity of call and put options are

$$C(T) = \max(S(T) - K, 0) = \begin{cases} S(T) - K, & \text{if } S(T) > K; \\ 0, & \text{if } S(T) \leq K; \end{cases} \qquad (1.60)$$

$$P(T) = \max(K - S(T), 0) = \begin{cases} 0, & \text{if } S(T) \geq K; \\ K - S(T), & \text{if } S(T) < K. \end{cases} \qquad (1.61)$$

From (1.60), we obtain that the values of the six months ATM call with strike \$40 on the asset are given by

$$C(1/2) = \max(S(1/2) - 40, 0),$$

where $S(1/2)$ denotes the value of the asset in six months, and are as follows:

state ω^1 : $S(1/2) = 30$ and $C(1/2) = \max(30 - 40, 0) = 0$;
state ω^2 : $S(1/2) = 35$ and $C(1/2) = \max(35 - 40, 0) = 0$;
state ω^3 : $S(1/2) = 40$ and $C(1/2) = \max(40 - 40, 0) = 0$;
state ω^4 : $S(1/2) = 42$ and $C(1/2) = \max(42 - 40, 0) = 2$;
state ω^5 : $S(1/2) = 45$ and $C(1/2) = \max(45 - 40, 0) = 5$;
state ω^6 : $S(1/2) = 50$ and $C(1/2) = \max(50 - 40, 0) = 10$.

Thus, the vector $S_{3,1/2}$ of the six possible prices of the six months ATM call with strike \$40 on the asset is

$$S_{3,1/2} = (0 \ 0 \ 0 \ 2 \ 5 \ 10). \qquad (1.62)$$

From (1.61), we obtain that the values of the six months ATM put with strike \$40 on the asset are given by

$$P(1/2) = \max(40 - S(1/2), 0),$$

and are as follows:

state ω^1 : $S(1/2) = 30$ and $P(1/2) = \max(40 - 30, 0) = 10$;
state ω^2 : $S(1/2) = 35$ and $P(1/2) = \max(40 - 35, 0) = 5$;
state ω^3 : $S(1/2) = 40$ and $P(1/2) = \max(40 - 40, 0) = 0$;
state ω^4 : $S(1/2) = 42$ and $P(1/2) = \max(40 - 42, 0) = 0$;
state ω^5 : $S(1/2) = 45$ and $P(1/2) = \max(40 - 45, 0) = 0$;
state ω^6 : $S(1/2) = 50$ and $P(1/2) = \max(40 - 50, 0) = 0$.

Thus, the vector $S_{4,1/2}$ of the six possible prices of the six months ATM put with strike \$40 on the asset is

$$S_{4,1/2} = (10 \ 5 \ 0 \ 0 \ 0 \ 0). \qquad (1.63)$$

From (1.58), (1.59), (1.62), and (1.63), we conclude that the payoff matrix $M_{1/2}$ of the market model is the following 4×6 matrix:

$$M_{1/2} = \begin{pmatrix} S_{1,1/2} \\ S_{2,1/2} \\ S_{3,1/2} \\ S_{4,1/2} \end{pmatrix} = \begin{pmatrix} e^{0.03} & e^{0.03} & e^{0.03} & e^{0.03} & e^{0.03} & e^{0.03} \\ 30 & 35 & 40 & 42 & 45 & 50 \\ 0 & 0 & 0 & 2 & 5 & 10 \\ 10 & 5 & 0 & 0 & 0 & 0 \end{pmatrix}. \quad (1.64)$$

(ii) This market model is not complete since it has fewer (four) securities than market states (six), and a necessary condition for a one period market model to be complete is that the model has at least as many securities as market states.

(iii) From (1.58), (1.59), (1.62), and (1.63), we obtain that

$$S_{4,1/2} + S_{2,1/2} - S_{3,1/2} = (40 \ \ 40 \ \ 40 \ \ 40 \ \ 40 \ \ 40) = \frac{40}{e^{0.03}} S_{1,1/2}. \quad (1.65)$$

In other words, the price vectors $S_{1,1/2}$, $S_{2,1/2}$, $S_{3,1/2}$, and $S_{4,1/2}$ are not linearly independent. We conclude that, e.g., the ATM put on the asset can be replicated using cash and positions on the asset and on ATM calls on the asset, since

$$S_{4,1/2} = \frac{40}{e^{0.03}} S_{1,1/2} - S_{2,1/2} + S_{3,1/2},$$

and therefore that the ATM put on the asset is a redundant security.[6]   □

---

[6]Note that the redundancy in this model is due to the Put–Call parity; see section 1.2.1 from Stefanica [3] for more details.

Chapter 2

LU decomposition and linear systems solutions. Discount factors computation. Cubic spline interpolation.

2.1 Exercises

1. Let

$$
L_1 = \begin{pmatrix} 1 & 0 & 0 & 0 \\ -1 & 2 & 0 & 0 \\ 2 & -2 & 3 & 0 \\ 2 & 2 & -3 & 4 \end{pmatrix}; \quad U_1 = \begin{pmatrix} 2 & -1 & 0 & 1 \\ 0 & -1/2 & 1/2 & 0 \\ 0 & 0 & 1/3 & -1/3 \\ 0 & 0 & 0 & -1/4 \end{pmatrix}.
$$

$$
L_2 = \begin{pmatrix} 1 & 0 & 0 & 0 \\ -1 & 1 & 0 & 0 \\ 2 & -1 & 1 & 0 \\ 2 & 1 & -1 & 1 \end{pmatrix}; \quad U_2 = \begin{pmatrix} 2 & -1 & 0 & 1 \\ 0 & -1 & 1 & 0 \\ 0 & 0 & 1 & -1 \\ 0 & 0 & 0 & -1 \end{pmatrix}.
$$

(i) Show that $L_1 U_1 = L_2 U_2$.

(ii) Explain why this does not contradict the uniqueness of the LU decomposition of a matrix.

(iii) Show that

$$
L_1 = L_2 \begin{pmatrix} 1 & 0 & 0 & 0 \\ 0 & 2 & 0 & 0 \\ 0 & 0 & 3 & 0 \\ 0 & 0 & 0 & 4 \end{pmatrix} \quad \text{and} \quad U_1 = \begin{pmatrix} 1 & 0 & 0 & 0 \\ 0 & 1/2 & 0 & 0 \\ 0 & 0 & 1/3 & 0 \\ 0 & 0 & 0 & 1/4 \end{pmatrix} U_2.
$$

2. Let L_1 and L_2 be nonsingular lower triangular matrices and let U_1 and U_2 be nonsingular upper triangular matrices. If $L_1 U_1 = L_2 U_2$, show that there exists a nonsingular diagonal matrix D such that

$$
L_1 = L_2 D \quad \text{and} \quad U_1 = D^{-1} U_2.
$$

31

3. Let

$$A = \begin{pmatrix} 2 & -1 & 0 & 1 \\ -2 & 0 & 1 & -1 \\ 4 & -1 & 0 & 1 \\ 4 & -3 & 0 & 2 \end{pmatrix}.$$

(i) Show that the LU decomposition with row pivoting of the matrix A is given by $PA = LU$, where

$$P = \begin{pmatrix} 0 & 0 & 1 & 0 \\ 0 & 0 & 0 & 1 \\ 0 & 1 & 0 & 0 \\ 1 & 0 & 0 & 0 \end{pmatrix};$$

$$L = \begin{pmatrix} 1 & 0 & 0 & 0 \\ 1 & 1 & 0 & 0 \\ -0.5 & 0.25 & 1 & 0 \\ 0.5 & 0.25 & 0 & 1 \end{pmatrix}; \quad U = \begin{pmatrix} 4 & -1 & 0 & 1 \\ 0 & -2 & 0 & 1 \\ 0 & 0 & 1 & -0.75 \\ 0 & 0 & 0 & 0.25 \end{pmatrix}.$$

(ii) Solve $Ax = b$, where $b = \begin{pmatrix} 3 \\ -1 \\ 0 \\ 2 \end{pmatrix}$.

(iii) Find A^{-1}, the inverse matrix of A.

4. Let $A = \begin{pmatrix} 2 & -1 & 3 & -1 \\ 1 & 0 & -2 & -4 \\ 3 & 1 & 1 & -2 \\ -4 & 1 & 0 & 2 \end{pmatrix}$ and $b = \begin{pmatrix} -1 \\ 0 \\ 1 \\ 2 \end{pmatrix}$.

(i) Find the LU decomposition with row pivoting of the matrix A;

(ii) Use the linear solver linear_solve_lu_row_pivoting to solve the linear system $Ax = b$.

(iii) Let $P_1 = \begin{pmatrix} 1 & 0 & 0 & 0 \\ 0 & 0 & 0 & 1 \\ 0 & 1 & 0 & 0 \\ 0 & 0 & 1 & 0 \end{pmatrix}$ and $P_2 = \begin{pmatrix} 0 & 1 & 0 & 0 \\ 0 & 0 & 0 & 1 \\ 1 & 0 & 0 & 0 \\ 0 & 0 & 1 & 0 \end{pmatrix}$ be permutation

matrices, and let

$$L_1 = \begin{pmatrix} 1 & 0 & 0 & 0 \\ 0 & 1 & 0 & 0 \\ -0.6667 & -0.5833 & 1 & 0 \\ 0.3333 & -0.5833 & 0.1429 & 1 \end{pmatrix};$$

$$U_1 = \begin{pmatrix} 3 & 2 & -1 & -1 \\ 0 & -4 & 2 & 1 \\ 0 & 0 & -3.5 & -0.0833 \\ 0 & 0 & 0 & 1.9286 \end{pmatrix}.$$

Show that

$$P_1 \, A \, P_2 \; = \; L_1 \, U_1.$$

(iv) Use forward substitution to solve $L_1 y = P_1 b$, and use backward substitution to solve $U_1 x_1 = y$. Let $x_2 = P_2 x_1$. Show that x_2 is the same as the solution x to $Ax = b$ obtained at (ii), and explain why this happens.

Note: The LU decomposition with full pivoting of A is of the form $P_1 A P_2 = L_1 U_1$.

5. Find the LU decomposition without pivoting of the matrix

$$\begin{pmatrix} 1 & 0 & 0 & 0 & 1 \\ -1 & 1 & 0 & 0 & 1 \\ -1 & -1 & 1 & 0 & 1 \\ -1 & -1 & -1 & 1 & 1 \\ -1 & -1 & -1 & -1 & 1 \end{pmatrix}.$$

Note: This is the 5×5 version of the classic example of a matrix whose LU decomposition is unstable.

6. Let

$$A = \begin{pmatrix} 2 & -1 & 1 \\ -2 & 1 & 3 \\ 4 & 0 & -1 \end{pmatrix}.$$

(i) Show that the 2×2 leading principal minor of A is 0, i.e., show that

$$\det \begin{pmatrix} 2 & -1 \\ -2 & 1 \end{pmatrix} = 0.$$

(ii) Attempt to do the LU decomposition without pivoting of the matrix A, and show that the division by $U(2, 2)$ cannot be performed when trying to compute the second row of L.

(iii) Show that the matrix A is nonsingular, and compute the LU decomposition with row pivoting of A.

7. The LU decomposition with column pivoting of an $n \times n$ nonsingular matrix A is $AP = LU$, where P is an $n \times n$ permutation matrix, L is an $n \times n$ lower triangular matrix with all entries on the main diagonal equal to 1, and U is an $n \times n$ upper triangular matrix. Write a pseudocode for solving linear systems of the form $Ax = b$ by using the LU decomposition with column pivoting of A.

8. Write the pseudocode for the forward substitution corresponding to a lower triangular banded matrix of band m, i.e., for solving $Lx = b$ where b is an $n \times 1$ vector and L is an $n \times n$ lower triangular matrix such that

$$L(j, k) = 0, \quad \forall \, 1 \le j, k \le n \quad \text{with} \quad j - k > m.$$

What is the corresponding operation count?

9. Write the pseudocode for the backward substitution corresponding to an upper triangular banded matrix of band m, i.e., for solving $Ux = b$ where b is an $n \times 1$ vector and U is an $n \times n$ upper triangular matrix such that

$$U(j, k) = 0, \quad \forall \, 1 \le j, k \le n \quad \text{with} \quad k - j > m.$$

What is the corresponding operation count?

10. Write the pseudocode for the LU decomposition without pivoting for banded matrices of band m. What is the corresponding operation count?

Use the fact that the L and U factors from the LU decomposition without pivoting of a banded matrix of band m are a banded lower triangular matrix of band m and a banded upper triangular matrix of band m, respectively.

11. What is the operation count for solving a linear system corresponding to a banded matrix of band m using a linear solver based on the LU decomposition without pivoting of the matrix?

12. The values of the following coupon bonds with face value $100 are given:

Bond Type	Coupon Rate	Bond Price
10 months semiannual	3%	$101.30
16 months semiannual	4%	$102.95
22 months annual	6%	$107.35
22 months semiannual	5%	$105.45

(i) List the cash flows and cash flow dates for each bond.

(ii) Identify the 4×4 matrix and the 4×1 right hand side vector corresponding to the linear system whose solution are the 4 months, 10 months, 16 months, and 22 months discount factors.

(iii) Find the 4 months, 10 months, 16 months, and 22 months discount factors.

13. The values of the following coupon bonds with face value $100 are given:

Bond Type	Coupon Rate	Bond Price
6 months semiannual	0	$98.50
1 year semiannual	3%	$101.00
18 months semiannual	5%	$102.00
2 years semiannual	3%	$103.50

(i) List the cash flows and cash flow dates for each bond.

(ii) Find the 6 months, 1 year, 18 months, and 2 years discount factors.

14. The values of the following coupon bonds with face value $100 are given:

Bond Type	Coupon Rate	Bond Price
9 months semiannual	2%	$100.80
15 months semiannual	4%	$103.50
15 months annual	5%	$107.50
21 months semiannual	5%	$110.50

(i) List the cash flows and cash flow dates for each bond.

(ii) Find the 3 months, 9 months, 15 months, and 21 months discount factors.

15. The following discount factors are obtained by fitting market data:

Date	Discount Factor
2 months	99.80
5 months	99.35
11 months	98.20
15 months	97.75

The overnight rate is 1%.

(i) What is the linear system that has to be solved for the cubic spline interpolation of the zero rate curve?

(ii) Use cubic spline interpolation to find a zero rate curve for all times less than 15 months matching the discount factors above.

(iii) Find the value of a 13 months quarterly bond with 2.5% coupon rate.

Note: A quarterly coupon bond with face value $100, coupon rate C, and maturity T pays the holder of the bond a coupon payment equal to $\frac{C}{4} \cdot 100$ every three months, except at maturity. The final payment at maturity T is equal to the face value of the bond plus one coupon payment, i.e., $100 + \frac{C}{4}100$.

16. Consider three assets with the following expected values, standard deviations, and correlations of their returns:

$$\mu_1 = 0.10; \quad \sigma_1 = 0.15; \quad \rho_{1,2} = -0.25;$$
$$\mu_2 = 0.15; \quad \sigma_2 = 0.30; \quad \rho_{2,3} = 0.20;$$
$$\mu_3 = 0.20; \quad \sigma_3 = 0.35; \quad \rho_{1,3} = 0.30.$$

(i) Find the covariance matrix M of the returns of the three assets.

(ii) A minimum variance portfolio with 16% expected rate of return and fully invested in the three assets (i.e., with no cash position) can be set up by investing a percentage w_i of the total value of the portfolio in asset i, with $i = 1 : 3$, where w_1, w_2, w_3 can be found by solving the following linear system:

$$\begin{pmatrix} 2M & 1 & \mu \\ 1^t & 0 & 0 \\ \mu^t & 0 & 0 \end{pmatrix} \begin{pmatrix} w \\ \lambda_1 \\ \lambda_2 \end{pmatrix} = \begin{pmatrix} 0 \\ 1 \\ \mu_P \end{pmatrix}, \tag{2.1}$$

where $w = \begin{pmatrix} w_1 \\ w_2 \\ w_3 \end{pmatrix}$ and

$$\mu_P = 0.16; \quad \mu = \begin{pmatrix} 0.1 \\ 0.15 \\ 0.2 \end{pmatrix}; \quad \mathbf{1} = \begin{pmatrix} 1 \\ 1 \\ 1 \end{pmatrix}.$$

Show that the matrices from the LU decomposition with row pivoting of the matrix on the left hand side of (2.1) are

$$P = \begin{pmatrix} 0 & 0 & 0 & 1 & 0 \\ 0 & 1 & 0 & 0 & 0 \\ 0 & 0 & 1 & 0 & 0 \\ 1 & 0 & 0 & 0 & 0 \\ 0 & 0 & 0 & 0 & 1 \end{pmatrix};$$

$$L = \begin{pmatrix} 1 & 0 & 0 & 0 & 0 \\ -0.0225 & 1 & 0 & 0 & 0 \\ 0.0315 & 0.051852 & 1 & 0 & 0 \\ 0.045 & -0.333333 & 0.038067 & 1 & 0 \\ 0.1 & 0.246914 & 0.400056 & -0.482738 & 1 \end{pmatrix};$$

$$U = \begin{pmatrix} 1 & 1 & 1 & 0 & 0 \\ 0 & 0.2025 & 0.0645 & 1 & 0.15 \\ 0 & 0 & 0.210555 & 0.948148 & 0.192222 \\ 0 & 0 & 0 & 1.297240 & 0.142683 \\ 0 & 0 & 0 & 0 & -0.045059 \end{pmatrix}.$$

(iii) Find the weights of each asset in the minimum variance portfolio with 16% expected return. Find the standard deviation of the return of this portfolio.

(iv) Show that the two portfolios below have 16% expected return, and compute the standard deviation of the returns of each portfolio:

• 30% invested in asset 1, 20% invested in asset 2, 50% invested in asset 3;

• 50% invested in asset 1, 70% invested in asset 3, and short an amount equal to 20% of the value of the portfolio of asset 2.

Hint: If w_1, w_2, and w_3 denote the weights of asset 1, of asset 2, and of asset 3 in the portfolio, respectively, then the expected return of the portfolio is

$$w_1\mu_1 + w_2\mu_2 + w_3\mu_3.$$

2.2 Solutions to Chapter 2 Exercises

Problem 1: Let

$$
L_1 = \begin{pmatrix} 1 & 0 & 0 & 0 \\ -1 & 2 & 0 & 0 \\ 2 & -2 & 3 & 0 \\ 2 & 2 & -3 & 4 \end{pmatrix}; \quad U_1 = \begin{pmatrix} 2 & -1 & 0 & 1 \\ 0 & -1/2 & 1/2 & 0 \\ 0 & 0 & 1/3 & -1/3 \\ 0 & 0 & 0 & -1/4 \end{pmatrix}.
$$

$$
L_2 = \begin{pmatrix} 1 & 0 & 0 & 0 \\ -1 & 1 & 0 & 0 \\ 2 & -1 & 1 & 0 \\ 2 & 1 & -1 & 1 \end{pmatrix}; \quad U_2 = \begin{pmatrix} 2 & -1 & 0 & 1 \\ 0 & -1 & 1 & 0 \\ 0 & 0 & 1 & -1 \\ 0 & 0 & 0 & -1 \end{pmatrix}.
$$

(i) Show that $L_1 U_1 = L_2 U_2$.

(ii) Explain why this does not contradict the uniqueness of the LU decomposition of a matrix.

(iii) Show that

$$
L_1 = L_2 \begin{pmatrix} 1 & 0 & 0 & 0 \\ 0 & 2 & 0 & 0 \\ 0 & 0 & 3 & 0 \\ 0 & 0 & 0 & 4 \end{pmatrix}; \quad U_1 = \begin{pmatrix} 1 & 0 & 0 & 0 \\ 0 & 1/2 & 0 & 0 \\ 0 & 0 & 1/3 & 0 \\ 0 & 0 & 0 & 1/4 \end{pmatrix} U_2. \tag{2.2}
$$

Solution: (i) By matrix multiplication, we find that

$$
L_1 U_1 = L_2 U_2 = \begin{pmatrix} 2 & -1 & 0 & 1 \\ -2 & 0 & 1 & -1 \\ 4 & -1 & 0 & 1 \\ 4 & -3 & 0 & 2 \end{pmatrix}. \tag{2.3}
$$

(ii) Let

$$
A = \begin{pmatrix} 2 & -1 & 0 & 1 \\ -2 & 0 & 1 & -1 \\ 4 & -1 & 0 & 1 \\ 4 & -3 & 0 & 2 \end{pmatrix}.
$$

The LU decomposition without pivoting of the matrix A is $A = L_2 U_2$; see (2.3). Note that $A = L_1 U_1$ is not an LU decomposition without pivoting of A, since the lower triangular matrix L_1 does not have all the main diagonal entries equal to 1.

(iii) The relationships from (2.2) can be obtained directly by multiplication.

Alternatively, note that multiplying the matrix L_2 to the right by a diagonal matrix multiplies the columns of L_2 by the corresponding diagonal entries, and therefore

$$
L_2 \begin{pmatrix} 1 & 0 & 0 & 0 \\ 0 & 2 & 0 & 0 \\ 0 & 0 & 3 & 0 \\ 0 & 0 & 0 & 4 \end{pmatrix} = \begin{pmatrix} 1 & 0 & 0 & 0 \\ -1 & 1 & 0 & 0 \\ 2 & -1 & 1 & 0 \\ 2 & 1 & -1 & 1 \end{pmatrix} \begin{pmatrix} 1 & 0 & 0 & 0 \\ 0 & 2 & 0 & 0 \\ 0 & 0 & 3 & 0 \\ 0 & 0 & 0 & 4 \end{pmatrix}
$$

$$
= \begin{pmatrix} 1 & 0 & 0 & 0 \\ -1 & 2 \cdot 1 & 0 & 0 \\ 2 & 2 \cdot (-1) & 3 \cdot 1 & 0 \\ 2 & 2 \cdot 1 & 3 \cdot (-1) & 4 \cdot 1 \end{pmatrix}
$$

$$= \begin{pmatrix} 1 & 0 & 0 & 0 \\ -1 & 2 & 0 & 0 \\ 2 & -2 & 3 & 0 \\ 2 & 2 & -3 & 4 \end{pmatrix}$$

$$= L_1.$$

Also, note that multiplying the matrix U_2 to the left by a diagonal matrix multiplies the rows of U_2 by the corresponding diagonal entries, and therefore

$$\begin{pmatrix} 1 & 0 & 0 & 0 \\ 0 & 1/2 & 0 & 0 \\ 0 & 0 & 1/3 & 0 \\ 0 & 0 & 0 & 1/4 \end{pmatrix} U_2 = \begin{pmatrix} 1 & 0 & 0 & 0 \\ 0 & 1/2 & 0 & 0 \\ 0 & 0 & 1/3 & 0 \\ 0 & 0 & 0 & 1/4 \end{pmatrix} \begin{pmatrix} 2 & -1 & 0 & 1 \\ 0 & -1 & 1 & 0 \\ 0 & 0 & 1 & -1 \\ 0 & 0 & 0 & -1 \end{pmatrix}$$

$$= \begin{pmatrix} 2 & -1 & 0 & 1 \\ 0 & 1/2 \cdot (-1) & 1/2 \cdot 1 & 0 \\ 0 & 0 & 1/3 \cdot 1 & 1/3 \cdot (-1) \\ 0 & 0 & 0 & 1/4 \cdot (-1) \end{pmatrix}$$

$$= \begin{pmatrix} 2 & -1 & 0 & 1 \\ 0 & -1/2 & 1/2 & 0 \\ 0 & 0 & 1/3 & -1/3 \\ 0 & 0 & 0 & -1/4 \end{pmatrix}$$

$$= U_1. \quad \square$$

Problem 2: Let L_1 and L_2 be nonsingular lower triangular matrices and let U_1 and U_2 be nonsingular upper triangular matrices. If $L_1U_1 = L_2U_2$, show that there exists a nonsingular diagonal matrix D such that

$$L_1 = L_2D \quad \text{and} \quad U_1 = D^{-1}U_2.$$

Solution: If

$$L_1U_1 = L_2U_2, \tag{2.4}$$

and since the matrices L_2 and U_1 are nonsingular, we can multiply (2.4) by L_2^{-1} on the left and by U_1^{-1} on the right to obtain that

$$L_1U_1 = L_2U_2 \iff L_2^{-1} \cdot (L_1U_1) \cdot U_1^{-1} = L_2^{-1} \cdot (L_2U_2) \cdot U_1^{-1}$$
$$\iff (L_2^{-1}L_1) \cdot (U_1U_1^{-1}) = (L_2^{-1}L_2) \cdot (U_2U_1^{-1})$$
$$\iff L_2^{-1}L_1 = U_2U_1^{-1}, \tag{2.5}$$

since $U_1U_1^{-1} = I$ and $L_2^{-1}L_2 = I$.

Recall that the inverse of a lower triangular matrix is lower triangular and the inverse of an upper triangular matrix is upper triangular. Also, recall that the product of two lower triangular matrices is lower triangular, and the product of two upper triangular matrices is upper triangular. Thus, the matrix L_2^{-1} is lower triangular and therefore the matrix $L_2^{-1}L_1$ is also lower triangular. Similarly, the matrix U_1^{-1} is upper triangular and therefore the matrix $U_2U_1^{-1}$ is also upper triangular. Since the matrices $L_2^{-1}L_1$ and $U_2U_1^{-1}$ are equal, see (2.5), it follows that they must be diagonal matrices.

Let D be a diagonal matrix such that

$$D = L_2^{-1}L_1 = U_2 U_1^{-1}. \tag{2.6}$$

Note that the matrix D is nonsingular, since it is the product of two nonsingular matrices.

Since the matrices L_1, L_2, U_1, and U_1 are nonsingular, it follows from (2.6) that

$$
\begin{aligned}
D = L_2^{-1}L_1 \quad &\Longleftrightarrow \quad L_2 \cdot D = L_2 \cdot (L_2^{-1}L_1) \\
&\Longleftrightarrow \quad L_2 D = (L_2 L_2^{-1}) \cdot L_1 \\
&\Longleftrightarrow \quad L_2 D = L_1; \tag{2.7} \\
D = U_2 U_1^{-1} \quad &\Longleftrightarrow \quad D^{-1} \cdot D \cdot U_1 = D^{-1} \cdot (U_2 U_1^{-1}) \cdot U_1 \\
&\Longleftrightarrow \quad (D^{-1}D) \cdot U_1 = D^{-1}U_2 \cdot (U_1^{-1}U_1) \\
&\Longleftrightarrow \quad U_1 = D^{-1}U_2, \tag{2.8}
\end{aligned}
$$

since $L_2 L_2^{-1} = I$, $D^{-1}D = I$, and $U_1^{-1}U_1 = I$.

From (2.7) and (2.8), we conclude that $L_1 = L_2 D$ and $U_1 = D^{-1}U_2$. $\square$

Problem 3: Let

$$A = \begin{pmatrix} 2 & -1 & 0 & 1 \\ -2 & 0 & 1 & -1 \\ 4 & -1 & 0 & 1 \\ 4 & -3 & 0 & 2 \end{pmatrix}. \tag{2.9}$$

(i) Show that the LU decomposition with row pivoting of the matrix A is given by $PA = LU$, where

$$P = \begin{pmatrix} 0 & 0 & 1 & 0 \\ 0 & 0 & 0 & 1 \\ 0 & 1 & 0 & 0 \\ 1 & 0 & 0 & 0 \end{pmatrix}; \tag{2.10}$$

$$L = \begin{pmatrix} 1 & 0 & 0 & 0 \\ 1 & 1 & 0 & 0 \\ -0.5 & 0.25 & 1 & 0 \\ 0.5 & 0.25 & 0 & 1 \end{pmatrix}; \quad U = \begin{pmatrix} 4 & -1 & 0 & 1 \\ 0 & -2 & 0 & 1 \\ 0 & 0 & 1 & -0.75 \\ 0 & 0 & 0 & 0.25 \end{pmatrix}. \tag{2.11}$$

(ii) Solve $Ax = b$, where $b = \begin{pmatrix} 3 \\ -1 \\ 0 \\ 2 \end{pmatrix}$.

(iii) Find A^{-1}, the inverse matrix of A.

Solution: (i) The matrix P given by (2.10) is a permutation matrix, the matrix L given by (2.11) is a lower triangular matrix with entries equal to 1 on the main diagonal, and the matrix U given by (2.11) is an upper triangular matrix. By direct computation, we find that $PA = LU$.

(ii) The solution to the linear system $Ax = b$ is

$$x = \text{linear\_solve\_lu\_row\_pivoting}(A, b) = \begin{pmatrix} -1.5 \\ 4 \\ 6 \\ 10 \end{pmatrix}. \qquad (2.12)$$

(iii) Let $A^{-1} = \text{col}\,(c_k)_{k=1:4}$ be the column form of the inverse matrix of the nonsingular matrix A given by (2.9). The columns c_k, $k = 1 : 4$ of A^{-1} can be found by solving the linear systems $Ac_k = e_k$ for $k = 1 : 4$, where e_k denotes the k-th column of the 4×4 identity matrix,[1] i.e.,

$$Ac_1 = \begin{pmatrix} 1 \\ 0 \\ 0 \\ 0 \end{pmatrix}; \quad Ac_2 = \begin{pmatrix} 0 \\ 1 \\ 0 \\ 0 \end{pmatrix}; \quad Ac_3 = \begin{pmatrix} 0 \\ 0 \\ 1 \\ 0 \end{pmatrix}; \quad Ac_4 = \begin{pmatrix} 0 \\ 0 \\ 0 \\ 1 \end{pmatrix}.$$

We find that

$$c_1 = \begin{pmatrix} -0.5 \\ 2 \\ 3 \\ 4 \end{pmatrix}; \quad c_1 = \begin{pmatrix} 0 \\ 0 \\ 1 \\ 0 \end{pmatrix}; \quad c_1 = \begin{pmatrix} 0.5 \\ 0 \\ 0 \\ -1 \end{pmatrix}; \quad c_1 = \begin{pmatrix} 0 \\ -1 \\ -1 \\ -1 \end{pmatrix},$$

and therefore

$$A^{-1} = \text{col}\,(c_1 \quad c_2 \quad c_3 \quad c_4) = \begin{pmatrix} -0.5 & 0 & 0.5 & 0 \\ 2 & 0 & 0 & -1 \\ 3 & 1 & 0 & -1 \\ 4 & 0 & -1 & -1 \end{pmatrix}. \quad \Box$$

Problem 4: Let $A = \begin{pmatrix} 2 & -1 & 3 & -1 \\ 1 & 0 & -2 & -4 \\ 3 & 1 & 1 & -2 \\ -4 & 1 & 0 & 2 \end{pmatrix}$ and $b = \begin{pmatrix} -1 \\ 0 \\ 1 \\ 2 \end{pmatrix}.$

(i) Find the LU decomposition with row pivoting of the matrix A;

(ii) Use the linear solver linear_solve_lu_row_pivoting to solve the linear system $Ax = b$.

---

[1]Solving $Ac_k = e_k$ for $k = 1 : 4$ can be done efficiently by first computing the LU decomposition with row pivoting of A, i.e., $[P, L, U] = \text{lu\_row\_pivoting}(A)$, which gives the matrices P, L, and U from (2.10) and (2.11), followed by the "for" loop

$$\text{for } k = 1 : 4$$
$$y = \text{forward\_subst}\,(L, Pe_k);$$
$$c_k = \text{backward\_subst}\,(U, y);$$
$$\text{end}$$

(iii) Let $P_1 = \begin{pmatrix} 1 & 0 & 0 & 0 \\ 0 & 0 & 0 & 1 \\ 0 & 1 & 0 & 0 \\ 0 & 0 & 1 & 0 \end{pmatrix}$ and $P_2 = \begin{pmatrix} 0 & 1 & 0 & 0 \\ 0 & 0 & 0 & 1 \\ 1 & 0 & 0 & 0 \\ 0 & 0 & 1 & 0 \end{pmatrix}$ be permutation ma-

trices, and let

$$L_1 = \begin{pmatrix} 1 & 0 & 0 & 0 \\ 0 & 1 & 0 & 0 \\ -0.6667 & -0.5833 & 1 & 0 \\ 0.3333 & -0.5833 & 0.1429 & 1 \end{pmatrix};$$

$$U_1 = \begin{pmatrix} 3 & 2 & -1 & -1 \\ 0 & -4 & 2 & 1 \\ 0 & 0 & -3.5 & -0.0833 \\ 0 & 0 & 0 & 1.9286 \end{pmatrix}.$$

Show that

$$P_1\, A\, P_2 \;=\; L_1\, U_1.$$

(iv) Use forward substitution to solve $L_1 y = P_1 b$, and use backward substitution to solve $U_1 x_1 = y$. Let $x_2 = P_2 x_1$. Show that x_2 is the same as the solution x to $Ax = b$ obtained at (ii), and explain why this happens.

Note: The LU decomposition with full pivoting of A is of the form $P_1 A P_2 = L_1 U_1$.

Solution: (i) The LU decomposition with row pivoting of the matrix A is $PA = LU$ where the permutation matrix P, the lower triangular matrix L, and the upper triangular matrix U are given by

$$P = \;=\; \begin{pmatrix} 0 & 0 & 0 & 1 \\ 0 & 0 & 1 & 0 \\ 1 & 0 & 0 & 0 \\ 0 & 1 & 0 & 0 \end{pmatrix};$$

$$L \;=\; \begin{pmatrix} 1 & 0 & 0 & 0 \\ -0.75 & 1 & 0 & 0 \\ -0.5 & -0.2857 & 1 & 0 \\ -0.25 & 0.1429 & -0.6522 & 1 \end{pmatrix};$$

$$U \;=\; \begin{pmatrix} -4 & 1 & 0 & 2 \\ 0 & 1.75 & 1 & -0.5 \\ 0 & 0 & 3.2857 & -0.1429 \\ 0 & 0 & 0 & -3.5217 \end{pmatrix}.$$

(ii) The solution to the linear system $Ax = b$ is

$$x \;=\; \text{linear\_solve\_lu\_row\_pivoting}(A, b) \;=\; \begin{pmatrix} -0.2716 \\ 1.2593 \\ 0.2099 \\ -0.1728 \end{pmatrix}. \tag{2.13}$$

(iii) By direct computation, we obtain that

$$P_1\, A\, P_2 \;=\; \begin{pmatrix} 3 & 2 & -1 & -1 \\ 0 & -4 & 2 & 1 \\ -2 & 1 & -4 & 0 \\ 1 & 3 & -2 & 1 \end{pmatrix} \;=\; L_1\, U_1.$$

(iv) Using forward substitution, we find that the solution y to $L_1 y = P_1 b$ is

$$ y \;=\; \text{forward\_subst}(L_1, P_1 b) \;=\; \begin{pmatrix} -1 \\ 2 \\ 0.5 \\ 2.4285 \end{pmatrix}, $$

and, using backward substitution, we find that the solution x_1 to $U_1 x_1 = y$ is

$$ x_1 \;=\; \text{backward\_subst}(U_1, y) \;=\; \begin{pmatrix} 0.2099 \\ -0.2716 \\ -0.1728 \\ 1.2592 \end{pmatrix}. $$

Then,

$$ x_2 \;=\; P_2 x_1 \;=\; \begin{pmatrix} -0.2716 \\ 1.2592 \\ 0.2099 \\ -0.1728 \end{pmatrix}, $$

which is the same (up to the roundoff error at the fourth decimal) as the solution x to $Ax = b$ given by (2.13).

To see why x_2 is a solution to $Ax = b$, note that solving $Ax = b$ is the same as solving

$$ P_1 A x \;=\; P_1 b, \tag{2.14} $$

since P_1 is a nonsingular matrix.

Denote by x_1 the vector such that

$$ x \;=\; P_2 x_1; \tag{2.15} $$

such a vector x_1 exists since P_2 is a nonsingular matrix.[2] From (2.14) and (2.15), it follows that

$$ P_1 A P_2 x_1 \;=\; P_1 b, \tag{2.16} $$

and, since

$$ P_1 \, A \, P_2 \;=\; L_1 \, U_1, \tag{2.17} $$

we find from (2.16) and (2.17) that

$$ L_1 U_1 x_1 \;=\; P_1 b. \tag{2.18} $$

Solving (2.18) is the same as solving

$$ L_1 y \;=\; P_1 b $$

for y, and then solving

$$ U_1 x_1 = y $$

for x_1.

In other words, the solution x to the linear system $Ax = b$ can be computed as $x = P_2 x_1$, where x_1 is obtained by solving $U_1 x_1 = y$ where y is the solution to $L_1 y = P_1 b$. □

---

[2]In fact, since P_2 is an orthogonal matrix and therefore $P_2^{-1} = P_2^t$, the vector x_1 is given by $x_1 = P_2^{-1} x = P_2^t x$.

Problem 5: Find the LU decomposition without pivoting of the matrix

$$\begin{pmatrix} 1 & 0 & 0 & 0 & 1 \\ -1 & 1 & 0 & 0 & 1 \\ -1 & -1 & 1 & 0 & 1 \\ -1 & -1 & -1 & 1 & 1 \\ -1 & -1 & -1 & -1 & 1 \end{pmatrix}.$$

Solution: Let

$$A = \begin{pmatrix} 1 & 0 & 0 & 0 & 1 \\ -1 & 1 & 0 & 0 & 1 \\ -1 & -1 & 1 & 0 & 1 \\ -1 & -1 & -1 & 1 & 1 \\ -1 & -1 & -1 & -1 & 1 \end{pmatrix},$$

and let L and U be the LU factors of the matrix A. Rather than proving that the matrix A has an LU decomposition without pivoting by showing that all the leading principal minors of A are nonzero, we will apply the LU decomposition without pivoting algorithm and obtain that the algorithm will yield matrices L and U such that $A = LU$.

We compute one row of the matrix U and one column of the matrix L at a time, followed by computing an update of the remaining part of the matrix A, and then continue the process recursively.

– By computing the first column of L and the first row of U, we find that

$$L = \begin{pmatrix} 1 & 0 & 0 & 0 & 0 \\ -1 & 1 & 0 & 0 & 0 \\ -1 & \cdot & 1 & 0 & 0 \\ -1 & \cdot & \cdot & 1 & 0 \\ -1 & \cdot & \cdot & \cdot & 1 \end{pmatrix}; \quad U = \begin{pmatrix} 1 & 0 & 0 & 0 & 1 \\ 0 & \cdot & \cdot & \cdot & \cdot \\ 0 & 0 & \cdot & \cdot & \cdot \\ 0 & 0 & 0 & \cdot & \cdot \\ 0 & 0 & 0 & 0 & \cdot \end{pmatrix}$$

and the updated form of the 4×4 matrix $A(2:5, 2:5)$ is

$$\begin{pmatrix} 1 & 0 & 0 & 1 \\ -1 & 1 & 0 & 1 \\ -1 & -1 & 1 & 1 \\ -1 & -1 & -1 & 1 \end{pmatrix} - \begin{pmatrix} -1 \\ -1 \\ -1 \\ -1 \end{pmatrix} (0\ 0\ 0\ 1)$$

$$= \begin{pmatrix} 1 & 0 & 0 & 2 \\ -1 & 1 & 0 & 2 \\ -1 & -1 & 1 & 2 \\ -1 & -1 & -1 & 2 \end{pmatrix}. \tag{2.19}$$

– By computing the second column of L and the second row of U using (2.19), we find that

$$L = \begin{pmatrix} 1 & 0 & 0 & 0 & 0 \\ -1 & 1 & 0 & 0 & 0 \\ -1 & -1 & 1 & 0 & 0 \\ -1 & -1 & \cdot & 1 & 0 \\ -1 & -1 & \cdot & \cdot & 1 \end{pmatrix}; \quad U = \begin{pmatrix} 1 & 0 & 0 & 0 & 1 \\ 0 & 1 & 0 & 0 & 2 \\ 0 & 0 & \cdot & \cdot & \cdot \\ 0 & 0 & 0 & \cdot & \cdot \\ 0 & 0 & 0 & 0 & \cdot \end{pmatrix}$$

and the updated form of the 3×3 matrix $A(3:5,3:5)$ is

$$\begin{pmatrix} 1 & 0 & 2 \\ -1 & 1 & 2 \\ -1 & -1 & 2 \end{pmatrix} - \begin{pmatrix} -1 \\ -1 \\ -1 \end{pmatrix} (0\ 0\ 2) \; = \; \begin{pmatrix} 1 & 0 & 4 \\ -1 & 1 & 4 \\ -1 & -1 & 4 \end{pmatrix}. \qquad (2.20)$$

– By computing the third column of L and the third row of U using (2.20), we find that

$$L = \begin{pmatrix} 1 & 0 & 0 & 0 & 0 \\ -1 & 1 & 0 & 0 & 0 \\ -1 & -1 & 1 & 0 & 0 \\ -1 & -1 & -1 & 1 & 0 \\ -1 & -1 & -1 & \cdot & 1 \end{pmatrix}; \quad U = \begin{pmatrix} 1 & 0 & 0 & 0 & 1 \\ 0 & 1 & 0 & 0 & 2 \\ 0 & 0 & 1 & 0 & 4 \\ 0 & 0 & 0 & \cdot & \cdot \\ 0 & 0 & 0 & 0 & \cdot \end{pmatrix}$$

and the updated form of the 2×2 matrix $A(4:5,4:5)$ is

$$\begin{pmatrix} 1 & 4 \\ -1 & 4 \end{pmatrix} - \begin{pmatrix} -1 \\ -1 \end{pmatrix} (0\ 4) \; = \; \begin{pmatrix} 1 & 8 \\ -1 & 8 \end{pmatrix}. \qquad (2.21)$$

– By computing the fourth column of L and the fourth row of U using (2.21), we find that

$$L = \begin{pmatrix} 1 & 0 & 0 & 0 & 0 \\ -1 & 1 & 0 & 0 & 0 \\ -1 & -1 & 1 & 0 & 0 \\ -1 & -1 & -1 & 1 & 0 \\ -1 & -1 & -1 & -1 & 1 \end{pmatrix}; \quad U = \begin{pmatrix} 1 & 0 & 0 & 0 & 1 \\ 0 & 1 & 0 & 0 & 2 \\ 0 & 0 & 1 & 0 & 4 \\ 0 & 0 & 0 & 1 & 8 \\ 0 & 0 & 0 & 0 & \cdot \end{pmatrix}$$

and the updated form of $A(5,5)$, which is a number, is

$$8 - (-1) \cdot 8 \; = \; 16, \qquad (2.22)$$

From (2.22), it follows that $U(5,5) = 16$, and we conclude that the matrix A has an LU decomposition without pivoting and the L and U factors of A are

$$L = \begin{pmatrix} 1 & 0 & 0 & 0 & 0 \\ -1 & 1 & 0 & 0 & 0 \\ -1 & -1 & 1 & 0 & 0 \\ -1 & -1 & -1 & 1 & 0 \\ -1 & -1 & -1 & -1 & 1 \end{pmatrix}; \quad U = \begin{pmatrix} 1 & 0 & 0 & 0 & 1 \\ 0 & 1 & 0 & 0 & 2 \\ 0 & 0 & 1 & 0 & 4 \\ 0 & 0 & 0 & 1 & 8 \\ 0 & 0 & 0 & 0 & 16 \end{pmatrix}$$

This is the 5×5 version of the classic example of a matrix whose LU decomposition is unstable. For an $n \times n$ matrix A similar to the 5×5 matrix from this example, i.e., with

$$\begin{aligned} A(i,i) &= 1, & \forall\, i = 1:n; \\ A(j,k) &= -1, & \forall\, 1 \le k < j \le n; \\ A(j,k) &= 0, & \forall\, 1 \le j < k \le n-1; \\ A(j,n) &= 1, & \forall\, j = 1:n, \end{aligned}$$

the upper triangular matrix U from the LU decomposition of A is given by

$$U(i,i) = 1, \qquad \forall\, i = 1:n;$$
$$U(j,n) = 2^{j-1}, \quad \forall\, j = 1:n;$$
$$U(j,k) = 0, \qquad \text{else.}$$

Thus, the last column of U $\begin{pmatrix} 1 \\ 2 \\ 4 \\ \vdots \\ 2^{n-1} \end{pmatrix}$. In other words, while all the entries of the

matrix A are uniformly bounded in absolute value by 1, the entries of the upper triangular factor U from the LU decomposition without pivoting of A are not uniformly bounded and grow exponentially with the size of the matrix n.   □

Problem 6: Let
$$A = \begin{pmatrix} 2 & -1 & 1 \\ -2 & 1 & 3 \\ 4 & 0 & -1 \end{pmatrix}.$$

(i) Show that the 2×2 leading principal minor of A is 0, i.e., show that

$$\det \begin{pmatrix} 2 & -1 \\ -2 & 1 \end{pmatrix} = 0. \tag{2.23}$$

(ii) Attempt to do the LU decomposition without pivoting of the matrix A, and show that the division by $U(2,2)$ cannot be performed when trying to compute the second row of L.

(iii) Show that the matrix A is nonsingular, and compute the LU decomposition with row pivoting of A.

Solution: (i) The 2×2 leading principal minor of A is the determinant of the 2×2 upper left corner matrix of A, i.e.,

$$\det \begin{pmatrix} 2 & -1 \\ -2 & 1 \end{pmatrix} = 2 \cdot 1 - (-1) \cdot (-2) = 2 - 2 = 0.$$

(ii) To attempt to do the LU decomposition without pivoting of the matrix A, let L and U be the LU factors of A. Then, $LU = A$, which can be written as

$$\begin{pmatrix} 1 & 0 & 0 \\ L(2,1) & 1 & 0 \\ L(3,1) & L(3,2) & 1 \end{pmatrix} \begin{pmatrix} U(1,1) & U(1,2) & U(1,3) \\ 0 & U(2,2) & U(2,3) \\ 0 & 0 & U(3,3) \end{pmatrix} = \begin{pmatrix} 2 & -1 & 1 \\ -2 & 1 & 3 \\ 4 & 0 & -1 \end{pmatrix}.$$

The entries of the first row of U are given by $U(1,k) = A(1,k)$ for $k = 1:3$, i.e.,

$$U(1,1) = 2; \quad U(1,2) = -1; \quad U(1,3) = 1.$$

The entries of the first column of L are given by $L(k,1) = \frac{A(k,1)}{U(1,1)}$ for $k = 2:3$, i.e.,

$$L(2,1) = \frac{-2}{U(1,1)} = \frac{-2}{2} = -1; \quad L(3,1) = \frac{4}{U(1,1)} = \frac{4}{2} = 2.$$

The current forms of L and U are

$$L = \begin{pmatrix} 1 & 0 & 0 \\ -1 & 1 & 0 \\ 2 & L(3,2) & 1 \end{pmatrix}; \quad U = \begin{pmatrix} 2 & -1 & 1 \\ 0 & U(2,2) & U(2,3) \\ 0 & 0 & U(3,3) \end{pmatrix}.$$

The updated form of the 2×2 matrix $A(2:3,2:3)$ is

$$
\begin{aligned}
&A(2:3,2:3) \\
=\ & A(2:3,2:3) - L(2:3,1)\, U(1,2:3) \\
=\ & \begin{pmatrix} 1 & 3 \\ 0 & -1 \end{pmatrix} - \begin{pmatrix} -1 \\ 2 \end{pmatrix} (-1\ \ 1) \\
=\ & \begin{pmatrix} 1 & 3 \\ 0 & -1 \end{pmatrix} - \begin{pmatrix} 1 & -1 \\ -2 & 2 \end{pmatrix} \\
=\ & \begin{pmatrix} 0 & 4 \\ 2 & -3 \end{pmatrix}.
\end{aligned}
$$

Then,

$$L(2:3,2:3)\, U(2:3,2:3) = \begin{pmatrix} 0 & 4 \\ 2 & -3 \end{pmatrix},$$

which can be written as

$$\begin{pmatrix} 1 & 0 \\ L(3,2) & 1 \end{pmatrix} \begin{pmatrix} U(2,2) & U(2,3) \\ 0 & U(3,3) \end{pmatrix} = \begin{pmatrix} 0 & 4 \\ 2 & -3 \end{pmatrix}. \tag{2.24}$$

From (2.24), we find that $U(2,2) = 0$ and $U(2,3) = 4$, however, solving for $L(3,2)$ is not possible since it would require division by 0, i.e.,

$$L(3,2) = \frac{2}{U(2,2)} = \frac{2}{0}.$$

We conclude that the matrix A does not have an LU decomposition without pivoting. This is due to the fact that a leading principal minor of A is 0; cf. (2.23).

(iii) By direct computation, we find that $\det(A) = -16 \neq 0$. In other words, the matrix A has nonzero determinant and is therefore nonsingular.

The LU decomposition with row pivoting of A is $PA = LU$, where the permutation matrix P, the lower triangular matrix L, and the upper triangular matrix U are given by

$$P = \begin{pmatrix} 0 & 0 & 1 \\ 0 & 1 & 0 \\ 1 & 0 & 0 \end{pmatrix}; \quad L = \begin{pmatrix} 1 & 0 & 0 \\ -0.5 & 1 & 0 \\ 0.5 & -1 & 1 \end{pmatrix}; \quad U = \begin{pmatrix} 4 & 0 & -1 \\ 0 & 1 & 2.5 \\ 0 & 0 & 4 \end{pmatrix}. \ \square$$

Problem 7: The LU decomposition with column pivoting of an $n \times n$ nonsingular matrix A is $AP = LU$, where P is an $n \times n$ permutation matrix, L is an $n \times n$ lower triangular matrix with all entries on the main diagonal equal to 1, and U is an $n \times n$ upper triangular matrix. Write a pseudocode for solving linear systems of the form $Ax = b$ by using the LU decomposition with column pivoting of A.

Solution: Let v be an $n \times 1$ vector such that $x = Pv$. (Note that $v = P^t x$ since the permutation matrix P is orthogonal and $PP^t = I$, and therefore $Pv = PP^t x = x$.)

Then, solving a linear system $Ax = b$ is equivalent to solving

$$APv = b.$$

Since $AP = LU$, this is equivalent to

$$LUv = b.$$

This is the same as solving

$$Ly = b$$

for y, which can be done using forward substitution since the matrix L is lower triangular, and then solving

$$Uv = y$$

for v, which can be done using backward substitution since the matrix U is upper triangular. Then, the solution x of $Ax = b$ is then given by

$$x = Pv.$$

The pseudocode for solving a linear system using the LU decomposition with column pivoting of a matrix can be found in Table 2.1.

Table 2.1: Linear solver using LU decomposition with column pivoting

Function Call:
 $x = $ linear_solve_lu_column_pivoting(A,b)
 	
 Input:
 $A = $ nonsingular square matrix of size n with LU decomposition
 $b = $ column vector of size n
 	
 Output:
 $x = $ solution to $Ax = b$
 	
 $[P, L, U] = $ lu_column_pivoting(A);   // LU decomposition of A
 $y = $ forward_subst(L, b);   // solve $Ly = b$
 $v = $ backward_subst(U, y);   // solve $Uv = y$
 $x = Pv$;

Problem 8: Write the pseudocode for the forward substitution corresponding to a lower triangular banded matrix of band m, i.e., for solving $Lx = b$ where b is an $n \times 1$ vector and L is an $n \times n$ lower triangular matrix such that

$$L(j, k) = 0, \quad \forall\, 1 \leq j, k \leq n \quad \text{with} \quad j - k > m. \tag{2.25}$$

What is the corresponding operation count?

Solution: If x is an $n \times 1$ vector such that $Lx = b$, then, by multiplying the j–th row of L by x, we obtain that

$$\sum_{k=1}^{n} L(j,k)x(k) = b(j). \tag{2.26}$$

Note that $L(j,k) = 0$ for all $k > j$, since L is a lower triangular matrix, and therefore (2.26) becomes

$$\sum_{k=1}^{j} L(j,k)x(k) = b(j). \tag{2.27}$$

Since $L(j,k) = 0$ for all $k < j - m$, see (2.25), it follows that

$$L(j,k) = 0, \quad \forall\, k = 1 : (j - m - 1). \tag{2.28}$$

Using (2.28), we find from (2.27) that

$$\sum_{k=max(1,j-m)}^{j} L(j,k)x(k) = b(j),$$

which can also be written as

$$L(j,j)x(j) + \sum_{k=max(1,j-m)}^{j-1} L(j,k)x(k) = b(j). \tag{2.29}$$

By solving (2.29) for $x(j)$, we find that

$$x(j) = \frac{1}{L(j,j)} \left(b(j) - \sum_{k=max(1,j-m)}^{j-1} L(j,k)x(k) \right); \tag{2.30}$$

note that $L(j,j) \neq 0$ since the lower triangular matrix L is nonsingular.

The Forward Substitution for banded matrices implements formula (2.30) for computing the value of $x(j)$ given the values of $x(1)$, $x(2)$, $\ldots$, $x(j-1)$; see the pseudocode from Table 2.2.

To find the operation count for Forward Substitution for banded matrices, note that the "for" loop from the pseudocode from Table 2.2 can also be written as

```
for j = 2 : n
    sum = 0;
    if j ≤ m
        for k = 1 : (j − 1)
            sum = sum + L(j, k)x(k);
        end
    else if j ≥ m + 1
        for k = (j − m) : (j − 1)
            sum = sum + L(j, k)x(k);
        end
    end
    x(j) = b(j)−sum / L(j,j);
end
```

Table 2.2: Pseudocode for Forward Substitution for banded matrices

```
Function Call:
x = forward_subst_banded(L,b)

Input:
L = n × n nonsingular lower triangular matrix of band m
b = column vector of size n

Output:
x = solution to Lx = b
```

$x(1) = \frac{b(1)}{L(1,1)};$
for $j = 2 : n$
  sum = 0;
  for $k = \max(1, j - m) : (j - 1)$
    sum = sum + $L(j, k)x(k);$
  end
  $x(j) = \frac{b(j) - \text{sum}}{L(j,j)};$
end

The "for" loop

```
for k = 1 : (j − 1)
    sum = sum + L(j, k)x(k);
end
```

requires $2(j - 1)$ operations and is executed if $2 \le j \le m$, for a total of

$$\sum_{j=2}^{m} 2(j-1) = 2\sum_{j=2}^{m}(j-1) = 2\sum_{l=1}^{m-1} l = 2 \cdot \frac{(m-1)m}{2} = m^2 - m \quad (2.31)$$

operations, since $\sum_{l=1}^{m-1} l = \frac{(m-1)m}{2}$; see, e.g., section 10.3.1 from Stefanica [2]. In (2.31), we used the notation $l = j - 1$.

The "for" loop

```
for k = (j − m) : (j − 1)
    sum = sum + L(j, k)x(k);
end
```

requires $2m$ operations and is executed if $m + 1 \le j \le n$ for a total of

$$\sum_{j=m+1}^{n} 2m = 2m(n-m) = 2mn - 2m^2 \quad (2.32)$$

operations.

Note that computing $x(1) = \frac{b(1)}{L(1,1)}$ requires 1 operation, and computing $x(j) = \frac{b(j) - \text{sum}}{L(j,j)}$ requires 2 operations, and it is done $n - 1$ times in the outer "for" loop from Table 2.2, for a total of

$$1 + 2(n-1) = 2n - 1 \quad (2.33)$$

operations.

We conclude from (2.31), (2.32), and (2.33) that the operations count for the Forward Substitution for banded matrices is

$$(m^2 - m) + (2mn - 2m^2) + (2n - 1)$$
$$= \quad 2mn + 2n - m^2 - m - 1. \quad \square \qquad (2.34)$$

Problem 9: Write the pseudocode for the backward substitution corresponding to an upper triangular banded matrix of band m, i.e., for solving $Ux = b$ where b is an $n \times 1$ vector and U is an $n \times n$ upper triangular matrix such that

$$U(j, k) = 0, \quad \forall\, 1 \le j, k \le n \quad \text{with} \quad k - j > m. \qquad (2.35)$$

What is the corresponding operation count?

Solution: If x is an $n \times 1$ vector such that $Ux = b$, then, by multiplying the j–th row of U by x, we obtain that

$$\sum_{k=1}^{n} U(j, k) x(k) = b(j). \qquad (2.36)$$

Note that $U(j, k) = 0$ for all $1 \le k < j$, since U is an upper triangular matrix, and therefore (2.36) becomes

$$\sum_{k=j}^{n} U(j, k) x(k) = b(j). \qquad (2.37)$$

Since $U(j, k) = 0$ for all $k > j + m$, see (2.35), it follows that

$$U(j, k) = 0, \quad \forall\, k = (j + m + 1) : n. \qquad (2.38)$$

Using (2.38), we find from (2.37) that

$$\sum_{k=j}^{min(j+m,n)} U(j, k) x(k) = b(k),$$

which can also be written as

$$U(j, j) x(j) + \sum_{k=j+1}^{min(j+m,n)} U(j, k) x(k) = b(k). \qquad (2.39)$$

By solving (2.39) for $x(j)$, we find that

$$x(j) = \frac{1}{U(j, j)} \left(b(j) - \sum_{k=j+1}^{min(j+m,n)} U(j, k) x(k) \right); \qquad (2.40)$$

note that $U(j, j) \ne 0$ since the upper triangular matrix U is nonsingular.

The Backward Substitution for banded matrices implements formula (2.40) for computing the value of $x(j)$ given the values of $x(n), x(n-1), \ldots, x(j+1)$; see the pseudocode from Table 2.3.

To find the operation count for the Forward Substitution for banded matrices, note that the "for" loop from the pseudocode from Table 2.3 can also be written as

Table 2.3: Pseudocode for Backward Substitution for banded matrices

Function Call:
x = backward_subst_banded(U,b)

Input:
$U = n \times n$ nonsingular upper triangular matrix of band m
b = column vector of size n

Output:
x = solution to $Ux = b$

$x(n) = \frac{b(n)}{U(n,n)}$;
for $j = (n-1) : 1$
  sum = 0;
  for $k = (j+1) : \min(j+m, n)$
    sum = sum + $U(j,k)x(k)$;
  end
  $x(j) = \frac{b(j)-\text{sum}}{U(j,j)}$;
end

---

for $j = (n-1) : 1$
  sum = 0;
  if $j > n - m$
    for $k = (j+1) : n$
      sum = sum + $U(j,k)x(k)$;
    end
  else if $j \leq n - m$
    for $k = (j+1) : (j+m)$
      sum = sum + $U(j,k)x(k)$;
    end
  end
  $x(j) = \frac{b(j)-\text{sum}}{U(j,j)}$;
end

The "for" loop

for $k = (j+1) : n$
  sum = sum + $U(j,k)x(k)$;
end

requires $2(n-j)$ operations and is executed if $j > n - m$, i.e., if $n - m + 1 \leq j \leq n$, for a total of

$$\sum_{j=n-m+1}^{n-1} 2(n-j) = \sum_{l=1}^{m-1} 2l = 2 \cdot \frac{(m-1)m}{2} = m^2 - m \qquad (2.41)$$

operations, since $\sum_{l=1}^{m-1} l = \frac{(m-1)m}{2}$; in (2.41), we used the notation $l = n - j$.
  The "for" loop

```
for k = (j + 1) : (j + m)
    sum = sum + U(j, k)x(k);
end
```

requires $2m$ operations and is executed if $1 \leq j \leq n - m$, i.e., it is executed $n - m$ times, for a total of

$$2m(n - m) \; = \; 2mn - 2m^2 \tag{2.42}$$

operations.

Note that computing $x(1)$ requires 1 operation, and computing $x(j) = \frac{b(j) - \text{sum}}{U(j,j)}$ requires 2 operations, and it is done $n-1$ times in the outer "for" loop from Table 2.3, for a total of

$$1 + 2(n - 1) \; = \; 2n - 1 \tag{2.43}$$

operations.

We conclude from (2.41), (2.42), and (2.43) that the operations count for Backward Substitution for banded matrices is

$$(m^2 - m) \; + \; (2mn - 2m^2) \; + \; (2n - 1)$$
$$= \; 2mn + 2n - m^2 - m - 1, \tag{2.44}$$

which is the same as the operations count for Forward Substitution for banded matrices; cf. (2.34).   □

Problem 10: Write the pseudocode for the LU decomposition without pivoting for banded matrices of band m. What is the corresponding operation count?

Solution: If L and U are the factors from the LU decomposition without pivoting of a banded matrix of band m, then L is a banded lower triangular matrix of band m and U is a banded upper triangular matrix of band m; see, e.g., [1] or [5]. Then, for a banded matrix of band m, computing the updated form of the matrix A after the row i of the matrix U and the column i of the matrix L are computed only requires updating the $m \times m$ matrix $A(i + 1 : i + m, i + 1 : i + m)$, if $i + m \leq n$.

The corresponding pseudocode for the banded LU decomposition without pivoting can be found in Table 2.4.

We now proceed to finding the operation count for the pseudocode from Table 2.4.

If $i + m \leq n$, i.e., for $i \leq n - m$, it follows that $\min(i + m, n) = i + m$ and the outside "for" loop from the pseudocode from Table 2.4 becomes

```
for i = 1 : (n - m)
    for k = i : (i + m)
        U(i, k) = A(i, k);           // compute row i of U
        L(k, i) = A(k, i)/U(i, i);   // compute column i of L
    end
    for j = (i + 1) : (i + m)
        for k = (i + 1) : (i + m)
            A(j, k) = A(j, k) - L(j, i)U(i, k);
        end
    end
end
```

Table 2.4: Pseudocode for banded LU decomposition without pivoting

```
Function Call:
[L, U] = lu_no_pivoting_banded(A)

Input:
A = banded matrix of size n and band m with LU decomposition

Output:
L = banded lower triangular matrix with band m
U = banded upper triangular matrix with band m
such that A = LU

for i = 1 : (n − 1)
    for k = i : min(i + m, n)
        U(i, k) = A(i, k);              // compute row i of U
        L(k, i) = A(k, i)/U(i, i);      // compute column i of L
    end
    for j = (i + 1) : min(i + m, n)
        for k = (i + 1) : min(i + m, n)
            A(j, k) = A(j, k) − L(j, i)U(i, k);
        end
    end
end
L(n, n) = 1; U(n, n) = A(n, n)
```

In this case, at step i, computing the row i of U and computing the column i of L, i.e., going through the "for" loop

```
for k = i : (i + m)
    U(i, k) = A(i, k);
    L(k, i) = A(k, i)/U(i, i);
end
```

requires $m + 1$ operations. Also at step i, the double "for" loop

```
for j = (i + 1) : (i + m)
    for k = (i + 1) : (i + m)
        A(j, k) = A(j, k) − L(j, i)U(i, k);
    end
end
```

to update the matrix $A(i + 1 : i + m), i + 1 : i + m)$ requires

$$\sum_{j=i+1}^{i+m} \sum_{k=i+1}^{i+m} 2 = \sum_{j=i+1}^{i+m} 2m = 2m^2, \tag{2.45}$$

and therefore the number of operations required to go through the outside "for" loop for $i = 1 : (n − m)$ is

$$\sum_{i=1}^{n-m} \left(m + 1 + 2m^2 \right) = (n − m)(2m^2 + m + 1). \tag{2.46}$$

If $i + m > n$, i.e., for $i \geq n - m + 1$, it follows that $\min(i + m, n) = n$ and the outside "for" loop from the pseudocode from Table 2.4 becomes

```
for i = (n − m + 1) : (n − 1)
    for k = i : n
        U(i, k) = A(i, k);           // compute row i of U
        L(k, i) = A(k, i)/U(i, i);   // compute column i of L
    end
    for j = (i + 1) : n
        for k = (i + 1) : n
            A(j, k) = A(j, k) − L(j, i)U(i, k);
        end
    end
end
```

In this case, at step i, computing the row i of U and computing the column i of L, i.e., going through the "for" loop

```
for k = i : n
    U(i, k) = A(i, k);
    L(k, i) = A(k, i)/U(i, i);
end
```

requires

$$n - i + 1 \tag{2.47}$$

operations. Also at step i, the double "for" loop

```
for j = (i + 1) : n
    for k = (i + 1) : n
        A(j, k) = A(j, k) − L(j, i)U(i, k);
    end
end
```

requires

$$\sum_{j=i+1}^{n} \sum_{k=i+1}^{n} 2 = 2(n - i + 1)^2 \tag{2.48}$$

operations.

From (2.47) and (2.48), we obtain that the number of operations required to go through the outside "for" loop for $i = (n - m + 1) : (n - 1)$ is

$$\sum_{i=n-m+1}^{n-1} \left(2(n - i + 1)^2 + (n - i + 1) \right). \tag{2.49}$$

By letting $l = n - i + 1$ in (6.56), we obtain that

$$\sum_{i=n-m+1}^{n-1} \left(2(n - i + 1)^2 + (n - i + 1) \right)$$

$$= \sum_{l=2}^{m} (2l^2 + l) = \sum_{l=1}^{m} (2l^2 + l) - 3$$

$$= 2\sum_{l=1}^{m} l^2 + \sum_{l=1}^{m} l - 3$$

$$= 2\frac{m(m+1)(2m+1)}{6} + \frac{m(m+1)}{2} - 3$$

$$= \frac{2m^3}{3} + \frac{3m^2}{2} + \frac{5m}{6} - 3, \tag{2.50}$$

where we used the facts that $\sum_{l=1}^{m} l = \frac{m(m+1)}{2}$ and $\sum_{l=1}^{m} l^2 = \frac{m(m+1)(2m+1)}{6}$.

Thus, when accounting for the outside "for" loop "for $i = 1 : (n-1)$" from Table 2.4 separately for $i = 1 : (n-m)$ and for $i = (n-m+1) : n$, we obtain using (2.46) and (2.50) that the operation count for the banded LU decomposition pseudocode from Table 2.4 is

$$(n-m)(2m^2 + m + 1) + \frac{2m^3}{3} + \frac{3m^2}{2} + \frac{5m}{6} - 3$$

$$= n(2m^2 + m + 1) - \frac{4m^3}{3} + \frac{m^2}{2} - \frac{m}{6} - 3. \quad \square \tag{2.51}$$

Problem 11: What is the operation count for solving a linear system corresponding to a banded matrix of band m using a linear solver based on the LU decomposition without pivoting of the matrix?

Solution: Let A be a banded nonsingular square matrix of size n and band m, and let b be a column vector of size n. If the matrix A has an LU decomposition without pivoting $A = LU$, then solving a linear system $Ax = b$ is equivalent to solving $LUx = b$. This can be done by solving $Ly = b$ for y using the banded forward substitution routine from Table 2.2, and then solving $Ux = y$ for x using the banded backward substitution routine from Table 2.3.

The pseudocode for solving a linear system corresponding to a banded matrix with LU decomposition without pivoting can be found in Table 2.5.

Table 2.5: Banded LU linear solver without pivoting

Function Call:
$x =$ linear_solve_LU_no_pivoting_banded(A,b)

Input:
$A =$ nonsingular banded matrix of size n and band m with LU decomposition
$b =$ column vector of size n

Output:
$x =$ solution to $Ax = b$

$[L, U] =$ lu_no_pivoting_banded(A);
$y =$ forward_subst_banded(L, b);
$x =$ backward_subst_banded(U, y);

The operation count for the pseudocode from Table 2.5 is as follows:

• $n(2m^2 + m + 1) - \frac{4m^3}{3} + \frac{m^2}{2} - \frac{m}{6} - 3$ for $[L, U] = $ lu_no_pivoting_banded(A), the banded LU decomposition of A; cf. (2.51);

• $2mn + 2n - m^2 - m - 1$ for $y = $ forward_subst_banded(L, b), the banded forward substitution for solving $Ly = b$; cf. (2.34);

• $2mn + 2n - m^2 - m - 1$ for $x = $ backward_subst_banded(U, y), the banded backward substitution for solving $Ux = y$; cf. (2.44),

for a total operation count of

$$\left(n(2m^2 + m + 1) - \frac{4m^3}{3} + \frac{m^2}{2} - \frac{m}{6} - 3\right) + 2\left(2mn + 2n - m^2 - m - 1\right)$$

$$= \quad n(2m^2 + 5m + 5) - \frac{4m^3}{3} - \frac{3m^2}{2} - \frac{13m}{6} - 5. \quad \square$$

Problem 12: The values of the following coupon bonds with face value $100 are given:

Bond Type	Coupon Rate	Bond Price
10 months semiannual	3%	$101.30
16 months semiannual	4%	$102.95
22 months annual	6%	$107.35
22 months semiannual	5%	$105.45

(i) List the cash flows and cash flow dates for each bond.

(ii) Identify the 4×4 matrix and the 4×1 right hand side vector corresponding to the linear system whose solution are the 4 months, 10 months, 16 months, and 22 months discount factors.

(iii) Find the 4 months, 10 months, 16 months, and 22 months discount factors.

Solution: (i) Recall that a semiannual coupon bond with face value $100, coupon rate C, and maturity T pays the holder of the bond a coupon payment equal to $\frac{C}{2} \cdot 100$ every six months, except at maturity; the final payment at maturity T is equal to the face value of the bond plus one coupon payment, i.e., $100 + \frac{C}{2} \cdot 100$.

Also, an annual coupon bond with face value $100, coupon rate C, and maturity T pays the holder of the bond a coupon payment equal to $C \cdot 100$ every year, except at maturity; the final payment at maturity T is equal to the face value of the bond plus one coupon payment, i.e., $100 + C \cdot 100$.

The cash flows and the cash flow dates of the bonds above are recorded in Table 2.6.

(ii) The value of a bond is equal to the sum of the present value of all its future cash flows. In other words, if B is the value of a bond with future cash flows c_i to be paid at times t_i, and if $d_i = \text{Disc}(t_i)$ are the discount factors corresponding to time t_i, for $i = 1 : n$, then

$$B = \sum_{i=1}^{n} c_i \text{Disc}(t_i) = \sum_{i=1}^{n} c_i d_i. \qquad (2.52)$$

The four bonds considered here have exactly four cash flow dates, i.e., 4 months, 10 months, 16 months, and 22 months. Denote by d_1 the 4 months discount factor,

Table 2.6: Bonds cash flows

Maturity	Cash Flow & Date
10 months	$1.50 in 4 months
	$101.50 in 10 months
16 months	$2 in 4 months
	$2 in 10 months
	$102 in 16 months
22 months	$6 in 10 months
	$106 in 22 months
22 months	$2.50 in 4 months
	$2.50 in 10 months
	$2.50 in 16 months
	$102.50 in 22 months

by d_2 the 10 months discount factor, by d_3 the 16 months discount factor, and by d_4 the 22 months discount factor. Using the formula (2.52) for the value of a bond, we find that

$$
\begin{aligned}
101.30 &= 1.50d_1 + 101.50d_2; \\
102.95 &= 2d_1 + 2d_2 + 102d_3; \\
107.35 &= 6d_2 + 106d_4; \\
105.45 &= 2.50d_1 + 2.50d_2 + 2.50d_3 + 102.50d_4.
\end{aligned}
$$

This linear system can be written in matrix notation as $Md = b$, where M is a 4×4 matrix and d and b are column vectors given by

$$
M = \begin{pmatrix} 1.50 & 101.50 & 0 & 0 \\ 2 & 2 & 102 & 0 \\ 0 & 6 & 0 & 106 \\ 2.50 & 2.50 & 2.50 & 102.50 \end{pmatrix}; \quad d = \begin{pmatrix} d_1 \\ d_2 \\ d_3 \\ d_4 \end{pmatrix}; \quad b = \begin{pmatrix} 101.30 \\ 102.95 \\ 107.35 \\ 105.45 \end{pmatrix}.
$$

(iii) The solution of the linear system $Md = b$ is found by using the LU decomposition with row pivoting linear solver, i.e.,

$$
d = \text{linear\_solve\_lu\_row\_pivoting}(M,b) = \begin{pmatrix} 0.9860 \\ 0.9835 \\ 0.9707 \\ 0.9571 \end{pmatrix}.
$$

Thus, the 4 months, 10 months, 16 months, and 22 months discount factors are:

$$
\text{Disc}\left(\frac{4}{12}\right) = 0.9860; \quad \text{Disc}\left(\frac{10}{12}\right) = 0.9835;
$$

$$
\text{Disc}\left(\frac{16}{12}\right) = 0.9707; \quad \text{Disc}\left(\frac{22}{12}\right) = 0.9571. \quad \square
$$

Problem 13: The values of the following coupon bonds with face value $100 are given:

Bond Type	Coupon Rate	Bond Price
6 months semiannual	0	$98.50
1 year semiannual	3%	$101.00
18 months semiannual	5%	$102.00
2 years semiannual	3%	$103.50

(i) List the cash flows and cash flow dates for each bond.

(ii) Find the 6 months, 1 year, 18 months, and 2 years discount factors.

Solution: (i) A semiannual coupon bond with face value $100, coupon rate C, and maturity T pays the holder of the bond a coupon payment equal to $\frac{C}{2} \cdot 100$ every six months, except at maturity. The final payment at maturity T is equal to the face value of the bond plus one coupon payment, i.e., $100 + \frac{C}{2} \cdot 100$.

The cash flows and the cash flow dates of the bonds above are recorded in Table 2.7.

Table 2.7: Bonds cash flows

Maturity	Cash Flow & Date
6 months	$100 in 6 months
1 year	$1.50 in 6 months
	$101.50 in 1 year
18 months	$2.50 in 6 months
	$2.50 in 1 year
	$102.50 in 18 months
2 years	$1.50 in 6 months
	$1.50 in 1 year
	$1.50 in 18 months
	$101.50 in 2 years

(ii) The four bonds considered here have exactly four cash flow dates, i.e., 6 months, 1 year, 18 months, and 2 years. Denote by d_1 the 6 months discount factor, by d_2 the 1 year discount factor, by d_3 the 18 months discount factor, and by d_4 the 2 years discount factor. Using the formula (2.52) for the value of a bond as the sum of the present values of all the future cash flows of the bond, we find that

$$
\begin{aligned}
98.50 &= 100 d_1; \\
101 &= 1.50 d_1 + 101.50 d_2; \\
102 &= 2.50 d_1 + 2.50 d_2 + 102.50 d_3; \\
103.50 &= 1.50 d_1 + 1.50 d_2 + 1.50 d_3 + 101.50 d_4.
\end{aligned}
$$

This linear system can be written in matrix notation as $Ld = b$, where L is a lower triangular matrix, and d and b are column vectors given by

$$
L = \begin{pmatrix} 100 & 0 & 0 & 0 \\ 1.50 & 101.50 & 0 & 0 \\ 2.50 & 2.50 & 102.50 & 0 \\ 1.50 & 1.50 & 1.50 & 101.50 \end{pmatrix}; \quad d = \begin{pmatrix} d_1 \\ d_2 \\ d_3 \\ d_4 \end{pmatrix}; \quad b = \begin{pmatrix} 98.50 \\ 101 \\ 102 \\ 103.50 \end{pmatrix}.
$$

The solution of the linear system $Ld = b$ is found by using forward substitution, i.e.,

$$d = \text{forward\_subst}(L, b) = \begin{pmatrix} 0.9850 \\ 0.9805 \\ 0.9472 \\ 0.9767 \end{pmatrix}.$$

Thus, the 6 months, 1 year, 18 months, and 2 years discount factors are:

$$\text{Disc}\left(\frac{6}{12}\right) = 0.9850; \quad \text{Disc}(1) = 0.9805;$$

$$\text{Disc}\left(\frac{18}{12}\right) = 0.9472; \quad \text{Disc}(2) = 0.9767.$$

Note that, for continuously compounded interest, the discount factor $\text{Disc}(t)$ at time t can be written in terms of the risk–free zero rate $r(0, t)$ at time t as follows:

$$\text{Disc}(t) = e^{-tr(0,t)} = \exp\left(-t \cdot r(0, t)\right), \quad \forall\, t > 0,$$

and therefore

$$r(0, t) = -\frac{1}{t} \ln(\text{Disc}(t)), \quad \forall\, t > 0. \tag{2.53}$$

Formula (2.53) can be used to find the 6 months, 1 year, 18 months, and 2 years continuously compounded zero rates:

$$r\left(\frac{1}{2}\right) = -2\ln\left(\text{Disc}\left(\frac{1}{2}\right)\right) = 0.0302 = 3.02\%;$$

$$r(0, 1) = -\ln(\text{Disc}(1)) = 0.0197 = 1.97\%;$$

$$r\left(\frac{3}{2}\right) = -\frac{2}{3}\ln\left(\text{Disc}\left(\frac{3}{2}\right)\right) = 0.0362 = 3.62\%;$$

$$r(0, 2) = -\frac{1}{2}\ln(\text{Disc}(2)) = 0.0118 = 1.18\%. \quad \square$$

Problem 14: The values of the following coupon bonds with face value $100 are given:

Bond Type	Coupon Rate	Bond Price
9 months semiannual	2%	$100.80
15 months semiannual	4%	$103.50
15 months annual	5%	$107.50
21 months semiannual	5%	$110.50

(i) List the cash flows and cash flow dates for each bond.

(ii) Find the 3 months, 9 months, 15 months, and 21 months discount factors.

Solution: (i) Recall that an annual coupon bond with face value $100, coupon rate C, and maturity T pays the holder of the bond a coupon payment equal to $C \cdot 100$ every year, except at maturity. The final payment at maturity T is equal to the face value of the bond plus one coupon payment, i.e., $(1 + C)100$. Also, a semiannual coupon

bond with face value $100, coupon rate C, and maturity T pays the holder of the
bond a coupon payment equal to $\frac{C}{2} \cdot 100$ every six months, except at maturity. The
final payment at maturity T is equal to the face value of the bond plus one coupon
payment, i.e., $100 + \frac{C}{2} \cdot 100$.

The cash flows and the cash flow dates of the bonds above are recorded in Table 2.8.

Table 2.8: Bonds cash flows

Maturity	Cash Flow & Date
9 months	$1 in 3 months
	$101 in 9 months
15 months	$2 in 3 months
	$2 in 9 months
	$102 in 15 months
15 months	$5 in 3 months
	$105 in 15 months
21 months	$2.50 in 3 months
	$2.50 in 9 months
	$2.50 in 15 months
	$102.50 in 21 months

(ii) The value of a bond is equal to the sum of the present value of all its future cash
flows; see (2.52). The four bonds considered here have four cash flow dates, i.e., 3
months, 9 months, 15 months, and 21 months. Denote by d_1 the 3 months discount
factor, by d_2 the 9 months discount factor, by d_3 the 15 months discount factor,
and by d_4 the 21 months discount factor. Using formula (2.52) for the values of the
bonds, we find that

$$
\begin{aligned}
100.80 &= d_1 + 101 d_2; \\
103.50 &= 2d_1 + 2d_2 + 102 d_3; \\
107.50 &= 5d_1 + 105 d_3; \\
110.50 &= 2.50 d_1 + 2.50 d_2 + 2.50 d_3 + 102.50 d_4.
\end{aligned}
$$

This linear system can be written in matrix notation as $Ax = b$, where

$$
A = \begin{pmatrix} 1 & 101 & 0 & 0 \\ 2 & 2 & 102 & 0 \\ 5 & 0 & 105 & 0 \\ 2.50 & 2.50 & 2.50 & 102.50 \end{pmatrix} ; \quad x = \begin{pmatrix} d_1 \\ d_2 \\ d_3 \\ d_4 \end{pmatrix} ; \quad b = \begin{pmatrix} 100.80 \\ 103.50 \\ 107.50 \\ 110.50 \end{pmatrix} .
$$

The solution of $Ax = b$ is found by using the LU decomposition with row pivoting
linear solver, i.e.,

$$
x = \text{linear\_solve\_lu\_row\_pivoting}(A, b) = \begin{pmatrix} 1.0166 \\ 0.9880 \\ 0.9754 \\ 1.0054 \end{pmatrix} .
$$

Thus, the 3 months, 9 months, 15 months, and 21 months discount factors are:[3]

$$\text{Disc}\left(\frac{3}{12}\right) = 1.0166; \quad \text{Disc}\left(\frac{9}{12}\right) = 0.9880;$$

$$\text{Disc}\left(\frac{15}{12}\right) = 0.9754; \quad \text{Disc}\left(\frac{21}{12}\right) = 1.0054. \quad \square$$

Problem 15: The following discount factors are obtained by fitting market data:

Date	Discount Factor
2 months	0.9980
5 months	0.9935
11 months	0.9820
15 months	0.9775

The overnight rate is 1%.

(i) What is the linear system that has to be solved for the cubic spline interpolation of the zero rate curve?

(ii) Use cubic spline interpolation to find a zero rate curve for all times less than 15 months matching the discount factors above.

(iii) Find the value of a 13 months quarterly bond with 2.5% coupon rate.

Solution: (i) Recall that, for continuously compounded interest, the discount factor at time t is

$$\text{Disc}(t) = e^{-tr(0,t)}, \quad \forall\, t > 0,$$

where $r(0, t)$ is the zero rate corresponding to time t. Thus,

$$r(0, t) = -\frac{1}{t}\ln\left(\text{Disc}(t)\right), \quad \forall\, t > 0,$$

and therefore the 2 months, 5 months, 11 months, and 15 months zero rates corresponding to the discount factors above are

$$r\left(0, \frac{2}{12}\right) = 0.012012; \quad r\left(0, \frac{5}{12}\right) = 0.015651;$$

$$r\left(0, \frac{11}{12}\right) = 0.019815; \quad r\left(0, \frac{15}{12}\right) = 0.018206.$$

Also, since the overnight rate is 1%, it follows that

$$r(0, 0) = 0.01.$$

(ii) We use cubic spline interpolation to find the zero rate curve $r(0, t)$ for all maturities up to 15 months, i.e., for all $0 \le t \le \frac{15}{12}$.

---

[3]Note that, since the 3 months and the 21 months discount factors are greater than 1, it follows from formula (2.53) that the corresponding zero rates will be negative. For example, $r\left(\frac{3}{12}\right) = -0.0657 = -6.57\%$.

Thus, $r(0,t)$ is assumed to be a cubic polynomial on each of the intervals $\left[0, \frac{2}{12}\right]$, $\left[\frac{2}{12}, \frac{5}{12}\right]$, $\left[\frac{5}{12}, \frac{11}{12}\right]$, and $\left[\frac{11}{12}, \frac{15}{12}\right]$, i.e.,

$$
r(0,t) = \begin{cases}
a_1 + b_1 t + c_1 t^2 + d_1 t^3, & \text{if } 0 \le t \le \frac{2}{12}; \\
a_2 + b_2 t + c_2 t^2 + d_2 t^3, & \text{if } \frac{2}{12} \le t \le \frac{5}{12}; \\
a_3 + b_3 t + c_3 t^2 + d_3 t^3, & \text{if } \frac{5}{12} \le t \le \frac{11}{12}; \\
a_4 + b_4 t + c_4 t^2 + d_4 t^3, & \text{if } \frac{11}{12} \le t \le \frac{15}{12}.
\end{cases}
\tag{2.54}
$$

The coefficients of these cubic polynomials are obtained by solving the linear system

$$
\overline{M}\overline{x} = \overline{b}, \tag{2.55}
$$

where $\overline{x}$ is the 16×1 vector of the unknowns a_i, b_i, c_i, d_i, $i = 1:4$, given by

$$
\overline{x}(4i-3) = a_i; \ \overline{x}(4i-2) = b_i; \ \overline{x}(4i-1) = c_i; \ \overline{x}(4i) = d_i, \ \forall\, i = 1:4, \tag{2.56}
$$

the matrix $\overline{M}$ is the following 16×16 banded matrix given by (2.57–2.58):

$$
\overline{M} =
\begin{pmatrix}
0 & 0 & 2 & 0 & 0 & 0 & 0 & 0 \\
1 & 0 & 0 & 0 & 0 & 0 & 0 & 0 \\
1 & 0.1667 & 0.0278 & 0.0046 & 0 & 0 & 0 & 0 \\
0 & 1 & 0.3333 & 0.0833 & 0 & -1 & -0.3333 & -0.0833 \\
0 & 0 & 2 & 1 & 0 & 0 & -2 & -1 \\
0 & 0 & 0 & 0 & 1 & 0.1667 & 0.0278 & 0.0046 \\
0 & 0 & 0 & 0 & 1 & 0.4167 & 0.1736 & 0.0723 \\
0 & 0 & 0 & 0 & 0 & 1 & 0.8333 & 0.5208 \\
0 & 0 & 0 & 0 & 0 & 0 & 2 & 2.5 \\
0 & 0 & 0 & 0 & 0 & 0 & 0 & 0 \\
0 & 0 & 0 & 0 & 0 & 0 & 0 & 0 \\
0 & 0 & 0 & 0 & 0 & 0 & 0 & 0 \\
0 & 0 & 0 & 0 & 0 & 0 & 0 & 0 \\
0 & 0 & 0 & 0 & 0 & 0 & 0 & 0 \\
0 & 0 & 0 & 0 & 0 & 0 & 0 & 0 \\
0 & 0 & 0 & 0 & 0 & 0 & 0 & 0
\end{pmatrix}
\tag{2.57}
$$

$$
\begin{pmatrix}
0 & 0 & 0 & 0 & 0 & 0 & 0 & 0 \\
0 & 0 & 0 & 0 & 0 & 0 & 0 & 0 \\
0 & 0 & 0 & 0 & 0 & 0 & 0 & 0 \\
0 & 0 & 0 & 0 & 0 & 0 & 0 & 0 \\
0 & 0 & 0 & 0 & 0 & 0 & 0 & 0 \\
0 & 0 & 0 & 0 & 0 & 0 & 0 & 0 \\
0 & 0 & 0 & 0 & 0 & 0 & 0 & 0 \\
0 & -1 & -0.8333 & -0.5208 & 0 & 0 & 0 & 0 \\
0 & 0 & -2 & -2.5 & 0 & 0 & 0 & 0 \\
1 & 0.4167 & 0.1736 & 0.0723 & 0 & 0 & 0 & 0 \\
1 & 0.9167 & 0.8403 & 0.7703 & 0 & 0 & 0 & 0 \\
0 & 1 & 1.8333 & 2.5208 & 0 & -1 & -1.8333 & -2.5208 \\
0 & 0 & 2 & 5.5 & 0 & 0 & -2 & -5.5 \\
0 & 0 & 0 & 0 & 1 & 0.9167 & 0.8403 & 0.7703 \\
0 & 0 & 0 & 0 & 1 & 1.25 & 1.5625 & 1.9531 \\
0 & 0 & 0 & 0 & 0 & 0 & 2 & 7.5
\end{pmatrix} .
\tag{2.58}
$$

and $\bar{b}$ is the following 16×1 vector:

$$\bar{b} = \begin{pmatrix} 0 \\ 0.01 \\ 0.012012 \\ 0 \\ 0 \\ 0.012012 \\ 0.015651 \\ 0 \\ 0 \\ 0.015651 \\ 0.019815 \\ 0 \\ 0 \\ 0.019815 \\ 0.018206 \\ 0. \end{pmatrix}.$$

(ii) By solving the linear system (2.55), we obtain that

$$\bar{x} = \begin{pmatrix} 0.01 \\ 0.011457 \\ 0 \\ 0.022152 \\ 0.010215 \\ 0.007589 \\ 0.023204 \\ -0.024257 \\ 0.009156 \\ 0.015212 \\ 0.004909 \\ -0.009621 \\ -0.014852 \\ 0.093786 \\ -0.080807 \\ 0.021549 \end{pmatrix}.$$

Using the relationship (2.56) between $\bar{x}$ and the coefficients a_i, b_i, c_i, d_i, $i = 1 : 4$, we find from (2.54) that the resulting zero rate curve is

$$r(0,t) = \begin{cases} 0.01 + 0.011457t + 0.022152t^3, & \text{if } 0 \le t \le \frac{2}{12}; \\ 0.010215 + 0.007589t + 0.023204t^2 - 0.024257t^3, & \text{if } \frac{2}{12} \le t \le \frac{5}{12}; \\ 0.009156 + 0.015212t + 0.004909t^2 - 0.009621t^3, & \text{if } \frac{5}{12} \le t \le \frac{11}{12}; \\ -0.014852 + 0.093786t - 0.080807t^2 + 0.021549t^3, & \text{if } \frac{11}{12} \le t \le \frac{13}{12}. \end{cases}$$
$$(2.59)$$

(iii) Recall that a quarterly coupon bond with face value \$100, coupon rate C, and maturity T pays the holder of the bond a coupon payment equal to $\frac{C}{4} \cdot 100$ every three months, except at maturity. The final payment at maturity T is equal to the face value of the bond plus one coupon payment, i.e., $100 + \frac{C}{4}100$.

Then, a 13 months quarterly bond with 2.5% coupon rate and face value \$100 has the following cash flows:

Using the zero rate curve $r(0,t)$ given by (2.59) and continuous compounding, we obtain that the value of the 13 months quarterly bond with 2.5% coupon rate and

Date	Cash flow
1 month	$0.625
4 months	$0.625
7 months	$0.625
10 months	$0.625
13 months	$100.625

face value $100 is

$$
B = 0.625 \exp\left(-\frac{1}{12} r\left(0, \frac{1}{12}\right)\right) + 0.625 \exp\left(-\frac{4}{12} r\left(0, \frac{4}{12}\right)\right)
$$

$$
+ 0.625 \exp\left(-\frac{7}{12} r\left(0, \frac{7}{12}\right)\right) + 0.625 \exp\left(-\frac{10}{12} r\left(0, \frac{10}{12}\right)\right)
$$

$$
+ 100.625 \exp\left(-\frac{13}{12} r\left(0, \frac{13}{12}\right)\right).
$$

From (2.59), we find that

$$
r\left(0, \frac{1}{12}\right) = 0.010968; \quad r\left(0, \frac{4}{12}\right) = 0.014424; \quad r\left(0, \frac{7}{12}\right) = 0.017790;
$$

$$
r\left(0, \frac{10}{12}\right) = 0.019674; \quad r\left(0, \frac{13}{12}\right) = 0.019311
$$

Thus, $B = 101.0216$, i.e., the value of the bond is $101.02. □

Problem 16: Consider three assets with the following expected values, standard deviations, and correlations of their returns:

$$
\begin{array}{llll}
\mu_1 = 0.10; & \sigma_1 = 0.15; & \rho_{1,2} = -0.25; \\
\mu_2 = 0.15; & \sigma_2 = 0.30; & \rho_{2,3} = 0.20; \\
\mu_3 = 0.20; & \sigma_3 = 0.35; & \rho_{1,3} = 0.30.
\end{array}
$$

(i) Find the covariance matrix M of the returns of the three assets.

(ii) A minimum variance portfolio with 16% expected rate of return and fully invested in the three assets (i.e., with no cash position) can be set up by investing a percentage w_i of the total value of the portfolio in asset i, with $i = 1 : 3$, where w_1, w_2, w_3 can be found by solving the following linear system:

$$
\begin{pmatrix} 2M & 1 & \mu \\ 1^t & 0 & 0 \\ \mu^t & 0 & 0 \end{pmatrix} \begin{pmatrix} w \\ \lambda_1 \\ \lambda_2 \end{pmatrix} = \begin{pmatrix} 0 \\ 1 \\ \mu_P \end{pmatrix}, \tag{2.60}
$$

where $w = \begin{pmatrix} w_1 \\ w_2 \\ w_3 \end{pmatrix}$ and

$$
\mu_P = 0.16; \quad \mu = \begin{pmatrix} 0.1 \\ 0.15 \\ 0.2 \end{pmatrix}; \quad 1 = \begin{pmatrix} 1 \\ 1 \\ 1 \end{pmatrix}; \quad 0 = \begin{pmatrix} 0 \\ 0 \\ 0 \end{pmatrix}.
$$

Show that the matrices from the LU decomposition with row pivoting of the matrix on the left hand side of (2.60) are

$$P = \begin{pmatrix} 0 & 0 & 0 & 1 & 0 \\ 0 & 1 & 0 & 0 & 0 \\ 0 & 0 & 1 & 0 & 0 \\ 1 & 0 & 0 & 0 & 0 \\ 0 & 0 & 0 & 0 & 1 \end{pmatrix}; \qquad (2.61)$$

$$L = \begin{pmatrix} 1 & 0 & 0 & 0 & 0 \\ -0.0225 & 1 & 0 & 0 & 0 \\ 0.0315 & 0.051852 & 1 & 0 & 0 \\ 0.045 & -0.333333 & 0.038067 & 1 & 0 \\ 0.1 & 0.246914 & 0.400056 & -0.482738 & 1 \end{pmatrix}; \qquad (2.62)$$

$$U = \begin{pmatrix} 1 & 1 & 1 & 0 & 0 \\ 0 & 0.2025 & 0.0645 & 1 & 0.15 \\ 0 & 0 & 0.210555 & 0.948148 & 0.192222 \\ 0 & 0 & 0 & 1.297240 & 0.142683 \\ 0 & 0 & 0 & 0 & -0.045059 \end{pmatrix}. \qquad (2.63)$$

(iii) Find the weights of each asset in the minimum variance portfolio with 16% expected return. Find the standard deviation of the return of this portfolio.

(iv) Show that the two portfolios below have 16% expected return, and compute the standard deviation of the returns of each portfolio:
• 30% invested in asset 1, 20% invested in asset 2, 50% invested in asset 3;
• 50% invested in asset 1, 70% invested in asset 3, and short an amount equal to 20% of the value of the portfolio of asset 2.

Solution: (i) The covariance matrix M of the returns of the three assets is given by

$$\begin{aligned} M &= \begin{pmatrix} \sigma_1^2 & \sigma_1\sigma_2\rho_{1,2} & \sigma_1\sigma_3\rho_{1,3} \\ \sigma_1\sigma_2\rho_{1,2} & \sigma_2^2 & \sigma_2\sigma_3\rho_{2,3} \\ \sigma_1\sigma_3\rho_{1,3} & \sigma_2\sigma_3\rho_{2,3} & \sigma_3^2 \end{pmatrix} \\ &= \begin{pmatrix} 0.0225 & -0.0113 & 0.0158 \\ -0.0113 & 0.0900 & 0.0210 \\ 0.0158 & 0.0210 & 0.1225 \end{pmatrix}. \end{aligned} \qquad (2.64)$$

(ii) The linear system (2.60) used to find the minimum variance portfolio with 16% expected rate of return fully invested in the three assets and with assets weights w_1, w_2, w_3 can be written as

$$Ax = b, \qquad (2.65)$$

where

$$A = \begin{pmatrix} 2M & 1 & \mu \\ 1^t & 0 & 0 \\ \mu^t & 0 & 0 \end{pmatrix} = \begin{pmatrix} 0.0450 & -0.0225 & 0.0315 & 1 & 0.10 \\ -0.0225 & 0.18 & 0.0420 & 1 & 0.15 \\ 0.0315 & 0.0420 & 0.2450 & 1 & 0.20 \\ 1 & 1 & 1 & 0 & 0 \\ 0.10 & 0.15 & 0.20 & 0 & 0 \end{pmatrix}; \qquad (2.66)$$

$$x = \begin{pmatrix} w_1 \\ w_2 \\ w_3 \\ \lambda_1 \\ \lambda_2 \end{pmatrix} ; \quad b = \begin{pmatrix} 0 \\ 0 \\ 0 \\ 1 \\ 0.16 \end{pmatrix}. \tag{2.67}$$

The LU decomposition with row pivoting of the matrix A from (2.66) is $PA = LU$ where $[P, L, U] = $ lu_row_pivoting(A), and the matrices P, L, and U are given by (2.61–2.63).

(iii) By solving the linear system (2.65), we obtain that

$$x = \begin{pmatrix} w_1 \\ w_2 \\ w_3 \\ \lambda_1 \\ \lambda_2 \end{pmatrix} = \text{linear\_solve\_lu\_row\_pivoting}(A,b) = \begin{pmatrix} 0.2351 \\ 0.3298 \\ 0.4351 \\ 0.0941 \\ -1.1099 \end{pmatrix},$$

and therefore the weights of asset 1, asset 2, and asset 3 are $w_1 = 0.2351$, $w_2 = 0.3298$, and $w_3 = 0.4351$, respectively.

In other words, the minimum variance portfolio with 16% expected return is obtained by investing 23.51% of the portfolio in the first asset, 32.98% of the portfolio in the second asset, and 43.51% of the portfolio in the third asset.

The standard deviation of the return of this portfolio is

$$\sigma_m = (w_1 \; w_2 \; w_3) \; M \begin{pmatrix} w_1 \\ w_2 \\ w_3 \end{pmatrix} = 0.2043 = 20.43\%,$$

where the covariance matrix M of the returns of the assets is given by (2.64).

(iv) Recall that, if w_1, w_2, and w_3 denote the weights of asset 1, of asset 2, and of asset 3 in the portfolio, respectively, the expected return of the portfolio is

$$\mu_R = w_1\mu_1 + w_2\mu_2 + w_3\mu_3.$$

Then, a portfolio which is invested 30% in asset 1, 20% in asset 2, and 50% in asset 3, i.e., with $w_1 = 0.30$, $w_2 = 0.20$, and $w_3 = 0.50$, has expected return

$$0.30 \cdot 0.10 + 0.20 \cdot 0.15 + 0.50 \cdot 0.20 = 0.16 = 16\%,$$

and the standard deviation of the return of this portfolio is

$$(0.30 \; 0.20 \; 0.50) \; M \begin{pmatrix} 0.30 \\ 0.20 \\ 0.50 \end{pmatrix} = 0.2093 = 20.93\%,$$

where the matrix M is given by (2.64).

A portfolio which is invested 50% invested in asset 1, 70% invested in asset 3, and has a short position in asset 2 equal to 20% of the value of the portfolio, i.e., with $w_1 = 0.50$, $w_2 = -0.20$, and $w_3 = 0.70$, has expected return

$$0.50 \cdot 0.10 - 0.20 \cdot 0.15 + 0.70 \cdot 0.20 = 0.16 = 16\%,$$

and the standard deviation of the return of this portfolio is

$$(0.50 \ -0.20 \ 0.70) \ M \begin{pmatrix} 0.50 \\ -0.20 \\ 0.70 \end{pmatrix} = 0.2768 = 27.68\%,$$

where the matrix M is given by (2.64).

As expected, the standard deviations (20.93% and 27.68%) of the returns of both portfolios considered here are greater than 20.43%, the standard deviation of the return the minimum variance portfolio with 16% expected return. $\square$

Chapter 3

The Arrow–Debreu one period market model

3.1 Exercises

1. In three months, the value of an asset with spot price $50 will be either $60 or $40, with probability one half. What is the value of a three months at–the–money put on this asset? Assume the risk–free interest rate is zero.

2. A one period market with four securities and four states at time $\tau > t_0$ has the following price vector of spot prices at the initial time t_0:

$$S_{t_0} = \begin{pmatrix} 4 \\ 100 \\ 1 \\ -2 \end{pmatrix},$$

and the following payoff matrix at time τ:

$$M_\tau = \begin{pmatrix} 1 & 1 & 1 & 1 \\ 20 & 22 & 24 & 26 \\ -5 & -3 & -1 & 2 \\ 4 & 2 & 0 & 0 \end{pmatrix}.$$

Find an arbitrage opportunity.

3. A one period market model is made of four securities and has four states at time $\tau > t_0$. Assume that the market model is complete and that the state prices are -1, 1, 2, and 4, respectively. Find an arbitrage opportunity.

4. At time $\tau > t_0$, an asset with spot price S_0 at time t_0 will be worth either uS_0, in which case the value of $1 today would be FV_1, or dS_0, in which case the value of $1 today would be FV_2, with $d < u$. Consider the one period market model with two securities, i.e., cash and the asset, and two states, i.e., asset value at time τ equal to uS_0 and asset value at time τ equal to dS_0.

(i) Show that the payoff matrix at time τ of this one period market model is

$$\begin{pmatrix} FV_1 & FV_2 \\ uS_0 & dS_0 \end{pmatrix}.$$

(ii) Find necessary and sufficient conditions for the model to be complete.

(iii) Show that this model is arbitrage–free if and only if

$$\min\left(\frac{u}{FV_1}, \frac{d}{FV_2}\right) < 1 < \max\left(\frac{u}{FV_1}, \frac{d}{FV_2}\right). \tag{3.1}$$

(iv) Show that, if $FV_1 = FV_2 = e^{r\delta t}$, the condition (3.1) is equivalent to the no–arbitrage condition $d < e^{r\delta t} < u$ for the classical one period binomial model.

5. In a one period trinomial model, it is assumed that the price at time $\tau > t_0$ of an asset with price S_0 at time t_0 will be either dS_0, mS_0, or uS_0, where $d < m < u$.

   Consider a one period market model with two securities, i.e., the asset described above and cash, and three states, i.e., state w^1, when the price of the asset at time τ is dS_0, state w^2, when the price of the asset at time τ is mS_0, and state w^3, when the price of the asset at time τ is uS_0.

   (i) Show that the one period trinomial model is incomplete.

   (ii) Show that the one period trinomial model is arbitrage–free if and only if

$$d < e^{r\delta t} < u,$$

   where $\delta t = \tau - t_0$ and r is the continuously compounded interest rate between t_0 and τ.

6. In three months, the value of an asset with spot price $40 will be either $32, $38, $42, or $44. The value of a three months European call option with strike $36 on this asset is $8 and the value of a three months European put option with strike $40 on this asset is $5. For simplicity, assume zero risk free rates. Consider the one period market model with the following four securities and the following four states in three months:

   Securities:
   • cash;
   • asset;
   • three months call with strike $36;
   • three months put with strike $40;

   Market states:
   • asset price $32;
   • asset price $38;
   • asset price $42;
   • asset price $44.

   (i) Find the payoff matrix of this model, and show that the four securities are non–redundant.

(ii) Show that the one period market model is complete.

(iii) How do you replicate a bull spread made of a long position in a three months call with strike $34 and a short position in a three months call with strike $40 in this one period market?

(iv) Show that this model is not arbitrage–free, and find an arbitrage opportunity.

7. Two assets have spot prices $20 and $30, respectively. Assume that, in five months, the first asset will be worth either $18 or $22, and the second asset will be worth either $28 or $32. Also, assume that the risk–free rate for five months cash deposits is 0. Consider the one period market model with the following three securities and the following four states in five months:

Securities:

• cash;

• first asset;

• second asset.

Market states:

• first asset at $22 and second asset at $32;

• first asset at $22 and second asset at $28;

• first asset at $18 and second asset at $32;

• first asset at $18 and second asset at $28.

(i) Show that this one period market model is arbitrage–free.

(ii) Show that, in this one period market model, it is not possible to replicate a derivative security that pays $1 if the first state occurs (i.e., if, in five months, the first asset is worth $22 and the second asset is worth $32), and does not pay anything if any other state occurs. Conclude that the model is not complete.

Hint: For (i), consider the case when all state prices are equal to $\frac{1}{4}$.

8. This exercise is related to the example from Chapter 1.

Consider two assets with spot prices $30 and $50, respectively. Assume that, in three months, the first asset will be worth either $34 or $24, and the second asset will be worth either $56, $51, or $46. The value of a three months at–the–money European call option with strike $30 on the first asset is $2.5, the value of a three months at–the–money European call option with strike $50 on the second asset is $2.7, and the value of a three months European put option with strike $52 on the second asset is $4.1. Assume that the future value in three months of $1 today is $1.01. Consider the one period market model with the following six securities and the following six states in three months:

Securities:

• cash;

- first asset;
- second asset;
- three months ATM call on the first asset;
- three months ATM call on the second asset;
- three months put with strike $52 on the second asset;

Market states:

- first asset at $34 and second asset at $56;
- first asset at $34 and second asset at $51;
- first asset at $34 and second asset at $46;
- first asset at $24 and second asset at $56;
- first asset at $24 and second asset at $51;
- first asset at $24 and second asset at $46.

(i) Show that the payoff matrix for this market model is nonsingular, and conclude that the market is complete.

(ii) Compute the state prices for this model, and show that the market is arbitrage–free.

(iii) Use risk–neutral pricing to find the value of a three months call option with strike $52 on the second asset.

(iv) Use Put–Call parity to find the value of a three months call option with strike $52 on the second asset, and compare it to the value computed at (iii).

(v) What is the value of a bear spread made of a long position in a three months put option with strike $35 on the first asset and a short position in a three months put option with strike $28 on the first asset?

9. This exercise refers to the S&P 500 options prices from Table 3.1.

Consider a one period market model with the following nine securities:

P1200; P1275; P1350; P1375; C1375; C1400; C1450; C1550; C1600.

The nine states of the index price at maturity are as follows: seven states correspond to the midpoints between the strikes of the options above, i.e.,

$$\omega^2 : \{S(\tau) = 1237.50\}; \qquad \omega^6 : \{S(\tau) = 1425\};$$
$$\omega^3 : \{S(\tau) = 1312.50\}; \qquad \omega^7 : \{S(\tau) = 1500\};$$
$$\omega^4 : \{S(\tau) = 1362.50\}; \qquad \omega^8 : \{S(\tau) = 1575\};$$
$$\omega^5 : \{S(\tau) = 1387.50\};$$

the first and last state are

$$\omega^1 : \{S(\tau) = 950\}; \qquad \omega^9 : \{S(\tau) = 1675\}.$$

Table 3.1: Dec 2012 SPX option prices on 3/9/2012

Call Strike	Price	Volume	Put Strike	Price	Volume
C1175	225.40	250	P1175	46.60	1
C1200	205.55	215	P1200	51.55	3204
C1225	186.20	1	P1225	57.15	1401
C1250	167.50	650	P1250	63.30	104
C1275	149.15	163	P1275	70.15	56
C1300	131.70	1	P1300	77.70	150
C1325	115.25	40	P1325	86.20	200
C1350	99.55	320	P1350	95.30	10118
C1375	84.90	1002	P1375	105.30	1250
C1400	71.10	5300	P1400	116.55	1250
C1425	58.70	4	P1425	129.00	200
C1450	47.25	9050	P1450	143.20	1
C1500	29.25	1000	P1500	173.95	6
C1550	15.80	1000	P1550	210.80	9
C1575	11.10	200	P1575	230.90	0
C1600	7.90	546	P1600	252.40	9

(i) Find the payoff matrix M_τ of this one period market model, and show that the model is complete.

(ii) Find the state prices vector Q and show that the model is arbitrage–free.

(iii) Compute the root–mean–squared error (RMSE) of this model. Comment on the precision of this nine securities model compared to the seven securities model from section 3.5 in Stefanica [3].

10. This problem refers to the S&P 500 options prices from Table 3.1.

Consider a one period market model with the following same seven securities:

$$P1200; \ P1300; \ P1400; \ C1400; \ C1450; \ C1550; \ C1600.$$

The states of this market are the midpoints of the strikes, i.e.,

$$\omega^2 : \{S(\tau) = 1250\}; \qquad \omega^5 : \{S(\tau) = 1500\};$$
$$\omega^3 : \{S(\tau) = 1350\}; \qquad \omega^6 : \{S(\tau) = 1575\};$$
$$\omega^4 : \{S(\tau) = 1425\};$$

and the first and last state are

$$\omega^1 : \{S(\tau) = 1100\}; \quad \omega^7 : \{S(\tau) = 1700\}.$$

Show that this market model is not arbitrage–free.

3.2   Solutions to Chapter 3 Exercises

Problem 1: In three months, the value of an asset with spot price $50 will be either $60 or $40, with probability one half. What is the value of a three months at–the–money put on this asset? Assume the risk–free interest rate is zero.

Solution: This is a one period binomial model with up and down factors $u = \frac{60}{50} = 1.2$ and $d = \frac{40}{50} = 0.8$, respectively, over a time period $\tau = \delta t = \frac{3}{12} = 0.25$. The interest rate was assumed to be $r = 0$.

Note that real world probabilities do not play any role in valuing an option in a (one period) binomial tree model. Thus, the fact that the probability the asset goes to either $60 or $40 is one half does not play any role in solving the problem.

The one period binomial model is arbitrage free and complete, since $d < e^{r\delta t} < u$:

$$d = 0.8 \; < \; e^0 = 1 \; < \; 1.2 = u.$$

Recall that any derivative security in an arbitrage free and complete one period binomial model with payoffs at time $\tau = \delta t$ equal to $V_\tau(1)$, if the "up" state occurs, and equal to $V_\tau(2)$, if the "down" state occurs, can be priced using the following risk–neutral pricing formula:

$$V(0) \; = \; e^{-r\delta t}\left(\frac{e^{r\delta t} - d}{u - d}V_\tau(1) \; + \; \frac{u - e^{r\delta t}}{u - d}V_\tau(2)\right),$$

which, for $r = 0$, becomes

$$V(0) \; = \; \frac{1 - d}{u - d}V_\tau(1) \; + \; \frac{u - 1}{u - d}V_\tau(2). \tag{3.2}$$

The payoff at maturity of a put option is $P(T) = \max(K - S(T), 0)$. Thus, the value of the at–the–money put with strike $50 on this asset in three months, i.e., at maturity, is

$$V_\tau(1) \; = \; \max(50 - 60, 0) = 0, \quad \text{for } S(T) = 60; \tag{3.3}$$
$$V_\tau(2) \; = \; \max(50 - 40, 0) = 10, \quad \text{for } S(T) = 40. \tag{3.4}$$

From (3.2), (3.3), and (3.4), we find that the value of the put option is

$$V(0) \; = \; \frac{1 - 0.8}{1.2 - 0.8}V_\tau(1) \; + \; \frac{1.2 - 1}{1.2 - 0.8}V_\tau(2) \; = \; 5. \quad \square$$

Problem 2: A one period market with four securities and four states at time $\tau > t_0$ has the following price vector of spot prices at the initial time t_0:

$$S_{t_0} \; = \; \begin{pmatrix} 4 \\ 100 \\ 1 \\ -2 \end{pmatrix},$$

and the following payoff matrix at time τ:

$$M_\tau = \begin{pmatrix} 1 & 1 & 1 & 1 \\ 20 & 22 & 24 & 26 \\ -5 & -3 & -1 & 2 \\ 4 & 2 & 0 & 0 \end{pmatrix}.$$

Find an arbitrage opportunity.

Solution: A straightforward arbitrage exists in this market: security four has negative price -2 at time t_0 and has a nonnegative payoff vector $(4\ 2\ 0\ 0)$ at time τ. This constitutes an arbitrage opportunity. Taking a long position in the fourth security generates a cash flow of $\$2$ and will not lose money at time τ: the security pays $\$4$ if the first state occurs at time τ, pays $\$2$ if the second state occurs at time τ, and does not lose money in any of the other states.

Alternatively, while the one period market model is not complete since $\det(M_\tau) = 0$ and the matrix M_τ is singular, note that

$$M_\tau \begin{pmatrix} -2 \\ 3 \\ 2 \\ 1 \end{pmatrix} = \begin{pmatrix} 1 & 1 & 1 & 1 \\ 20 & 22 & 24 & 26 \\ -5 & -3 & -1 & 2 \\ 4 & 2 & 0 & 0 \end{pmatrix} \begin{pmatrix} -2 \\ 3 \\ 2 \\ 1 \end{pmatrix} = \begin{pmatrix} 4 \\ 100 \\ 1 \\ -2 \end{pmatrix} = S_{t_0}.$$

In other words, if $Q = \begin{pmatrix} -2 \\ 3 \\ 2 \\ 1 \end{pmatrix}$, then $M_\tau Q = S_{t_0}$. Since $Q_1 = -2 < 0$, this

contradicts the no–arbitrage condition that a one period market model is arbitrage–free if and only if there exists a vector $Q = (Q_k)_{k=1:n}$ of positive state prices $Q_k > 0$, $k = 1:n$, such that $S_{t_0} = M_\tau Q$.

An arbitrage opportunity is found by considering a portfolio with positions vector

$$\Theta = \begin{pmatrix} 0 \\ 0 \\ 0 \\ 1 \end{pmatrix}.$$ The value of this portfolio at time t_0 is

$$V_{t_0} = \Theta^t S_{t_0} = -2,$$

and the positions vector of the portfolio at time τ is

$$V_\tau = \Theta^t M_\tau = (4\ 2\ 0\ 0).$$

Thus, we found a portfolio with negative initial value $V_{t_0} = -2$ (in other words, the portfolio generates a positive cash flow of $\$2$ when set up at time t_0) which does not lose money at time τ, when its values correspond to $V_\tau = (4\ 2\ 0\ 0)$, i.e., the portfolio will generate $\$4$ if the first state occurs, $\$2$ if the second state occurs, but will not lose any money if any of the other states occur. □

Problem 3: A one period market model is made of four securities and has four states at time $\tau > t_0$. Assume that the market model is complete and that the state prices are -1, 1, 2, and 4, respectively. Find an arbitrage opportunity.

Solution: Since the market is complete, it follows that the 4×4 payoff matrix M_τ corresponding to this one period market model is nonsingular. Also, if $Q = \begin{pmatrix} -1 \\ 1 \\ 2 \\ 4 \end{pmatrix}$

is the vector of state prices and S_{t_0} is the price vector of spot prices at the initial time t_0, then

$$S_{t_0} = M_\tau Q. \tag{3.5}$$

The market model is not arbitrage–free since $Q_1 = -1 < 0$. An arbitrage opportunity can be obtained by setting up a portfolio paying \$1 if the first state (i.e., the state corresponding to the negative state price $Q_1 = -1$) occurs, and paying nothing if any other state occurs. In other words, the payoff of this portfolio at time τ is

$$V_\tau = (1 \ 0 \ 0 \ 0).$$

The positions vector $\Theta = (\Theta_k)_{k=1:4}$ of this portfolio is obtained by solving

$$V_\tau = \Theta^t M_\tau. \tag{3.6}$$

By taking the transpose on both sides of (3.6), it follows that

$$(M_\tau)^t \Theta = (V_\tau)^t,$$

and therefore

$$\Theta = \text{linear\_solve\_lu\_row\_pivoting}((M_\tau)^t, (V_\tau)^t). \tag{3.7}$$

Recall that M_τ is nonsingular, and therefore the matrix $(M_\tau)^t$ is nonsingular as well since $\det((M_\tau)^t) = \det(M_\tau) \neq 0$.

The value at time t_0 of the portfolio with positions vector Θ is

$$V_{t_0} = \Theta^t S_{t_0}. \tag{3.8}$$

Then, from (3.8) and using (3.5) and (3.6), we obtain that

$$
\begin{aligned}
V_{t_0} &= \Theta^t S_{t_0} = \Theta^t \cdot (M_\tau Q) = (\Theta^t M_\tau) \cdot Q \\
&= V_\tau \cdot Q \\
&= (1 \ 0 \ 0 \ 0) \begin{pmatrix} -1 \\ 1 \\ 2 \\ 4 \end{pmatrix} \\
&= -1.
\end{aligned}
$$

Thus, the portfolio with positions vector Θ given by (3.7) has negative initial value $V_{t_0} = -1$ (in other words, the portfolio generates a positive cash flow of \$1 when set up at time t_0) and does not lose money at time τ, when its values correspond to $V_\tau = (1 \ 0 \ 0 \ 0)$, i.e., the portfolio will receive \$1 if the first state occurs, but will not lose any money if any of the other states occurs.

This constitutes an arbitrage opportunity. $\square$

Problem 4: At time $\tau > t_0$, an asset with spot price S_0 at time t_0 will be worth either uS_0, in which case the value of \$1 today would be FV_1, or dS_0, in which case the value of \$1 today would be FV_2, with $d < u$. Consider the one period market model with two securities, i.e., cash and the asset, and two states, i.e., asset value at time τ equal to uS_0 and asset value at time τ equal to dS_0.

(i) Show that the payoff matrix at time τ of this one period market model is

$$\begin{pmatrix} FV_1 & FV_2 \\ uS_0 & dS_0 \end{pmatrix}.$$

(ii) Find necessary and sufficient conditions for the model to be complete.

(iii) Show that this model is arbitrage–free if and only if

$$\min\left(\frac{u}{FV_1}, \frac{d}{FV_2}\right) < 1 < \max\left(\frac{u}{FV_1}, \frac{d}{FV_2}\right). \tag{3.9}$$

(iv) Show that, if $FV_1 = FV_2 = e^{r\delta t}$, the condition (3.9) is equivalent to the no–arbitrage condition $d < e^{r\delta t} < u$ for the classical one period binomial model.

Solution: (i) The price vector of the securities at time t_0 corresponding to a \$1 cash position and to one unit of the asset is

$$S_{t_0} = \begin{pmatrix} 1 \\ S_0 \end{pmatrix}. \tag{3.10}$$

At time $\tau > t_0$, let the state ω^1 correspond to the asset being worth uS_0, in which case the value of \$1 at time t_0 is FV_1, and let the state ω^2 correspond to the asset being worth dS_0, in which case the value of \$1 at time t_0 is FV_2.

Then, the payoff matrix at time τ of this one period market model is

$$M_\tau = \begin{pmatrix} FV_1 & FV_2 \\ uS_0 & dS_0 \end{pmatrix}, \tag{3.11}$$

where the first column of M_τ corresponds to state ω^1 and the second column of M_τ corresponds to state ω^2.

(ii) Note that this model has two states and two securities, and recall that a one period market model with the same number of states as the number of securities is complete if and only if the payoff matrix M_τ is nonsingular.

From (3.11), it follows that

$$\begin{aligned} \det(M_\tau) &= FV_1 \cdot dS_0 - FV_2 \cdot uS_0 = S_0 \left(d\, FV_1 - u\, FV_2\right) \\ &= FV_1\, FV_2\, S_0 \left(\frac{d}{FV_2} - \frac{u}{FV_1}\right). \end{aligned} \tag{3.12}$$

Thus, we find from (3.12) that

$$M_\tau \text{ nonsingular matrix} \iff \det(M_\tau) \neq 0 \iff \frac{u}{FV_1} \neq \frac{d}{FV_2}. \tag{3.13}$$

Since this one period market model is complete if and only if the payoff matrix M_τ is nonsingular, we conclude from (3.13) that a necessary and sufficient condition for the model to be complete is

$$\frac{u}{FV_1} \neq \frac{d}{FV_2}. \tag{3.14}$$

(iii) This one period model is arbitrage–free if and only if there exists $Q = \begin{pmatrix} Q_1 \\ Q_2 \end{pmatrix}$ with $Q_1 > 0$ and $Q_2 > 0$ such that

$$M_\tau Q = S_{t_0},$$

where S_{t_0} is the price vector of the securities at time t_0 given by (3.10) and M_τ is the payoff matrix at time τ given by (3.11).

Note that

$$M_\tau Q = S_{t_0} \iff \begin{pmatrix} FV_1 & FV_2 \\ uS_0 & dS_0 \end{pmatrix} \begin{pmatrix} Q_1 \\ Q_2 \end{pmatrix} = \begin{pmatrix} 1 \\ S_0 \end{pmatrix}$$

$$\iff \begin{cases} FV_1\, Q_1 + FV_2\, Q_2 = 1 \\ uS_0\, Q_1 + dS_0\, Q_2 = S_0 \end{cases}$$

which can be written as

$$FV_1\, Q_1 + FV_2\, Q_2 = 1; \tag{3.15}$$
$$u\, Q_1 + d\, Q_2 = 1. \tag{3.16}$$

By multiplying (3.15) by d and by multiplying (3.16) by FV_2, we obtain that

$$d\, FV_1\, Q_1 + d\, FV_2\, Q_2 = d; \tag{3.17}$$
$$u\, FV_2\, Q_1 + d\, FV_2\, Q_2 = FV_2. \tag{3.18}$$

By subtracting (3.17) from (3.18) and solving for Q_1, we obtain that

$$Q_1 = \frac{FV_2 - d}{u\, FV_2 - d\, FV_1}. \tag{3.19}$$

From (3.16) and (3.19), we find that

$$Q_2 = \frac{u - FV_1}{u\, FV_2 - d\, FV_1}. \tag{3.20}$$

Note that (3.19) and (3.20) can also be written as follows:

$$Q_1 = \frac{FV_2\left(1 - \frac{d}{FV_2}\right)}{FV_1 FV_2\left(\frac{u}{FV_1} - \frac{d}{FV_2}\right)} = \frac{1}{FV_1} \cdot \frac{1 - \frac{d}{FV_2}}{\frac{u}{FV_1} - \frac{d}{FV_2}}; \tag{3.21}$$

$$Q_2 = \frac{FV_1\left(\frac{u}{FV_1} - 1\right)}{FV_1 FV_2\left(\frac{u}{FV_1} - \frac{d}{FV_2}\right)} = \frac{1}{FV_2} \cdot \frac{\frac{u}{FV_1} - 1}{\frac{u}{FV_1} - \frac{d}{FV_2}} \tag{3.22}$$

Then, from (3.21) and (3.22), it follows that $Q_1 > 0$ and $Q_2 > 0$ if and only if

$$\frac{1 - \frac{d}{FV_2}}{\frac{u}{FV_1} - \frac{d}{FV_2}} > 0 \quad \text{and} \quad \frac{\frac{u}{FV_1} - 1}{\frac{u}{FV_1} - \frac{d}{FV_2}} > 0. \tag{3.23}$$

Depending on whether $\frac{u}{FV_1} - \frac{d}{FV_2}$ is positive or negative, we consider the following cases:

• If $\frac{d}{FV_2} < \frac{u}{FV_1}$, then $\frac{u}{FV_1} - \frac{d}{FV_2} > 0$, and it follows from (3.23) that $Q_1 > 0$ and $Q_2 > 0$ if and only if

$$1 > \frac{d}{FV_2} \quad \text{and} \quad \frac{u}{FV_1} > 1,$$

which can be written compactly as the following double inequality:

$$\frac{d}{FV_2} < 1 < \frac{u}{FV_1}. \tag{3.24}$$

Note that, if $\frac{d}{FV_2} < \frac{u}{FV_1}$, then

$$\min\left(\frac{u}{FV_1}, \frac{d}{FV_2}\right) = \frac{d}{FV_2}; \quad \max\left(\frac{u}{FV_1}, \frac{d}{FV_2}\right) = \frac{u}{FV_1}. \tag{3.25}$$

From (3.24) and (3.25), we obtain that, if $\frac{d}{FV_2} < \frac{u}{FV_1}$, then $Q_1 > 0$ and $Q_2 > 0$ if and only if

$$\min\left(\frac{u}{FV_1}, \frac{d}{FV_2}\right) < 1 < \max\left(\frac{u}{FV_1}, \frac{d}{FV_2}\right). \tag{3.26}$$

• If $\frac{u}{FV_1} < \frac{d}{FV_2}$, then $\frac{u}{FV_1} - \frac{d}{FV_2} < 0$, and it follows from (3.23) that $Q_1 > 0$ and $Q_2 > 0$ if and only if

$$1 < \frac{d}{FV_2} \quad \text{and} \quad \frac{u}{FV_1} < 1,$$

which can be written compactly as the following double inequality:

$$\frac{u}{FV_1} < 1 < \frac{d}{FV_2}. \tag{3.27}$$

Note that, if $\frac{u}{FV_1} < \frac{d}{FV_2}$, then

$$\min\left(\frac{u}{FV_1}, \frac{d}{FV_2}\right) = \frac{u}{FV_1}; \quad \max\left(\frac{u}{FV_1}, \frac{d}{FV_2}\right) = \frac{d}{FV_2}. \tag{3.28}$$

From (3.27) and (3.28), we obtain that, if $\frac{u}{FV_1} < \frac{d}{FV_2}$, then $Q_1 > 0$ and $Q_2 > 0$ if and only if

$$\min\left(\frac{u}{FV_1}, \frac{d}{FV_2}\right) < 1 < \max\left(\frac{u}{FV_1}, \frac{d}{FV_2}\right). \tag{3.29}$$

Recall that the one period model is arbitrage–free if and only if $Q_1 > 0$ and $Q_2 > 0$. Thus, from (3.26) and (3.29), we conclude that the necessary and sufficient condition for this one period model to be arbitrage–free is

$$\min\left(\frac{u}{FV_1}, \frac{d}{FV_2}\right) < 1 < \max\left(\frac{u}{FV_1}, \frac{d}{FV_2}\right). \tag{3.30}$$

(iv) If $FV_1 = FV_2 = e^{r\delta t}$, and since $d < u$, we obtain that

$$\min\left(\frac{u}{FV_1}, \frac{d}{FV_2}\right) = \min\left(\frac{u}{e^{r\delta t}}, \frac{d}{e^{r\delta t}}\right) = e^{-r\delta t}\min(u, d)$$

$$= e^{-r\delta t}d; \tag{3.31}$$

$$\max\left(\frac{u}{FV_1}, \frac{d}{FV_2}\right) = \max\left(\frac{u}{e^{r\delta t}}, \frac{d}{e^{r\delta t}}\right) = e^{-r\delta t}\max(u, d)$$

$$= e^{-r\delta t}u. \tag{3.32}$$

From (3.31) and (3.32), we find that, if $FV_1 = FV_2 = e^{r\delta t}$, the no–arbitrage condition (3.30) reduces to

$$e^{-r\delta t}d < 1 < e^{-r\delta t}u. \tag{3.33}$$

Multiplying (3.33) by $e^{r\delta t}$, the no–arbitrage condition (3.33) becomes

$$d < e^{r\delta t} < u,$$

which is the same as the no–arbitrage condition for the classical one period binomial model. □

Problem 5: In a one period trinomial model, it is assumed that the price at time $\tau > t_0$ of an asset with price S_0 at time t_0 will be either dS_0, mS_0, or uS_0, where $d < m < u$.

Consider a one period market model with two securities, i.e., the asset described above and cash, and three states, i.e., state w^1, when the price of the asset at time τ is dS_0, state w^2, when the price of the asset at time τ is mS_0, and state w^3, when the price of the asset at time τ is uS_0.

(i) Show that the one period trinomial model is incomplete.

(ii) Show that the one period trinomial model is arbitrage–free if and only if

$$d < e^{r\delta t} < u,$$

where $\delta t = \tau - t_0$ and r is the continuously compounded interest rate between t_0 and τ.

Solution: (i) For a market to be complete a necessary (but not sufficient) condition is for the market to have at least as many securities as states. The trinomial model has fewer securities (two) than states (three) and therefore it is an incomplete market model.

(ii) The price vector at time t_0 corresponding to a \$1 cash position and one unit of the asset is $S_{t_0} = \begin{pmatrix} 1 \\ S_0 \end{pmatrix}$. The payoff matrix of the model is

$$M_\tau = \begin{pmatrix} e^{r\delta t} & e^{r\delta t} & e^{r\delta t} \\ uS_0 & mS_0 & dS_0 \end{pmatrix},$$

since the future value at time τ of \$1 at time t_0 is $e^{r(\tau-t_0)} = e^{r\delta t}$, where $\delta t = \tau - t_0$.

The one period trinomial model is arbitrage–free if and only if there exists a vector $Q = \begin{pmatrix} Q_1 \\ Q_2 \\ Q_3 \end{pmatrix}$ with $Q_1 > 0$, $Q_2 > 0$, $Q_3 > 0$ such that $M_\tau Q = S_{t_0}$. Note that

$$M_\tau Q = S_{t_0} \quad \Longleftrightarrow \quad \begin{pmatrix} e^{r\delta t} & e^{r\delta t} & e^{r\delta t} \\ uS_0 & mS_0 & dS_0 \end{pmatrix} \begin{pmatrix} Q_1 \\ Q_2 \\ Q_3 \end{pmatrix} = \begin{pmatrix} 1 \\ S_0 \end{pmatrix}$$

$$\Longleftrightarrow \quad \begin{cases} e^{r\delta t}Q_1 + e^{r\delta t}Q_2 + e^{r\delta t}Q_3 &= 1 \\ uS_0Q_1 + mS_0Q_2 + dS_0Q_3 &= S_0 \end{cases}$$

$$\Longleftrightarrow \quad \left\{ \begin{array}{rcl} Q_1 + Q_2 + Q_3 & = & e^{-r\delta t} \\ uQ_1 + mQ_2 + dQ_3 & = & 1 \end{array} \right.$$

$$\Longleftrightarrow \quad \left\{ \begin{array}{rcl} Q_1 + Q_3 & = & e^{-r\delta t} - Q_2 \\ uQ_1 + dQ_3 & = & 1 - mQ_2 \end{array} \right. \qquad (3.34)$$

By solving (3.34) for Q_1 and Q_3 in terms of Q_2, we obtain that

$$Q_1 = \frac{1 - mQ_2 - d(e^{-r\delta t} - Q_2)}{u - d} = e^{-r\delta t}\frac{e^{r\delta t} - d}{u - d} - Q_2\frac{m - d}{u - d}; \qquad (3.35)$$

$$Q_3 = \frac{u(e^{-r\delta t} - Q_2) - (1 - mQ_2)}{u - d} = e^{-r\delta t}\frac{u - e^{r\delta t}}{u - d} - Q_2\frac{u - m}{u - d}. \qquad (3.36)$$

We will use (3.35–3.36) to show that the one period trinomial model is arbitrage–free if and only if $d < e^{r\delta t} < u$ by double implication as follows:

- "If the one period trinomial model is arbitrage–free, then $d < e^{r\delta t} < u$"

If the one period trinomial model is arbitrage–free, then there exists a solution to (3.35–3.36) with $Q_1 > 0$, $Q_2 > 0$, $Q_3 > 0$.
If $e^{r\delta t} \leq d$, then $e^{r\delta t} - d \leq 0$ and from (3.35) we obtain that

$$Q_1 = e^{-r\delta t}\frac{e^{r\delta t} - d}{u - d} - Q_2\frac{m - d}{u - d} \leq -Q_2\frac{m - d}{u - d} < 0,$$

since $Q_2 > 0$ and $d < m < u$, which is a contradiction. Thus, $d < e^{r\delta t}$.
If $u \leq e^{-r\delta t}$, then $u - e^{-r\delta t} \leq 0$ and from (3.36) we obtain that

$$Q_3 = e^{-r\delta t}\frac{u - e^{r\delta t}}{u - d} - Q_2\frac{u - m}{u - d} \leq -Q_2\frac{u - m}{u - d} < 0,$$

since $Q_2 > 0$ and $d < m < u$, which is a contradiction. Thus, $e^{r\delta t} < u$.
We conclude that, if the one period trinomial model is arbitrage–free, then $d < e^{r\delta t} < u$.

- "If $d < e^{r\delta t} < u$, then the one period trinomial model is arbitrage–free"

If $d < e^{r\delta t} < u$, then $\frac{e^{r\delta t} - d}{u - d} > 0$ and $\frac{u - e^{r\delta t}}{u - d} > 0$. From (3.35) and (3.36), it follows that we can choose $Q_2 > 0$ small enough[1] such that

$$Q_1 = e^{-r\delta t}\frac{e^{r\delta t} - d}{u - d} - Q_2\frac{m - d}{u - d} > 0;$$

$$Q_3 = e^{-r\delta t}\frac{u - e^{r\delta t}}{u - d} - Q_2\frac{u - m}{u - d} > 0.$$

Thus, a solution $Q = \begin{pmatrix} Q_1 \\ Q_2 \\ Q_3 \end{pmatrix}$ with $Q_1 > 0$, $Q_2 > 0$, $Q_3 > 0$ can be found for $M_\tau Q = S_{t_0}$, and therefore the one period model is arbitrage–free. $\quad\Box$

---

[1] For example, we can choose

$$Q_2 = \frac{1}{2}\min\left(e^{-r\delta t}\frac{e^{r\delta t} - d}{m - d}, e^{-r\delta t}\frac{u - e^{r\delta t}}{u - m}\right) > 0.$$

Problem 6: In three months, the value of an asset with spot price $40 will be either $32, $38, $42, or $44. The value of a three months European call option with strike $36 on this asset is $8 and the value of a three months European put option with strike $40 on this asset is $5. For simplicity, assume zero risk free rates. Consider the one period market model with the following four securities and the following four states in three months:

Securities:
- cash;
- asset;
- three months call with strike $36;
- three months put with strike $40;

Market states:
- asset price $32;
- asset price $38;
- asset price $42;
- asset price $44.

(i) Find the payoff matrix of this model, and show that the four securities are non–redundant.

(ii) Show that the one period market model is complete.

(iii) How do you replicate a bull spread made of a long position in a three months call with strike $34 and a short position in a three months call with strike $40 in this one period market?

(iv) Show that this model is not arbitrage–free, and find an arbitrage opportunity.

Solution: (i) This is a one period market model with $m = 4$ securities and with $n = 4$ states. The price vector of the securities at time 0, corresponding to a $1 cash position and to positions equal to one unit of each of the other securities, is

$$ S_0 = \begin{pmatrix} 1 \\ 40 \\ 8 \\ 5 \end{pmatrix}. \tag{3.37}$$

For $j = 1 : 4$, let $S_{j,1/4}$ be the vector of the four possible prices of asset j in three months. The price vectors $S_{1,1/4}$ of cash and $S_{2,1/4}$ of the first asset are

$$ S_{1,1/4} = (1\ 1\ 1\ 1); \tag{3.38}$$
$$ S_{2,1/4} = (32\ 38\ 42\ 44). \tag{3.39}$$

The value at maturity T of a call option with strike 36 is $\max(S(T) - 36, 0)$. Then, the vector $S_{3,1/4}$ of the four possible prices in three months of the three months call with strike $36 on the asset is

$$ S_{3,1/4} = \max(S_{2,1/4} - 36, 0) = (0\ 2\ 6\ 8), \tag{3.40}$$

where $S_{2,1/4}$ is the price vector of the asset in three months given by (3.39).

Since the value at maturity T of a put option with strike 40 is $\max(40 - S(T), 0)$, it follows that the vector $S_{4,1/4}$ of the four possible prices in three months of the three months European put option with strike $40 on the asset is

$$ S_{4,1/4} = \max(40 - S_{2,1/4}, 0) = (8\ 2\ 0\ 0), \tag{3.41}$$

where $S_{2,1/4}$ is the price vector of the asset in three months given by (3.39).

From (3.38–3.41), we conclude that the payoff matrix $M_{1/4}$ of this one period market model is the following 4×4 matrix:

$$M_{1/4} = \begin{pmatrix} S_{1,1/4} \\ S_{2,1/4} \\ S_{3,1/4} \\ S_{4,1/4} \end{pmatrix} = \begin{pmatrix} 1 & 1 & 1 & 1 \\ 32 & 38 & 42 & 44 \\ 0 & 2 & 6 & 8 \\ 8 & 2 & 0 & 0 \end{pmatrix}. \tag{3.42}$$

The four securities are non–redundant if their payoff vectors are linearly independent, i.e., if the row rank of the payoff matrix $M_{1/4}$ is 4. Note that the matrix $M_{1/4}$ is nonsingular since $\det(M_{1/4}) = 16 \neq 0$. Thus, the row rank of $M_{1/4}$ is 4 and therefore the four securities are non–redundant.

(ii) Recall that a one period market model with the same number of states as the number of securities is complete if and only if the payoff matrix of the market model is nonsingular. This is the case for this market model, since, as mentioned above, $\det(M_{1/4}) = 16 \neq 0$, and therefore the matrix $M_{1/4}$ is nonsingular,

(iii) The value at maturity T of a bull spread made of a long position in a three months call with strike \$34 and a short position in a three months call with strike \$40 is $\max(S(T) - 34, 0) - \max(S(T) - 40, 0)$. Then, the vector $V_{1/4}$ of the four possible prices in three months of the three months bull spread on the asset is

$$\begin{aligned} V_{1/4} &= \max(S_{2,1/4} - 34, 0) - \max(S_{2,1/4} - 40, 0) \\ &= (0\ 4\ 8\ 10) - (0\ 0\ 2\ 4) \\ &= (0\ 4\ 6\ 6), \end{aligned} \tag{3.43}$$

where $S_{2,1/4}$ is the price vector of the asset in three months given by (3.39).

The positions vector $\Theta = (\Theta_i)_{i=1:4}$ in a portfolio replicating the bull spread can be found by solving $V_{1/4} = \Theta^t M_{1/4}$, which is equivalent to solving

$$M_{1/4}^t \Theta = V_{1/4}^t.$$

By using an LU decomposition with row pivoting solver, we obtain that

$$\Theta = \text{linear\_solve\_lu\_row\_pivoting}\left(M_{1/4}^t, V_{1/4}^t\right) = \begin{pmatrix} 24 \\ -0.5 \\ 0.5 \\ -1 \end{pmatrix}.$$

We conclude that the replicating portfolio for the bull spread is made of a \$24 long cash position, short 0.5 units of the asset, long 0.5 three months calls with strike \$36, short 1 three months put with strike \$40.

(iv) By solving the linear system $M_{1/4}Q = S_0$, where $M_{1/4}$ and S_0 are given by (3.42) and (3.37), respectively, we obtain that

$$\Theta = \text{linear\_solve\_lu\_row\_pivoting}(M_{1/4}, S_0) = \begin{pmatrix} 1 \\ -1.5 \\ 0.5 \\ 1 \end{pmatrix}.$$

Since one of the entries of the state price vector Q is negative, we conclude that the one period market model is not arbitrage–free.

To find an arbitrage opportunity, we look for a portfolio with payoff $1 if state ω^2 occurs (i.e., the state corresponding to the negative entry $Q_2 = -1.5$ of the state vector Q) and payoff $0 if any of the other states occurs. In other words, if Θ_{arb} denotes the positions vector of this portfolio, we look for Θ_{arb} such that

$$V_{1/4} = \Theta_{arb}^t M_{1/4} = (0\ 1\ 0\ 0) \iff M_{1/4}^t \Theta_{arb} = \begin{pmatrix} 0 \\ 1 \\ 0 \\ 0 \end{pmatrix}.$$

Thus,

$$\Theta_{arb} = \text{linear\_solve\_lu\_row\_pivoting}\left(M_{1/4}^t, \begin{pmatrix} 0 \\ 1 \\ 0 \\ 0 \end{pmatrix} \right) = \begin{pmatrix} -36 \\ 1 \\ -1 \\ 0.5 \end{pmatrix}.$$

The value V_0 at time 0 of this portfolio is

$$V_0 = \Theta_{arb}^t S_0 = (-36\ \ 1\ \ -1\ \ 0.5) \begin{pmatrix} 1 \\ 40 \\ 8 \\ 5 \end{pmatrix} = -1.5.$$

In other words, we found a portfolio with negative value $V_0 = -1.5$ at time 0, i.e., a portfolio generating a positive cash flow of $1.50 to set up, which has nonnegative payoff $V_{1/4} = (0\ \ 1\ \ 0\ \ 0)$ in every state of the market in three months, i.e., the portfolio does not lose money regardless of the state of the market in three months. This constitutes an arbitrage opportunity. □

Problem 7: Two assets have spot prices $20 and $30, respectively. Assume that, in five months, the first asset will be worth either $18 or $22, and the second asset will be worth either $28 or $32. Also, assume that the risk–free rate for five months cash deposits is 0. Consider the one period market model with the following three securities and the following four states in five months:

Securities:
• cash;
• first asset;
• second asset.

Market states:
• first asset at $22 and second asset at $32;
• first asset at $22 and second asset at $28;
• first asset at $18 and second asset at $32;
• first asset at $18 and second asset at $28.

(i) Show that this one period market model is arbitrage–free.

(ii) Show that, in this one period market model, it is not possible to replicate a derivative security that pays $1 if the first state occurs (i.e., if, in five months, the first asset is worth $22 and the second asset is worth $32), and does not pay anything if any other state occurs. Conclude that the model is not complete.

Solution: (i) The price vector of the three securities at time 0 corresponding to a $1 cash position, one unit of the first asset, and one unit of the second asset is $S_0 = \begin{pmatrix} 1 \\ 20 \\ 30 \end{pmatrix}$. The 3×4 payoff matrix of this one period market model in five months, i.e., at time 5/12, is

$$M_{5/12} = \begin{pmatrix} 1 & 1 & 1 & 1 \\ 22 & 22 & 18 & 18 \\ 32 & 28 & 32 & 28 \end{pmatrix}.$$

Let $Q = \begin{pmatrix} \frac{1}{4} \\ \frac{1}{4} \\ \frac{1}{4} \\ \frac{1}{4} \end{pmatrix}$, and note that

$$M_{5/12}Q = \begin{pmatrix} 1 & 1 & 1 & 1 \\ 22 & 22 & 18 & 18 \\ 32 & 28 & 32 & 28 \end{pmatrix} \begin{pmatrix} \frac{1}{4} \\ \frac{1}{4} \\ \frac{1}{4} \\ \frac{1}{4} \end{pmatrix} = \begin{pmatrix} 1 \\ 20 \\ 30 \end{pmatrix} = S_0.$$

Thus, there exists a state price vector Q with all entries positive such that $M_{5/12}Q = S_0$, and therefore the one period market model is arbitrage–free.

(ii) Note that this market model has more states (four) than securities (three) and therefore is not complete.

The price vector at time 5/12 of a derivative security that pays $1 if the first state occurs and does not pay anything if any other state occurs is $V_{5/12} = (1 \ 0 \ 0 \ 0)$. This derivative security is replicable if and only if there exists a positions vector $\Theta = \begin{pmatrix} \Theta_1 \\ \Theta_2 \\ \Theta_3 \end{pmatrix}$ such that

$$\Theta^t M_{5/12} = V_{5/12} \iff \left(\Theta^t M_{5/12}\right)^t = V_{5/12}^t$$

$$\iff M_{5/12}^t \Theta = V_{5/12}^t \iff \begin{pmatrix} 1 & 22 & 32 \\ 1 & 22 & 28 \\ 1 & 18 & 32 \\ 1 & 18 & 28 \end{pmatrix} \begin{pmatrix} \Theta_1 \\ \Theta_2 \\ \Theta_3 \end{pmatrix} = \begin{pmatrix} 1 \\ 0 \\ 0 \\ 0 \end{pmatrix},$$

which can be written as

$$\begin{aligned}
\Theta_1 + 22\Theta_2 + 32\Theta_3 &= 1; & (3.44) \\
\Theta_1 + 22\Theta_2 + 28\Theta_3 &= 0; & (3.45) \\
\Theta_1 + 18\Theta_2 + 32\Theta_3 &= 0; & (3.46) \\
\Theta_1 + 18\Theta_2 + 28\Theta_3 &= 0. & (3.47)
\end{aligned}$$

By subtracting (3.45) from (3.44), we obtain that $4\Theta_3 = 1$, and therefore $\Theta_3 = \frac{1}{4}$. By subtracting (3.46) from (3.47), we obtain that $4\Theta_3 = 0$, and therefore $\Theta_3 = 0$, which is a contradiction.

The linear system (3.44–3.47) does not have a solution, and therefore there is no positions vector Θ such that $\Theta^t M_{5/12} = V_{5/12}$.

We conclude that the derivative security that pays \$1 if the first state occurs and does not pay anything if any other state occurs is not replicable. $\Box$

Problem 8: Consider two assets with spot prices \$30 and \$50, respectively. Assume that, in three months, the first asset will be worth either \$34 or \$24, and the second asset will be worth either \$56, \$51, or \$46. The value of a three months at–the–money European call option with strike \$30 on the first asset is \$2.5, the value of a three months at–the–money European call option with strike \$50 on the second asset is \$2.7, and the value of a three months European put option with strike \$52 on the second asset is \$4.1. Assume that the future value in three months of \$1 today is \$1.01. Consider the one period market model with the following six securities and the following six states in three months:

Securities:
- cash;
- first asset;
- second asset;
- three months ATM call on the first asset;
- three months ATM call on the second asset;
- three months put with strike \$52 on the second asset;

Market states:
- first asset at \$34 and second asset at \$56;
- first asset at \$34 and second asset at \$51;
- first asset at \$34 and second asset at \$46;
- first asset at \$24 and second asset at \$56;
- first asset at \$24 and second asset at \$51;
- first asset at \$24 and second asset at \$46.

(i) Show that the payoff matrix for this market model is nonsingular, and conclude that the market is complete.

(ii) Compute the state prices for this model, and show that the market is arbitrage–free.

(iii) Use risk–neutral pricing to find the value of a three months call option with strike \$52 on the second asset.

(iv) Use Put–Call parity to find the value of a three months call option with strike \$52 on the second asset, and compare it to the value computed at (iii).

(v) What is the value of a bear spread made of a long position in a three months put option with strike \$35 on the first asset and a short position in a three months put option with strike \$28 on the first asset?

Solution: This problem is, unfortunately, incorrect, since the one period market model is incomplete. We will establish this by proving that the payoff matrix of the model is singular.

To find the payoff matrix of this model, for $j = 1 : 6$, let $S_{j,1/4}$ be the vector of the six possible prices of asset j in three months, corresponding to a \$1 cash position and to positions equal to one unit of each of the other securities. The price vectors $S_{1,1/4}$ of cash, $S_{2,1/4}$ of the first asset, and $S_{3,1/4}$ of the second asset, respectively, are

$$S_{1,1/4} = (1.01 \; 1.01 \; 1.01 \; 1.01 \; 1.01 \; 1.01); \quad (3.48)$$
$$S_{2,1/4} = (34 \; 34 \; 34 \; 24 \; 24 \; 24); \quad (3.49)$$
$$S_{3,1/4} = (56 \; 51 \; 46 \; 56 \; 51 \; 46). \quad (3.50)$$

Since the value at maturity T of a call option with strike 30 is $\max(S(T) - 30, 0)$, it follows that the vector $S_{4,1/4}$ of the six possible prices in three months of the three months ATM call with strike \$30 on the first asset is

$$S_{4,1/4} = \max(S_{2,1/4} - 30, 0) = (4 \; 4 \; 4 \; 0 \; 0 \; 0), \quad (3.51)$$

where $S_{2,1/4}$ is the price vector of the first asset in three months given by (3.49).

The values at maturity T of a call option with strike \$50 and of a put option with strike \$52 are $\max(S(T) - 50, 0)$ and $\max(52 - S(T), 0)$, respectively. Thus, the vectors $S_{5,1/4}$ and $S_{6,1/4}$ of the six possible prices in three months of a three months ATM call with strike \$50 on the second asset and of a three months put with strike \$52 on the second asset are

$$S_{5,1/4} = \max(S_{3,1/4} - 50, 0) = (6 \; 1 \; 0 \; 6 \; 1 \; 0); \quad (3.52)$$
$$S_{6,1/4} = \max(52 - S_{3,1/4}, 0) = (0 \; 1 \; 6 \; 0 \; 1 \; 6), \quad (3.53)$$

where $S_{3,1/4}$ is the price vector of the second asset in three months given by (3.50).

From (3.48–3.53), we obtain that the 6×6 payoff matrix of this model is

$$M_{1/4} = \begin{pmatrix} S_{1,1/4} \\ S_{2,1/4} \\ S_{3,1/4} \\ S_{4,1/4} \\ S_{5,1/4} \\ S_{6,1/4} \end{pmatrix} = \begin{pmatrix} 1.01 & 1.01 & 1.01 & 1.01 & 1.01 & 1.01 \\ 34 & 34 & 34 & 24 & 24 & 24 \\ 56 & 51 & 46 & 56 & 51 & 46 \\ 4 & 4 & 4 & 0 & 0 & 0 \\ 6 & 1 & 0 & 6 & 1 & 0 \\ 0 & 1 & 6 & 0 & 1 & 6 \end{pmatrix}. \quad (3.54)$$

Note that $\det(M_{1/4}) = 0$. Thus, the payoff matrix $M_{1/4}$ is singular, and this one period market model is not complete. □

Problem 9: This exercise refers to the S&P 500 options prices from Table 3.2.

Consider a one period market model with the following nine securities:

P1200; P1275; P1350; P1375; C1375; C1400; C1450; C1550; C1600.

The nine states of the index price at maturity are as follows: seven states correspond to the midpoints between the strikes of the options above, i.e.,

$$\omega^2 : \{S(\tau) = 1237.50\}; \qquad \omega^6 : \{S(\tau) = 1425\};$$
$$\omega^3 : \{S(\tau) = 1312.50\}; \qquad \omega^7 : \{S(\tau) = 1500\};$$
$$\omega^4 : \{S(\tau) = 1362.50\}; \qquad \omega^8 : \{S(\tau) = 1575\};$$
$$\omega^5 : \{S(\tau) = 1387.50\};$$

Table 3.2: Dec 2012 SPX option prices on 3/9/2012

Call Strike	Price	Volume	Put Strike	Price	Volume
C1175	225.40	250	P1175	46.60	1
C1200	205.55	215	P1200	51.55	3204
C1225	186.20	1	P1225	57.15	1401
C1250	167.50	650	P1250	63.30	104
C1275	149.15	163	P1275	70.15	56
C1300	131.70	1	P1300	77.70	150
C1325	115.25	40	P1325	86.20	200
C1350	99.55	320	P1350	95.30	10118
C1375	84.90	1002	P1375	105.30	1250
C1400	71.10	5300	P1400	116.55	1250
C1425	58.70	4	P1425	129.00	200
C1450	47.25	9050	P1450	143.20	1
C1500	29.25	1000	P1500	173.95	6
C1550	15.80	1000	P1550	210.80	9
C1575	11.10	200	P1575	230.90	0
C1600	7.90	546	P1600	252.40	9

the first and last state are

$$\omega^1 : \{S(\tau) = 950\}; \quad \omega^9 : \{S(\tau) = 1675\}.$$

(i) Find the payoff matrix M_τ of this one period market model, and show that the model is complete.

(ii) Find the state prices vector Q and show that the model is arbitrage–free.

(iii) Compute the root–mean–squared error (RMSE) of this model. Comment on the precision of this nine securities model compared to the seven securities model from section 3.5 in Stefanica [3].

Solution: (i) The payoffs at maturity of call and put options are

$$C(T) = \max(S(T) - K, 0); \tag{3.55}$$
$$P(T) = \max(K - S(T), 0). \tag{3.56}$$

Then, the payoff matrix M_τ of this one period market model with nine securities and nine states is the following 9×9 matrix:

$$M_\tau = \begin{pmatrix} 250 & 0 & 0 & 0 & 0 & 0 & 0 & 0 & 0 \\ 325 & 37.50 & 0 & 0 & 0 & 0 & 0 & 0 & 0 \\ 400 & 112.50 & 37.50 & 0 & 0 & 0 & 0 & 0 & 0 \\ 425 & 137.50 & 62.50 & 12.5 & 0 & 0 & 0 & 0 & 0 \\ 0 & 0 & 0 & 0 & 12.50 & 50 & 125 & 200 & 300 \\ 0 & 0 & 0 & 0 & 0 & 25 & 100 & 175 & 275 \\ 0 & 0 & 0 & 0 & 0 & 0 & 50 & 125 & 225 \\ 0 & 0 & 0 & 0 & 0 & 0 & 0 & 25 & 125 \\ 0 & 0 & 0 & 0 & 0 & 0 & 0 & 0 & 75 \end{pmatrix}. \tag{3.57}$$

For example, $M_\tau(4,3)$ is the payoff of the fourth security P1375, i.e., of the put option with strike 1375, if state ω^3 occurs, i.e., if $S(\tau) = 1312.50$, and from (3.56) we find that

$$M_\tau(4,3) \;=\; \max(1375 - S(\tau), 0) \;=\; \max(1375 - 1312.50, 0) \;=\; 62.50.$$

Similarly, $M_\tau(6,8)$ is the payoff of the sixth security C1400, i.e., of the call option with strike 1400, if state ω^8 occurs, i.e., if $S(\tau) = 1575$, and from (3.55) we find that

$$M_\tau(6,8) \;=\; \max(S(\tau) - 1400, 0) \;=\; \max(1575 - 1400, 0) \;=\; 175.$$

The matrix M_τ is made of a 4×4 lower triangular block and a 5×5 upper triangular block. Thus, its determinant $\det(M_\tau)$ is the product of the diagonal entries of M_τ, and therefore nonzero. Then, the matrix M_τ is nonsingular and therefore the one period market model is complete.

(ii) The price vector of the nine securities at time t_0 is the vector of their mid prices from Table 3.2, i.e.,

$$S_{t_0} \;=\; \begin{pmatrix} 51.55 \\ 70.15 \\ 95.30 \\ 105.30 \\ 84.90 \\ 71.10 \\ 47.25 \\ 15.80 \\ 7.90 \end{pmatrix}. \tag{3.58}$$

By solving the linear system $M_\tau Q = S_{t_0}$ where M_τ is given by (3.57) and S_{t_0} is given by (3.58), we obtain that the state prices vector Q is

$$Q \;=\; \text{linear\_solve\_lu\_row\_pivoting}(M_\tau, S_{t_0}) \;=\; \begin{pmatrix} 0.2062 \\ 0.0836 \\ 0.0911 \\ 0.0383 \\ 0.0327 \\ 0.1173 \\ 0.2077 \\ 0.1053 \\ 0.1053 \end{pmatrix}.$$

Since all the entries of Q (i.e., the state prices) are positive, we conclude that the one period market model is arbitrage–free.

(iii) The one period options market model above is both arbitrage–free and complete and therefore all the options can be priced using risk–neutral pricing. The risk neutral values[2] of the options can be found in Table 3.3.

Denote by $V_{model}(i)$ for $i = 1 : 23$ the risk–neutral values of the $32 - 9 = 23$ options from Table 3.2 that were not chosen to be the securities in this model, and let $V_{mid}(i)$,

---

[2]Note that the risk–neutral values from Table 3.3 of the nine options that were chosen as securities in this one period model, i.e., P1200, P1275, P1350, P1375, C1375, C1400, C1450, C1550, and C1600, are the same as the given mid option prices.

Table 3.3: Risk-neutral option prices

Call Strike	Risk–Neutral Price	Put Strike	Risk–Neutral Price
C1175	223.49	P1175	46.39
C1200	203.96	P1200	51.55
C1225	184.42	P1225	56.70
C1250	165.94	P1250	62.90
C1275	148.50	P1275	70.15
C1300	131.05	P1300	77.39
C1325	114.75	P1325	85.78
C1350	99.59	P1350	95.30
C1375	84.90	P1375	105.30
C1400	71.10	P1400	116.19
C1425	57.71	P1425	127.48
C1450	47.25	P1450	141.71
C1500	26.33	P1500	170.17
C1550	15.80	P1550	209.01
C1575	10.53	P1575	228.42
C1600	7.90	P1600	250.48

$i = 1 : 23$, be the mid prices of the options. The percentage root–mean–squared error (RMSE) of the model is

$$\text{RMSE} = \sqrt{\frac{1}{23} \sum_{i=1}^{23} \frac{(V_{model}(i) - V_{mid}(i))^2}{V_{mid}(i)^2}} = 0.0250 = 2.50\%.$$

Recall that the RMSE error of the one period model with seven securities from section 3.5 in Stefanica [3] is 5.06%. The one period model with nine securities considered here has a smaller RMSE error of 2.50%. However, note that, generally speaking, there is no guarantee that the root–mean–squared error of a model with more securities would be smaller than the RMSE error of a model with fewer securities.   □

Problem 10: This problem refers to the S&P 500 options prices from Table 3.2.

Consider a one period market model with the following seven securities:

P1200; P1300; P1400; C1400; C1450; C1550; C1600.

The states of this market are the midpoints of the strikes, i.e.,

$$\omega^2 : \{S(\tau) = 1250\}; \qquad \omega^5 : \{S(\tau) = 1500\};$$
$$\omega^3 : \{S(\tau) = 1350\}; \qquad \omega^6 : \{S(\tau) = 1575\};$$
$$\omega^4 : \{S(\tau) = 1425\};$$

and the first and last state are

$$\omega^1 : \{S(\tau) = 1100\}; \qquad \omega^7 : \{S(\tau) = 1700\}.$$

Show that this market model is not arbitrage–free.

Solution: The initial price vector S_{t_0} of the seven securities and the payoff matrix M_τ of the model are

$$
S_{t_0} = \begin{pmatrix} 51.55 \\ 77.70 \\ 116.55 \\ 71.10 \\ 47.25 \\ 15.80 \\ 7.90 \end{pmatrix} ; \quad M_\tau = \begin{pmatrix} 100 & 0 & 0 & 0 & 0 & 0 & 0 \\ 200 & 50 & 0 & 0 & 0 & 0 & 0 \\ 300 & 150 & 50 & 0 & 0 & 0 & 0 \\ 0 & 0 & 0 & 25 & 100 & 175 & 300 \\ 0 & 0 & 0 & 0 & 50 & 125 & 250 \\ 0 & 0 & 0 & 0 & 0 & 25 & 150 \\ 0 & 0 & 0 & 0 & 0 & 0 & 100 \end{pmatrix}.
$$

By solving the linear system $M_\tau Q = S_{t_0}$, we obtain that

$$
Q = \begin{pmatrix} 0.5155 \\ -0.5080 \\ 0.7620 \\ 0.1700 \\ 0.1550 \\ 0.1580 \\ 0.0790 \end{pmatrix}.
$$

Since not all the entries of Q are positive, it follows that the market model is not arbitrage–free. $\square$

Chapter 4

Eigenvalues and eigenvectors.

4.1 Exercises

1. Let A and B be square matrices of the same size. Show that, if v is an eigenvector of both A and B, then v is also an eigenvector of the matrix $M = c_1 A + c_2 B$, where c_1 and c_2 are constants. What is the eigenvalue of M corresponding to the eigenvector v?

2. Let λ and v be an eigenvalue and the corresponding eigenvector of the matrix A. Let d be a constant number, and let I be the identity matrix.

   (i) Show that v is an eigenvector of the matrix $B = dI + A + A^2$, and find the corresponding eigenvalue.

   (ii) If A is a nonsingular matrix, show that v is an eigenvector of the matrix $M = dI + A + A^{-1}$, and find the corresponding eigenvalue.

3. Let A be a 3×3 nonsingular matrix with eigenvalues -2, 1, and 3. What are the eigenvalues of $A^2 + 2I - 3A^{-1}$?

4. (i) Show that the eigenvalues of the matrix

$$
A = \begin{pmatrix}
8 & -18 & -30 & -24 \\
18 & -37 & -60 & -48 \\
-9 & 18 & 29 & 24 \\
0 & 0 & 0 & -1
\end{pmatrix}
$$

are -1, with multiplicity 3, and 2, with multiplicity 1.

Show that $\begin{pmatrix} 0 \\ 1 \\ 1 \\ -2 \end{pmatrix}$, $\begin{pmatrix} 2 \\ 2 \\ -3 \\ 3 \end{pmatrix}$, $\begin{pmatrix} 2 \\ 1 \\ 0 \\ 0 \end{pmatrix}$ are three linearly independent eigenvectors corresponding to the eigenvalue -1, and show that there exists only one

93

linearly independent eigenvector of the matrix A corresponding to the eigen-

value 2, e.g., $\begin{pmatrix} 1 \\ 2 \\ -1 \\ 0 \end{pmatrix}$.

(ii) Show that the eigenvalues of the matrix

$$B = \begin{pmatrix} 10 & -20 & -32 & -26 \\ 18 & -41 & -68 & -54 \\ -14 & 19 & 26 & 23 \\ 7 & 1 & 9 & 4 \end{pmatrix}$$

are -1, with multiplicity 3, and 2, with multiplicity 1.

Show that there exists only one linearly independent eigenvector of the matrix

B corresponding to the eigenvalue -1, e.g., $\begin{pmatrix} 0 \\ -1 \\ -1 \\ 2 \end{pmatrix}$, and show that $\begin{pmatrix} 1 \\ 2 \\ -1 \\ 0 \end{pmatrix}$

is an eigenvector corresponding to the eigenvalue 2.

5. Let $A = \begin{pmatrix} 2 & -1 \\ 1 & 4 \end{pmatrix}$.

   (i) What is the characteristic polynomial of A?

   (ii) Find the eigenvalues and the eigenvectors of A.

6. (i) Find the eigenvalues and the eigenvectors of the lower triangular matrix

   $$L = \begin{pmatrix} 2 & 0 & 0 \\ 1 & -3 & 0 \\ -1 & 2 & -1 \end{pmatrix}.$$

   (ii) Find the eigenvalues and the eigenvectors of the upper triangular matrix

   $$U = \begin{pmatrix} -2 & -1 & 3 \\ 0 & 1 & 2 \\ 0 & 0 & 3 \end{pmatrix}.$$

7. Let A be a square matrix such that $A^2 = A$. Show that any eigenvalue of A is either 0 or 1.

   Note: A matrix A with the property that $A^2 = A$ is called an idempotent matrix.

8. Let A be a square matrix with the property that there exists a positive integer p such that $A^p = 0$. Show that any eigenvalue of A must be equal to 0.

   Note: A matrix A with the property that $A^p = 0$ for a positive integer p is called a nilpotent matrix.

9. Let $v \neq 0$ be a column vector of size n, and let $A = vv^t$ be an $n \times n$ matrix.

(i) Show that the matrix A has exactly one non-zero eigenvalue.

(ii) Find the eigenvalues and the eigenvectors of A.

10. Find the eigenvalues and the eigenvectors of the $n \times n$ matrix

$$\begin{pmatrix} d & 1 & \cdots & 1 \\ 1 & d & \cdots & 1 \\ \vdots & \vdots & \ddots & \vdots \\ 1 & 1 & \cdots & d \end{pmatrix},$$

where $d \in \mathbb{R}$ is a constant.

Hint: Note that

$$\begin{pmatrix} d & 1 & \cdots & 1 \\ 1 & d & \cdots & 1 \\ \vdots & \vdots & \ddots & \vdots \\ 1 & 1 & \cdots & d \end{pmatrix} = (d-1)I + \begin{pmatrix} 1 & 1 & \cdots & 1 \\ 1 & 1 & \cdots & 1 \\ \vdots & \vdots & \ddots & \vdots \\ 1 & 1 & \cdots & 1 \end{pmatrix}$$

$$= (d-1)I + \mathbf{1} \cdot \mathbf{1}^t,$$

where I is the identity matrix and $\mathbf{1} = \begin{pmatrix} 1 \\ 1 \\ \vdots \\ 1 \end{pmatrix}$.

11. (i) Let A be a symmetric matrix of size n, and let λ_i, $i = 1 : n$, be the eigenvalues of A, with corresponding eigenvectors v_i, $i = 1 : n$. What are the eigenvalues and the eigenvectors of A^2?

(ii) Let A be a symmetric matrix of size n. Let ϕ_i, $i = 1 : n$, be the eigenvalues of the matrix A^2, with corresponding eigenvectors w_i, $i = 1 : n$. What can you say about the eigenvalues and the eigenvectors of A?

12. Let A be a square matrix of size n, and let S be a nonsingular matrix of size n.

(i) Let λ and v be an eigenvalue and the corresponding eigenvector of A. Show that λ is also an eigenvalue of the matrix $S^{-1}AS$. What is the corresponding eigenvector?

(ii) Show that the matrix $S^{-1}AS$ has the same characteristic polynomial as A, i.e., show that
$$P_{S^{-1}AS}(t) = P_A(t), \quad \forall\, t \in \mathbb{R}.$$

13. Let A be a square matrix with real entries. If $\lambda = a + ib$ is a complex eigenvalue of A (i.e., with $b \neq 0$), show that $\bar{\lambda} = a - ib$, the complex conjugate of λ, is also an eigenvalue of A.

14. The 2×2 matrix A has eigenvalues 1 and -2 with corresponding eigenvectors
$\begin{pmatrix} -1 \\ 2 \end{pmatrix}$ and $\begin{pmatrix} 3 \\ 1 \end{pmatrix}$. If $v = \begin{pmatrix} 2 \\ -3 \end{pmatrix}$, find Av.

15. Let $A = \begin{pmatrix} -1 & 2 \\ 2 & 2 \end{pmatrix}$.

   (i) Find the eigenvalues and the eigenvectors of the matrix A.

   (ii) What is the diagonal form of A?

   (iii) Compute A^{12}.

16. Let $A = \begin{pmatrix} a & b \\ c & d \end{pmatrix}$ be a 2×2 matrix, and let

$$P_A(t) \;=\; t^2 - (a + d)t + (ad - bc)$$

   be the characteristic polynomial associated to A.
   Show that $P_A(A) = 0$, i.e., show that

$$A^2 - (a + d)A + (ad - bc)I = 0.$$

   Note: This is the 2×2 case of the Cayley-Hamilton theorem which states that $P_A(A) = 0$ for any square matrix A.

17. Let A and B be square matrices of the same size.

   (i) If B is nonsingular, show that the matrices AB and BA have the same characteristic polynomial, i.e., show that

$$\det(tI - AB) \;=\; \det(tI - BA).$$

   (ii) Show that the characteristic polynomial of AB is the same as the characteristic polynomial of BA even if B is a singular matrix.

   Hint: If B is a singular matrix, it is possible to find a number $\epsilon > 0$ as small as needed such that the matrix $B + \epsilon I$ is nonsingular.

   (iii) Show that the matrices AB and BA have the same eigenvalues.

18. (i) Let A and B be square matrices of the same size. Show that the traces of the matrices AB and BA are equal, i.e., show that $\text{tr}(AB) = \text{tr}(BA)$.

   (ii) Show that you cannot find two $n \times n$ matrices A and B such that

$$AB - BA \;=\; I,$$

   where I is the $n \times n$ identity matrix.

19. Recall that the matrix

$$
B_4 = \begin{pmatrix} 2 & -1 & 0 & 0 \\ -1 & 2 & -1 & 0 \\ 0 & -1 & 2 & -1 \\ 0 & 0 & -1 & 2 \end{pmatrix}
$$

has four eigenvectors v_1, v_2, v_3, v_4 given by

$$
v_j(i) = \sin\left(\frac{ij\pi}{5}\right), \quad \forall\, i = 1 : 4,
$$

for $j = 1 : 4$.

Are the vectors v_1, v_2, v_3, and v_4 orthogonal and of norm 1?

20. Show that the matrix $\begin{pmatrix} 3 & -1 & -1 \\ 0 & 2 & 2 \\ 0 & 1 & 1 \end{pmatrix}$ is a weakly diagonally dominant singular matrix.

21. Show that the matrix

$$
\begin{pmatrix} -5 & -1 & 0.25 & -1 \\ 2 & 4 & -1 & 3 \\ 0.5 & -2 & 3 & -1 \\ 1 & 0.5 & 0 & -6 \end{pmatrix}
$$

is nonsingular.

22. Let A be an $n \times n$ matrix. For every $j = 1 : n$, let

$$
R_j = \sum_{k=1:n,\,k \neq j} |A(j,k)|,
$$

and denote by D_j the disc of center $A(j,j)$ and radius R_j, i.e.,

$$
D_j = \{z \in \mathbb{C} \text{ such that } |z - A(j,j)| \leq R_j\};
$$

note that D_j is called a Gershgorin disk corresponding to the matrix A.

A more general form of Gershgorin's theorem states that, if a Gershgorin disk D_i is disjoint from the union of the other $n-1$ Gershgorin disks of A, then exactly one eigenvalue of the matrix A is in the disk D_i.

Use this result to show that all the eigenvalues of the matrix

$$
\begin{pmatrix} -4 & 1 & 0 & -0.5 \\ 0 & 0.1 & 0 & -0.2 \\ 1 & 2 & 5 & -1 \\ 0.25 & -0.15 & 0.1 & -1 \end{pmatrix}
$$

are real numbers.

23. Show that all the eigenvalues of the matrix

$$\begin{pmatrix} 2 & 0.0012 & -0.0003 & 0.0015 \\ -0.0002 & -1.25 & 0.0010 & -0.0001 \\ 0 & 0.0016 & 3 & 0.0009 \\ -0.0011 & -0.0008 & -0.0002 & -2.5 \end{pmatrix}$$

are real numbers, and find estimates for the values of the eigenvalues of the matrix with 0.005 accuracy.

24. Let A be an $n \times n$ matrix given by

$$\begin{aligned} A(i,i) &= 2, & \forall\, i = 1:n; \\ A(i,i-1) &= 1, & \forall\, i = 2:n; \\ A(j,k) &= 0, & \text{otherwise.} \end{aligned}$$

Find the eigenvalues and the eigenvectors of A.

25. Let J be an $n \times n$ matrix given by

$$\begin{aligned} A(i,i) &= a, & \forall\, i = 1:n; \\ A(i,i+1) &= b, & \forall\, i = 1:(n-1); \\ A(j,k) &= 0, & \text{otherwise,} \end{aligned}$$

where $a, b \in \mathbb{R}$ are constants. Find the eigenvalues and the eigenvectors of J.

Note: The matrix J is called a Jordan block if $b = 1$.

26. The 2–norm of a symmetric matrix A is given by

$$\|A\|_2 = \max_{\lambda \text{ eigenvalue of } A} |\lambda|, \tag{4.1}$$

i.e., $\|A\|_2$ is equal to the largest absolute value of all the eigenvalues of A.

The Forward Euler finite difference discretization of the heat PDE is convergent if and only if $\|A_N\|_2 \leq 1$ for all $N \geq 2$, where A_N is the $N \times N$ matrix given by

$$A_N = \begin{pmatrix} 1-2\alpha & -\alpha & \cdots & & 0 \\ -\alpha & \ddots & \ddots & & \vdots \\ \vdots & \ddots & \ddots & & -\alpha \\ 0 & \cdots & & -\alpha & 1-2\alpha \end{pmatrix},$$

with $\alpha > 0$ a positive constant.

Show that $\|A_N\|_2 \leq 1$ for all $N \geq 2$ if and only if $0 < \alpha \leq \frac{1}{2}$.

Hint: Recall that the matrix

$$T_N = \begin{pmatrix} d & -a & \cdots & & 0 \\ -a & \ddots & \ddots & & \vdots \\ \vdots & \ddots & \ddots & & -a \\ 0 & \cdots & & -a & d \end{pmatrix}$$

has the following N different eigenvalues:

$$\lambda_j \;=\; d - 2a \cos\left(\frac{\pi j}{N+1}\right), \quad \text{for} \quad j = 1 : N.$$

27. The Backward Euler finite difference discretization of the heat PDE is convergent if and only if $\|A_N^{-1}\|_2 \le 1$ for all $N \ge 2$, where

$$A_N \;=\; \begin{pmatrix} 1+2\alpha & \alpha & \cdots & 0 \\ \alpha & \ddots & \ddots & \vdots \\ \vdots & \ddots & \ddots & \alpha \\ 0 & \cdots & \alpha & 1+2\alpha \end{pmatrix},$$

with $\alpha > 0$ a positive constant.

Show that $\|A_N^{-1}\|_2 \le 1$ for all $N \ge 2$ for any $\alpha > 0$.

4.2 Solutions to Chapter 4 Exercises

Problem 1: Let A and B be square matrices of the same size. Show that, if v is an eigenvector of both A and B, then v is also an eigenvector of the matrix $M = c_1 A + c_2 B$, where c_1 and c_2 are constants. What is the eigenvalue of M corresponding to the eigenvector v?

Solution: Let λ_1 be the eigenvalue of the matrix A corresponding to the eigenvector v, and let λ_2 be the eigenvalue of the matrix B corresponding to the eigenvector v. Then, $Av = \lambda_1 v$ and $Bv = \lambda_2 v$, and therefore

$$
\begin{aligned}
Mv &= (c_1 A + c_2 B)v = c_1 Av + c_2 Bv = c_1 \lambda_1 v + c_2 \lambda_2 v \\
&= (c_1 \lambda_1 + c_2 \lambda_2)v.
\end{aligned}
$$

Thus, $Mv = (c_1 \lambda_1 + c_2 \lambda_2)v$, and we conclude that v is an eigenvector of the matrix M and $c_1 \lambda_1 v + c_2 \lambda_2$ is the corresponding eigenvalue. □

Problem 2: Let λ and v be an eigenvalue and the corresponding eigenvector of the matrix A. Let d be a constant number, and let I be the identity matrix.

(i) Show that v is an eigenvector of the matrix $B = dI + A + A^2$, and find the corresponding eigenvalue.

(ii) If A is a nonsingular matrix, show that v is an eigenvector of the matrix $M = dI + A + A^{-1}$, and find the corresponding eigenvalue.

Solution: (i) If λ and $v \neq 0$ are an eigenvalue and a corresponding eigenvector of the matrix A, then

$$Av = \lambda v \tag{4.2}$$

and

$$A^2 v = A \cdot (Av) = A \cdot (\lambda v) = \lambda \cdot Av = \lambda \cdot \lambda v = \lambda^2 v. \tag{4.3}$$

Then, from (4.2) and (4.3), we find that

$$
\begin{aligned}
Bv &= (dI + A + A^2)v = dv + Av + A^2 v = dv + \lambda v + \lambda^2 v \\
&= (d + \lambda + \lambda^2)v.
\end{aligned}
$$

We conclude that v is an eigenvector of the matrix B and $d + \lambda + \lambda^2$ is the corresponding eigenvalue.

(ii) Recall that, if A is a nonsingular matrix and λ and $v \neq 0$ are an eigenvalue and a corresponding eigenvector of A, then $\lambda \neq 0$ and

$$A^{-1}v = \frac{1}{\lambda}v. \tag{4.4}$$

From (4.2) and (4.4), we find that

$$
\begin{aligned}
Mv &= (dI + A + A^{-1})v = dv + Av + A^{-1}v = dv + \lambda v + \frac{1}{\lambda}v \\
&= \left(d + \lambda + \frac{1}{\lambda}\right)v.
\end{aligned}
$$

We conclude that v is an eigenvector of the matrix M and $d + \lambda + \frac{1}{\lambda}$ is the corresponding eigenvalue. $\square$

Problem 3: Let A be a 3×3 nonsingular matrix with eigenvalues -2, 1, and 3. What are the eigenvalues of $A^2 + 2I - 3A^{-1}$?

Solution: Recall that, if A is a nonsingular matrix and λ and v are an eigenvalue and a corresponding eigenvector of A, then $Av = \lambda v$ and

$$A^2 v = \lambda^2 v; \tag{4.5}$$

$$A^{-1} v = \frac{1}{\lambda} v. \tag{4.6}$$

From (4.5) and (4.6), we obtain that

$$(A^2 + 2I - 3A^{-1})v = \left(\lambda^2 + 2 - \frac{3}{\lambda} \right) v. \tag{4.7}$$

Let $\lambda_1 = -2$, $\lambda_2 = 1$, and $\lambda_3 = 3$ be the eigenvalues of the matrix A. From (4.7), we obtain the following eigenvalues of the matrix $A^2 + 2I - 3A^{-1}$ corresponding to λ_1, λ_2, and λ_3:

$$\lambda_1^2 + 2 - \frac{3}{\lambda_1} = 7.5;$$

$$\lambda_2^2 + 2 - \frac{3}{\lambda_2} = 0;$$

$$\lambda_3^2 + 2 - \frac{3}{\lambda_3} = 10.$$

In other words, 7.5, 0, and 10 are eigenvalues of $A^2 + 2I - 3A^{-1}$.
Since the 3×3 matrix $A^2 + 2I - 3A^{-1}$ has exactly three eigenvalues, we conclude that we found all the eigenvalues of $A^2 + 2I - 3A^{-1}$, which are 0, 7.5, and 10. $\square$

Problem 4: (i) Show that the eigenvalues of the matrix

$$A = \begin{pmatrix} 8 & -18 & -30 & -24 \\ 18 & -37 & -60 & -48 \\ -9 & 18 & 29 & 24 \\ 0 & 0 & 0 & -1 \end{pmatrix}$$

are -1, with multiplicity 3, and 2, with multiplicity 1.

Show that $\begin{pmatrix} 0 \\ 1 \\ 1 \\ -2 \end{pmatrix}$, $\begin{pmatrix} 2 \\ 2 \\ 3 \\ 3 \end{pmatrix}$, $\begin{pmatrix} 2 \\ 1 \\ 0 \\ 0 \end{pmatrix}$ are three linearly independent eigenvec-

tors corresponding to the eigenvalue -1, and show that there exists only one linearly independent eigenvector of the matrix A corresponding to the eigenvalue 2, e.g., $\begin{pmatrix} 1 \\ 2 \\ -1 \\ 0 \end{pmatrix}$.

(ii) Show that the eigenvalues of the matrix

$$B = \begin{pmatrix} 10 & -20 & -32 & -26 \\ 18 & -41 & -68 & -54 \\ -14 & 19 & 26 & 23 \\ 7 & 1 & 9 & 4 \end{pmatrix}$$

are -1, with multiplicity 3, and 2, with multiplicity 1.

Show that there exists only one linearly independent eigenvector of the matrix B

corresponding to the eigenvalue -1, e.g., $\begin{pmatrix} 0 \\ -1 \\ -1 \\ 2 \end{pmatrix}$, and show that $\begin{pmatrix} 1 \\ 2 \\ -1 \\ 0 \end{pmatrix}$ is an

eigenvector corresponding to the eigenvalue 2.

Solution: (i) The characteristic polynomial of the matrix A is

$$
\begin{aligned}
P_A(t) &= \det(tI - A) = \det \begin{pmatrix} t-8 & 18 & 30 & 24 \\ -18 & t+37 & 60 & 48 \\ 9 & -18 & t-29 & -24 \\ 0 & 0 & 0 & t+1 \end{pmatrix} \\
&= (t+1) \det \begin{pmatrix} t-8 & 18 & 30 \\ -18 & t+37 & 60 \\ 9 & -18 & t-29 \end{pmatrix} \\
&= (t+1) \left((t-8)(t+37)(t-29) + 18 \cdot 60 \cdot 9 \right. \\
&\qquad\quad +30 \cdot (-18) \cdot (-18) - 30 \cdot 9 \cdot (t+37) \\
&\qquad\quad \left. -18 \cdot (-18) \cdot (t-29) - (t-8) \cdot 60 \cdot (-18) \right) \\
&= (t+1)(t^3 - 3t + 2) = (t+1) \cdot (t+1)^2(t-2) \\
&= (t+1)^3(t-2).
\end{aligned}
$$

Since the eigenvalues of A are the roots of its characteristic polynomial $P_A(t)$, we solve $P_A(t) = 0$ and obtain that the eigenvalues of the matrix A are -1 with multiplicity 3, and 2 with multiplicity 1.

Let $v = \begin{pmatrix} v_1 \\ v_2 \\ v_3 \end{pmatrix}$ be an eigenvector corresponding to the eigenvalue $\lambda = -1$.

Then,

$$Av = \lambda v \iff Av = -v$$

$$
\iff \begin{cases}
8v_1 - 18v_2 - 30v_3 - 24v_4 &= -v_1 \\
18v_1 - 37v_2 - 60v_3 - 48v_4 &= -v_2 \\
-9v_1 + 18v_2 + 29v_3 + 24v_4 &= -v_3 \\
-v_4 &= -v_4
\end{cases}
$$

$$
\iff \begin{cases}
9v_1 - 18v_2 - 30v_3 - 24v_4 &= 0 \\
18v_1 - 36v_2 - 60v_3 - 48v_4 &= 0 \\
-9v_1 + 18v_2 + 30v_3 + 24v_4 &= 0
\end{cases}
$$

$$\iff 3v_1 - 6v_2 - 10v_3 - 8v_4 = 0$$

$$\iff v_1 = 2v_2 + \frac{10}{3}v_3 + \frac{8}{3}v_4.$$

Thus,

$$v = \begin{pmatrix} v_1 \\ v_2 \\ v_3 \end{pmatrix} = \begin{pmatrix} 2v_2 + \frac{10}{3}v_3 + \frac{8}{3}v_4 \\ v_2 \\ v_3 \end{pmatrix}$$

$$= \begin{pmatrix} 2 \\ 1 \\ 0 \\ 0 \end{pmatrix} v_2 + 3 \begin{pmatrix} 10 \\ 0 \\ 3 \\ 0 \end{pmatrix} v_3 + 3 \begin{pmatrix} 8 \\ 0 \\ 0 \\ 3 \end{pmatrix} v_4.$$

We conclude that the matrix A has three linearly independent eigenvectors corresponding to the eigenvalue -1, for example, $\begin{pmatrix} 2 \\ 1 \\ 0 \\ 0 \end{pmatrix}$, $\begin{pmatrix} 10 \\ 0 \\ 3 \\ 0 \end{pmatrix}$, and $\begin{pmatrix} 8 \\ 0 \\ 0 \\ 3 \end{pmatrix}$.

Note that

$$\begin{pmatrix} 0 \\ 1 \\ 1 \\ -2 \end{pmatrix} = \begin{pmatrix} 2 \\ 1 \\ 0 \\ 0 \end{pmatrix} + \frac{1}{3}\begin{pmatrix} 10 \\ 0 \\ 3 \\ 0 \end{pmatrix} - \frac{2}{3}\begin{pmatrix} 8 \\ 0 \\ 0 \\ 3 \end{pmatrix};$$

$$\begin{pmatrix} 2 \\ 2 \\ -3 \\ 3 \end{pmatrix} = 2\begin{pmatrix} 2 \\ 1 \\ 0 \\ 0 \end{pmatrix} - \begin{pmatrix} 10 \\ 0 \\ 3 \\ 0 \end{pmatrix} + \begin{pmatrix} 8 \\ 0 \\ 0 \\ 3 \end{pmatrix},$$

and therefore we can also choose $\begin{pmatrix} 0 \\ 1 \\ 1 \\ -2 \end{pmatrix}$, $\begin{pmatrix} 2 \\ 2 \\ -3 \\ 3 \end{pmatrix}$, $\begin{pmatrix} 2 \\ 1 \\ 0 \\ 0 \end{pmatrix}$ as three linearly independent eigenvectors corresponding to the eigenvalue -1.

Since the eigenvalue 2 of the matrix A has multiplicity 1, it follows that there is only one linearly independent eigenvector of the matrix A corresponding to the eigenvalue 2. By direct multiplication, we find that

$$A\begin{pmatrix} 1 \\ 2 \\ -1 \\ 0 \end{pmatrix} = \begin{pmatrix} 8 \\ 18 \\ -9 \\ 0 \end{pmatrix} + 2\begin{pmatrix} -18 \\ -37 \\ 18 \\ 0 \end{pmatrix} - \begin{pmatrix} -30 \\ -60 \\ 29 \\ 0 \end{pmatrix} = \begin{pmatrix} 2 \\ 4 \\ -2 \\ 0 \end{pmatrix} = 2\begin{pmatrix} 1 \\ 2 \\ -1 \\ 0 \end{pmatrix}.$$

Thus, $\begin{pmatrix} 1 \\ 2 \\ -1 \\ 0 \end{pmatrix}$ is an eigenvector corresponding to the eigenvalue 2 of A.

(ii) The characteristic polynomial of the matrix B is

$$P_B(t) = \det(tI - B) = \det\begin{pmatrix} t - 10 & 20 & 32 & 26 \\ -18 & t + 41 & 68 & 54 \\ 14 & -19 & t - 26 & -23 \\ -7 & -1 & -9 & t - 4 \end{pmatrix}$$

$$= (t - 10)\det\begin{pmatrix} t + 41 & 68 & 54 \\ -19 & t - 26 & -23 \\ -1 & -9 & t - 4 \end{pmatrix}$$

$$-20 \det \begin{pmatrix} -18 & 68 & 54 \\ 14 & t-26 & -23 \\ -7 & -9 & t-4 \end{pmatrix}$$

$$+32 \det \begin{pmatrix} -18 & t+41 & 54 \\ 14 & -19 & -23 \\ -7 & -1 & t-4 \end{pmatrix}.$$

$$-26 \det \begin{pmatrix} -18 & t+41 & 68 \\ 14 & -19 & t-26 \\ -7 & -1 & -9 \end{pmatrix}$$

$$= \quad (t+1)^3(t-2).$$

Since the eigenvalues of B are the roots of its characteristic polynomial $P_B(t)$, we solve $P_B(t) = 0$ and obtain that the eigenvalues of the matrix B are -1 with multiplicity 3, and 2 with multiplicity 1.

Let $v = \begin{pmatrix} v_1 \\ v_2 \\ v_3 \end{pmatrix}$ be an eigenvector corresponding to the eigenvalue $\lambda = -1$.

Then,

$$Bv = \lambda v \iff Bv = -v$$

$$\iff \begin{cases} 10v_1 - 20v_2 - 32v_3 - 26v_4 & = & -v_1 \\ 18v_1 - 41v_2 - 68v_3 - 54v_4 & = & -v_2 \\ -14v_1 + 19v_2 + 26v_3 + 23v_4 & = & -v_3 \\ 7v_1 + v_2 + 9v_3 + 4v_4 & = & -v_4 \end{cases}$$

$$\iff \begin{cases} 11v_1 - 20v_2 - 32v_3 - 26v_4 & = & 0 \quad (EV1.1) \\ 18v_1 - 40v_2 - 68v_3 - 54v_4 & = & 0 \quad (EV1.2) \\ -14v_1 + 19v_2 + 27v_3 + 23v_4 & = & 0 \quad (EV1.3) \\ 7v_1 + v_2 + 9v_3 + 5v_4 & = & 0 \quad (EV1.4) \end{cases}$$

Then,

$$\begin{array}{lrcl} (EV1.1) + (EV1.3): & -3v_1 - v_2 - 5v_3 - 3v_4 & = & 0; \\ 2(EV1.1) - (EV1.2): & 4v_1 + 4v_3 + 2v_4 & = & 0; \\ (EV1.3) + 2(EV1.4): & 21v_2 + 45v_3 + 33v_4 & = & 0; \\ (EV1.4): & 7v_1 + v_2 + 9v_3 + 5v_4 & = & 0, \end{array}$$

which can be simplified to

$$\begin{array}{rcll} -3v_1 - v_2 - 5v_3 - 3v_4 & = & 0; & (EV1.5) \\ 2v_1 + 2v_3 + v_4 & = & 0; & (EV1.6) \\ 7v_2 + 15v_3 + 11v_4 & = & 0; & (EV1.7) \\ 7v_1 + v_2 + 9v_3 + 5v_4 & = & 0. & (EV1.8) \end{array}$$

By solving (EV1.6) for v_4, we find that

$$v_4 = -2v_1 - 2v_3. \tag{4.8}$$

By solving (EV1.8) for v_2 and using (4.8), it follows that

$$\begin{aligned} v_2 &= -7v_1 - 9v_3 - 5v_4 = -7v_1 - 9v_3 - 5(-2v_1 - 2v_3) \\ &= 3v_1 + v_3. \end{aligned} \tag{4.9}$$

Using (4.8) and (4.9), the equations (EV1.5)and (EV1.7) become, respectively,

$$
\begin{aligned}
0 &= -3v_1 - v_2 - 5v_3 - 3v_4 \\
&= -3v_1 - (3v_1 + v_3) - 5v_3 - 3(-2v_1 - 2v_3) \\
&= 0; \\
0 &= 7v_2 + 15v_3 + 11v_4 \\
&= 7(3v_1 + v_3) + 15v_3 + 11(-2v_1 - 2v_3) \\
&= -v_1.
\end{aligned}
$$

Then, $v_1 = 0$, and we obtain from (4.8) and (4.9) that $v_4 = -2v_3$ and $v_2 = v_3$. Thus, the eigenvector $v = (v_i)_{i=1:4}$ is given by

$$
v = \begin{pmatrix} v_1 \\ v_2 \\ v_3 \\ v_4 \end{pmatrix} = \begin{pmatrix} 0 \\ v_3 \\ v_3 \\ -2v_3 \end{pmatrix} = -v_3 \begin{pmatrix} 0 \\ -1 \\ -1 \\ 2 \end{pmatrix}.
$$

We conclude that the matrix B has only one linearly independent eigenvectors corresponding to the eigenvalue -1 of multiplicity 3, e.g., $\begin{pmatrix} 0 \\ -1 \\ -1 \\ 2 \end{pmatrix}$.

Since the eigenvalue 2 of the matrix B has multiplicity 1, it follows that there is only one linearly independent eigenvector of the matrix B corresponding to the eigenvalue 2. By direct multiplication, we find that

$$
B \begin{pmatrix} 1 \\ 2 \\ -1 \\ 0 \end{pmatrix} = \begin{pmatrix} 2 \\ 4 \\ -2 \\ 0 \end{pmatrix} = 2 \begin{pmatrix} 1 \\ 2 \\ -1 \\ 0 \end{pmatrix}.
$$

Thus, $\begin{pmatrix} 1 \\ 2 \\ -1 \\ 0 \end{pmatrix}$ is an eigenvector corresponding to the eigenvalue 2 of B.   □

Problem 5: Let $A = \begin{pmatrix} 2 & -1 \\ 1 & 4 \end{pmatrix}$.

(i) What is the characteristic polynomial of A?

(ii) Find the eigenvalues and the eigenvectors of A.

Solution: (i) The characteristic polynomial of the matrix A is

$$
\begin{aligned}
P_A(t) &= \det(tI - A) \\
&= \det \begin{pmatrix} t-2 & 1 \\ -1 & t-4 \end{pmatrix} = (t-2)(t-4) - (-1) \\
&= t^2 - 6t + 9. \tag{4.10}
\end{aligned}
$$

Alternatively, since A is a 2×2 matrix, the characteristic polynomial of A can be computed as

$$P_A(t) = t^2 - \mathrm{tr}(A)t + \det(A) = t^2 - 6t + 9,$$

where the trace $\mathrm{tr}(A)$ and the determinant $\det(A)$ of the matrix $A = \begin{pmatrix} 2 & -1 \\ 1 & 4 \end{pmatrix}$ are given by

$$\begin{aligned} \mathrm{tr}(A) &= 2 + 4 = 6; \\ \det(A) &= 2 \cdot 4 - (-1) \cdot 1 = 9. \end{aligned}$$

(ii) From (4.10), we find that

$$P_A(t) = t^2 - 6t + 9 = (t-3)^2.$$

By solving

$$P_A(t) = 0 \iff (t-3)^2 = 0 \iff t = 3,$$

we obtain that $t = 3$ is the only root of $P_A(t)$ and has multiplicity 2.

Thus, $\lambda = 3$ is the only eigenvalue of the matrix A and has multiplicity 2.

Let $v = \begin{pmatrix} v_1 \\ v_2 \end{pmatrix}$ be an eigenvector corresponding to the eigenvalue $\lambda = 3$. Then,

$$Av = \lambda v \iff Av = 3v \iff \begin{pmatrix} 2 & -1 \\ 1 & 4 \end{pmatrix} \begin{pmatrix} v_1 \\ v_2 \end{pmatrix} = 3 \begin{pmatrix} v_1 \\ v_2 \end{pmatrix}$$

$$\iff \begin{cases} 2v_1 - v_2 &= 3v_1 \\ v_1 + 4v_2 &= 3v_2 \end{cases} \iff \begin{cases} -v_2 &= v_1 \\ v_1 &= -v_2 \end{cases}$$

$$\iff v_1 = -v_2.$$

Thus,

$$v = \begin{pmatrix} v_1 \\ v_2 \end{pmatrix} = \begin{pmatrix} -v_2 \\ v_2 \end{pmatrix} = v_2 \begin{pmatrix} -1 \\ 1 \end{pmatrix}.$$

We conclude that there exists only one linear independent eigenvector, for example, $\begin{pmatrix} -1 \\ 1 \end{pmatrix}$, corresponding to the eigenvalue $\lambda = 3$ of A. □

Problem 6: (i) Find the eigenvalues and the eigenvectors of the lower triangular matrix

$$L = \begin{pmatrix} 2 & 0 & 0 \\ 1 & -3 & 0 \\ -1 & 2 & -1 \end{pmatrix}.$$

(ii) Find the eigenvalues and the eigenvectors of the upper triangular matrix

$$U = \begin{pmatrix} -2 & -1 & 3 \\ 0 & 1 & 2 \\ 0 & 0 & 3 \end{pmatrix}.$$

Solution: (i) The characteristic polynomial of the matrix L is

$$P_L(t) \;=\; \det(tL - I) \;=\; \det \begin{pmatrix} t-2 & 0 & 0 \\ -1 & t+3 & 0 \\ 1 & -2 & t+1 \end{pmatrix}$$

$$= \;(t-2)(t+3)(t+1).$$

By solving

$$P_L(t) = 0 \iff (t-2)(t+3)(t+1) = 0 \iff t_1 = 2,\; t_2 = -3,\; t_3 = -1,$$

and since the eigenvalues of the matrix L are the roots of its characteristic polynomial $P_L(t)$, we obtain that the eigenvalues of the matrix L are

$$\lambda_1 = 2; \quad \lambda_2 = -3; \quad \lambda_3 = -1.$$

We proceed to identify the eigenvectors corresponding to each of these eigenvalues.

Let $v = \begin{pmatrix} v_1 \\ v_2 \\ v_3 \end{pmatrix}$ be an eigenvector corresponding to the eigenvalue $\lambda_1 = 2$. Then,

$$Lv = \lambda_1 v \iff Lv = 2v$$

$$\iff \begin{cases} 2v_1 &=& 2v_1 \\ v_1 - 3v_2 &=& 2v_2 \\ -v_1 + 2v_2 - v_3 &=& 2v_3 \end{cases} \iff \begin{cases} v_1 &=& 5v_2 \\ 3v_3 &=& -v_1 + 2v_2 \end{cases}$$

$$\iff \begin{cases} v_1 &=& 5v_2 \\ 3v_3 &=& -3v_2 \end{cases} \iff \begin{cases} v_1 &=& 5v_2 \\ v_3 &=& -v_2 \end{cases}$$

Thus, $v = \begin{pmatrix} v_1 \\ v_2 \\ v_3 \end{pmatrix} = \begin{pmatrix} 5v_2 \\ v_2 \\ -v_2 \end{pmatrix} = v_2 \begin{pmatrix} 5 \\ 1 \\ -1 \end{pmatrix}$. We conclude that the eigenvalue

$\lambda_1 = 2$ has one corresponding eigenvector, e.g., $\begin{pmatrix} 5 \\ 1 \\ -1 \end{pmatrix}$.

Let $v = \begin{pmatrix} v_1 \\ v_2 \\ v_3 \end{pmatrix}$ be an eigenvector corresponding to the eigenvalue $\lambda_2 = -3$. Then,

$$Lv = \lambda_2 v \iff Lv = -3v$$

$$\iff \begin{cases} 2v_1 &=& -3v_1 \\ v_1 - 3v_2 &=& -3v_2 \\ -v_1 + 2v_2 - v_3 &=& -3v_3 \end{cases} \iff \begin{cases} 5v_1 &=& 0 \\ v_1 &=& 0 \\ 2v_2 &=& v_1 - 2v_3 \end{cases}$$

$$\iff \begin{cases} v_1 &=& 0 \\ 2v_2 &=& -2v_3 \end{cases} \iff \begin{cases} v_1 &=& 0 \\ v_2 &=& -v_3 \end{cases}$$

Thus, $v = \begin{pmatrix} v_1 \\ v_2 \\ v_3 \end{pmatrix} = \begin{pmatrix} 0 \\ -v_3 \\ v_3 \end{pmatrix} = v_3 \begin{pmatrix} 0 \\ -1 \\ 1 \end{pmatrix}$. We conclude that the eigenvalue

$\lambda_2 = -3$ has one corresponding eigenvector, e.g., $\begin{pmatrix} 0 \\ -1 \\ 1 \end{pmatrix}$.

Let $v = \begin{pmatrix} v_1 \\ v_2 \\ v_3 \end{pmatrix}$ be an eigenvector corresponding to the eigenvalue $\lambda_3 = -1$. Then,

$$Lv = \lambda_3 v \iff Lv = -v$$

$$\iff \left\{ \begin{array}{rcl} 2v_1 & = & -v_1 \\ v_1 - 3v_2 & = & -v_2 \\ -v_1 + 2v_2 - v_3 & = & -v_3 \end{array} \right. \iff \left\{ \begin{array}{rcl} 3v_1 & = & 0 \\ v_1 & = & 2v_2 \\ -v_1 + 2v_2 & = & 0 \end{array} \right.$$

$$\iff \left\{ \begin{array}{rcl} v_1 & = & 0 \\ v_2 & = & 0 \end{array} \right.$$

Thus, $v = \begin{pmatrix} v_1 \\ v_2 \\ v_3 \end{pmatrix} = \begin{pmatrix} 0 \\ 0 \\ v_3 \end{pmatrix} = v_3 \begin{pmatrix} 0 \\ 0 \\ 1 \end{pmatrix}$. We conclude that the eigenvalue

$\lambda_3 = -1$ has one corresponding eigenvector, e.g., $\begin{pmatrix} 0 \\ 0 \\ 1 \end{pmatrix}$.

(ii) The characteristic polynomial of the matrix U is

$$P_U(t) = \det(tU - I) = \det \begin{pmatrix} t+2 & 1 & -3 \\ 0 & t-1 & -2 \\ 0 & 0 & t-3 \end{pmatrix}$$

$$= (t+2)(t-1)(t-3).$$

By solving

$$P_U(t) = 0 \iff (t+2)(t-1)(t-3) = 0 \iff t_1 = -2, \ t_2 = 1, \ t_3 = 3,$$

and since the eigenvalues of the matrix U are the roots of its characteristic polynomial $P_U(t)$, we obtain that the eigenvalues of the matrix U are

$$\lambda_1 = -2; \quad \lambda_2 = 1; \quad \lambda_3 = 3.$$

We proceed to identify the eigenvectors corresponding to each of these eigenvalues.

Let $v = \begin{pmatrix} v_1 \\ v_2 \\ v_3 \end{pmatrix}$ be an eigenvector corresponding to the eigenvalue $\lambda_1 = -2$. Then,

$$Uv = \lambda_1 v \iff Uv = -2v$$

$$\iff \left\{ \begin{array}{rcl} -2v_1 - v_2 + 3v_3 & = & -2v_1 \\ v_2 + 2v_3 & = & -2v_2 \\ 3v_3 & = & -2v_3 \end{array} \right. \iff \left\{ \begin{array}{rcl} 3v_3 & = & v_2 \\ 3v_2 & = & -2v_3 \\ 5v_3 & = & 0 \end{array} \right.$$

$$\iff \left\{ \begin{array}{rcl} v_2 & = & 0 \\ v_3 & = & 0 \end{array} \right.$$

Thus, $v = \begin{pmatrix} v_1 \\ v_2 \\ v_3 \end{pmatrix} = \begin{pmatrix} v_1 \\ 0 \\ 0 \end{pmatrix} = v_1 \begin{pmatrix} 1 \\ 0 \\ 0 \end{pmatrix}$. We conclude that the eigenvalue

$\lambda_1 = -2$ has one corresponding eigenvector, e.g., $\begin{pmatrix} 1 \\ 0 \\ 0 \end{pmatrix}$.

Let $v = \begin{pmatrix} v_1 \\ v_2 \\ v_3 \end{pmatrix}$ be an eigenvector corresponding to the eigenvalue $\lambda_2 = 1$. Then,

$$Uv = \lambda_2 v \quad \Longleftrightarrow \quad Uv = v$$

$$\Longleftrightarrow \quad \begin{cases} -2v_1 - v_2 + 3v_3 &= v_1 \\ v_2 + 2v_3 &= v_2 \\ 3v_3 &= v_3 \end{cases} \quad \Longleftrightarrow \quad \begin{cases} -v_2 + 3v_3 &= 3v_1 \\ 2v_3 &= 0 \\ 2v_3 &= 0 \end{cases}$$

$$\Longleftrightarrow \quad \begin{cases} v_2 &= -3v_1 \\ v_3 &= 0 \end{cases}$$

Thus, $v = \begin{pmatrix} v_1 \\ v_2 \\ v_3 \end{pmatrix} = \begin{pmatrix} v_1 \\ -3v_1 \\ 0 \end{pmatrix} = v_1 \begin{pmatrix} 1 \\ -3 \\ 0 \end{pmatrix}$. We conclude that the eigenvalue

$\lambda_2 = 1$ has one corresponding eigenvector, e.g., $\begin{pmatrix} 1 \\ -3 \\ 0 \end{pmatrix}$.

Let $v = \begin{pmatrix} v_1 \\ v_2 \\ v_3 \end{pmatrix}$ be an eigenvector corresponding to the eigenvalue $\lambda_3 = 3$. Then,

$$Uv = \lambda_3 v \quad \Longleftrightarrow \quad Uv = 3v$$

$$\Longleftrightarrow \quad \begin{cases} -2v_1 - v_2 + 3v_3 &= 3v_1 \\ v_2 + 2v_3 &= 3v_2 \\ 3v_3 &= 3v_3 \end{cases} \quad \Longleftrightarrow \quad \begin{cases} 5v_1 &= -v_2 + 3v_3 \\ 2v_2 &= 2v_3 \end{cases}$$

$$\Longleftrightarrow \quad \begin{cases} 5v_1 &= 2v_3 \\ v_2 &= v_3 \end{cases} \quad \Longleftrightarrow \quad \begin{cases} v_1 &= \frac{2v_3}{5} \\ v_2 &= v_3 \end{cases}$$

Thus, $v = \begin{pmatrix} v_1 \\ v_2 \\ v_3 \end{pmatrix} = \begin{pmatrix} \frac{2v_3}{5} \\ v_3 \\ v_3 \end{pmatrix} = \frac{v_3}{5} \begin{pmatrix} 2 \\ 5 \\ 5 \end{pmatrix}$. We conclude that the eigenvalue

$\lambda_3 = 3$ has one corresponding eigenvector, e.g., $\begin{pmatrix} 2 \\ 5 \\ 5 \end{pmatrix}$. $\square$

Problem 7: Let A be a square matrix such that $A^2 = A$. Show that any eigenvalue of A is either 0 or 1.

Solution: Let λ and $v \neq 0$ be an eigenvalue and a corresponding eigenvector of the matrix A. Then,

$$Av = \lambda v \qquad (4.11)$$

and

$$A^2 v = A \cdot (Av) = A \cdot (\lambda v) = \lambda \cdot Av = \lambda \cdot \lambda v = \lambda^2 v. \qquad (4.12)$$

Note that $A^2 v = Av$, since $A^2 = A$. Then, from (4.11) and (4.12), we obtain that

$$\lambda^2 v = \lambda v,$$

which can be written as

$$0 = \lambda^2 v - \lambda v = (\lambda^2 - \lambda)v = \lambda(\lambda - 1)v.$$

Thus, $\lambda(\lambda - 1)v = 0$, and, since $v \neq 0$, we find that either $\lambda = 0$ or $\lambda = 1$. We conclude that any eigenvalue of the matrix A is either 0 or 1.   □

Problem 8: Let A be a square matrix with the property that there exists a positive integer p such that $A^p = 0$. Show that any eigenvalue of A must be equal to 0.

Solution: Let λ and $v \neq 0$ be an eigenvalue and a corresponding eigenvector of the matrix A. Recall that, for any positive integer k, $A^k v = \lambda^k v$. Then, since $A^p = 0$, it follows that

$$A^p v = \lambda^p v = 0.$$

Thus, $\lambda^p v = 0$, and, since $v \neq 0$, we obtain that $\lambda^p = 0$, and therefore that $\lambda = 0$. We conclude that all the eigenvalues of the matrix A are equal to 0.   □

Problem 9: Let $v \neq 0$ be a column vector of size n, and let $A = vv^t$ be an $n \times n$ matrix.

(i) Show that the matrix A has exactly one non-zero eigenvalue.

(ii) Find the eigenvalues and the eigenvectors of A.

Solution: (i) Let λ and $w \neq 0$ be an eigenvalue and a corresponding eigenvector of the matrix $A = vv^t$. Then,

$$Aw = \lambda w \iff (vv^t)w = \lambda w \iff v \cdot (v^t w) = \lambda w$$
$$\iff (v^t w) \cdot v = \lambda w, \tag{4.13}$$

where (4.13) follows from the fact that $v^t w$ is a number since it is the product of the row vector v^t and the column vector w.[1]

If $\lambda \neq 0$ is a nonzero eigenvalue of A, we obtain from (4.13) that

$$w = \frac{v^t w}{\lambda} \cdot v = cv,$$

where $c = \frac{v^t w}{\lambda}$ is a real number.

Note that $v^t w \neq 0$. Otherwise, from (4.13), it would follow that $\lambda w = 0$ which is not possible since $w \neq 0$ and we assumed that $\lambda \neq 0$. Thus, $c = \frac{v^t w}{\lambda} \neq 0$.

Since $w = cv$ and $c \neq 0$, we find that

$$Aw = \lambda w \iff A(cv) = \lambda(cv) \iff cAv = c\lambda v \iff Av = \lambda v$$
$$\iff (vv^t)v = \lambda v \iff v \cdot (v^t v) = \lambda v \iff v \cdot ||v||^2 = \lambda v$$
$$\iff \lambda = ||v||^2,$$

---

[1]Note that $v^t w = (w, v)$ is also the inner product of the vectors v and w and therefore is a number.

since $v \neq 0$; recall that $||v||^2 = (v, v) = v^t v$.

We conclude that $\lambda = ||v||^2$ is the only nonzero eigenvalue of the matrix A.

(ii) Recall from (i) that $\lambda = ||v||^2$ is the only nonzero eigenvalue of the matrix A and note that v is an eigenvector corresponding to λ since

$$Av = (vv^t)v = v \cdot (v^t v) = ||v||^2 \cdot v = \lambda v.$$

We will show that $\lambda = 0$ is an eigenvalue of the matrix A which has $n - 1$ linearly independent eigenvectors. Therefore, $\lambda = 0$ will be an eigenvalue of A with multiplicity $n - 1$.

Note that $\lambda = 0$ is an eigenvalue of A with corresponding eigenvector $w \neq 0$ if and only if

$$Aw = 0 \iff (vv^t)w = 0 \iff v \cdot (v^t w) = 0 \iff v^t w = 0,$$

since $v \neq 0$. Thus, any vector $w \neq 0$ which is orthogonal to the vector v, i.e., with $v^t w = 0$, is an eigenvector of the matrix A corresponding to the eigenvalue A. The space of $n \times 1$ vectors orthogonal to the vector $v \neq 0$ has dimension $n - 1$, i.e., there are $n - 1$ linearly independent eigenvectors w such that $v^t w = 0$.

Summarizing, the matrix $A = vv^t$ has the following eigenvalues and eigenvectors:

• eigenvalue $\lambda = 0$ of multiplicity $n - 1$ and with $n - 1$ linearly independent corresponding eigenvectors w orthogonal to the vector v;
• eigenvalue $\lambda = ||v||^2$ of multiplicity 1 and corresponding eigenvector v.   □

Problem 10: Find the eigenvalues and the eigenvectors of the $n \times n$ matrix

$$\begin{pmatrix} d & 1 & \cdots & 1 \\ 1 & d & \cdots & 1 \\ \vdots & \vdots & \ddots & \vdots \\ 1 & 1 & \cdots & d \end{pmatrix},$$

where $d \in \mathbb{R}$ is a constant.

Solution: Let $A = \begin{pmatrix} d & 1 & \cdots & 1 \\ 1 & d & \cdots & 1 \\ \vdots & \vdots & \ddots & \vdots \\ 1 & 1 & \cdots & d \end{pmatrix}$, and note that

$$A = \begin{pmatrix} d-1 & 0 & \cdots & 0 \\ 0 & d-1 & \cdots & 0 \\ \vdots & \vdots & \ddots & \vdots \\ 0 & 0 & \cdots & d-1 \end{pmatrix} I + \begin{pmatrix} 1 & 1 & \cdots & 1 \\ 1 & 1 & \cdots & 1 \\ \vdots & \vdots & \ddots & \vdots \\ 1 & 1 & \cdots & 1 \end{pmatrix}$$

$$= (d-1)I + \mathbf{1} \cdot \mathbf{1}^t,$$

where I is the identity matrix and $\mathbf{1} = \begin{pmatrix} 1 \\ 1 \\ \vdots \\ 1 \end{pmatrix}$.

If λ and $w \neq 0$ are an eigenvalue and a corresponding eigenvector of the matrix $1 \cdot 1^t$, i.e., such that $(1 \cdot 1^t)w = \lambda w$, then

$$Aw = (d-1)w + (1 \cdot 1^t)w = (d-1)w + \lambda w = (d-1+\lambda)w.$$

In other words, $\mu = d - 1 + \lambda$ and w will be an eigenvalue and a corresponding eigenvector of the matrix A.

Recall that the matrix $1 \cdot 1^t$ has eigenvalue $\lambda = ||1||^2 = n$ of multiplicity 1 and corresponding eigenvector 1 and eigenvalue $\lambda = 0$ of multiplicity $n-1$ and with $n-1$ linearly independent corresponding eigenvectors w orthogonal to the vector 1. Note that a vector $w = (w_i)_{i=1:n}$ is orthogonal to the vector 1 if and only if $1^t w = \sum_{i=1}^{n} w_i = 0$. For example, $n-1$ linearly independent vectors orthogonal to the vector 1 could be

$$\begin{pmatrix} 1 \\ -1 \\ 0 \\ 0 \\ \vdots \\ 0 \end{pmatrix}, \begin{pmatrix} 1 \\ 0 \\ -1 \\ 0 \\ \vdots \\ 0 \end{pmatrix}, \dots, \begin{pmatrix} 1 \\ 0 \\ 0 \\ \vdots \\ 0 \\ -1 \end{pmatrix}. \tag{4.14}$$

We conclude that the matrix A has the following eigenvalues and eigenvectors:

• eigenvalue $\mu = d - 1 + n$ of multiplicity 1 and corresponding eigenvector 1;
• eigenvalue $\mu = d - 1$ of multiplicity $n - 1$ and with $n - 1$ linearly independent corresponding eigenvectors $w = (w_i)_{i=1:n}$ with $\sum_{i=1}^{n} w_i = 0$, for example, the $n - 1$ vectors from (4.14).   □

Problem 11: (i) Let A be a symmetric matrix of size n, and let λ_i, $i = 1 : n$, be the eigenvalues of A, with corresponding eigenvectors v_i, $i = 1 : n$. What are the eigenvalues and the eigenvectors of A^2?

(ii) Let A be a symmetric matrix of size n. Let ϕ_i, $i = 1 : n$, be the eigenvalues of the matrix A^2, with corresponding eigenvectors w_i, $i = 1 : n$. What can you say about the eigenvalues and the eigenvectors of A?

Solution: (i) The diagonal form of the symmetric matrix A is

$$A = Q\Lambda Q^t,$$

where $\Lambda = \text{diag}(\lambda_k)_{k=1:n}$ is the diagonal matrix whose diagonal entries are the eigenvalues of the matrix A and $Q = \text{col}(v_k)_{k=1:n}$ is the orthogonal matrix whose column vectors are the eigenvectors of A.

Then,

$$\begin{aligned} A^2 &= Q\Lambda Q^t \cdot Q\Lambda Q^t = Q\Lambda(Q^t Q)\Lambda Q^t = Q\Lambda \cdot \Lambda Q^t \\ &= Q\Lambda^2 Q^t, \end{aligned}$$

since Q is an orthogonal matrix and therefore $Q^t Q = I$. In other words, the diagonal form of the matrix A^2 is

$$A^2 = Q\Lambda^2 Q^t,$$

with $\Lambda^2 = \text{diag}(\lambda_k^2)_{k=1:n}$ and $Q = \text{col}(v_k)_{k=1:n}$.

We conclude that λ_k^2, $k = 1 : n$, are the eigenvalues of A with corresponding eigenvectors v_k, $i = k : n$.

(ii) Let ϕ_i, $i = 1 : n$, be the eigenvalues of the matrix A^2, with corresponding eigenvectors w_i, $i = 1 : n$. From (i), we know that the matrices A and A^2 have the same eigenvectors. Thus, if w_i, $i = 1 : n$, are the eigenvectors of A^2, then w_i, $i = 1 : n$, are the eigenvectors of the matrix A as well.

Let λ_i be the eigenvalue of the matrix A corresponding to the eigenvector w_i. Then, $Aw_i = \lambda_i w_i$ and therefore

$$
\begin{aligned}
A^2 w_i &= A \cdot A w_i = A \cdot \lambda_i w_i = \lambda_i \cdot A w_i = \lambda_i \cdot \lambda_i w_i \\
&= \lambda_i^2 w_i.
\end{aligned}
\tag{4.15}
$$

Since ϕ_i is the eigenvalue of A^2 corresponding to w_i, it follows that

$$
A^2 w_i = \phi_i w_i.
\tag{4.16}
$$

From (4.15) and (4.16), we find that

$$
\phi_i = \lambda_i^2,
\tag{4.17}
$$

since w_i is an eigenvector and therefore $w_i \neq 0$.

Then, from (4.17), we obtain that either $\lambda_i = \sqrt{\phi_i}$, or $\lambda_i = -\sqrt{\phi_i}$.

We conclude that the eigenvectors of the matrix A are w_i, $i = 1 : n$, and the eigenvalue λ_i corresponding to w_i is either $\sqrt{\phi_i}$ or $-\sqrt{\phi_i}$. $\quad\square$

Problem 12: Let A be a square matrix of size n, and let S be a nonsingular matrix of size n.

(i) Let λ and v be an eigenvalue and a corresponding eigenvector of A. Show that λ is also an eigenvalue of the matrix $S^{-1}AS$. What is a corresponding eigenvector?

(ii) Show that the matrix $S^{-1}AS$ has the same characteristic polynomial as A, i.e., show that

$$
P_{S^{-1}AS}(t) = P_A(t), \quad \forall\, t \in \mathbb{R}.
$$

Solution: (i) If λ and $v \neq 0$ are an eigenvalue and a corresponding eigenvector of A, then $Av = \lambda v$. Since S is a nonsingular matrix, and using the fact that $SS^{-1} = I$, we obtain that

$$
\begin{aligned}
Av = \lambda v &\iff S^{-1}Av = S^{-1}(\lambda v) & (4.18) \\
&\iff S^{-1}Av = \lambda S^{-1}v & (4.19) \\
&\iff S^{-1}A(SS^{-1})v = \lambda S^{-1}v & \\
&\iff S^{-1}AS \cdot S^{-1}v = \lambda S^{-1}v & \\
&\iff S^{-1}ASw = \lambda w, & (4.20)
\end{aligned}
$$

where $w = S^{-1}v \neq 0$.

From (4.20), we conclude that λ is an eigenvalue of $S^{-1}AS$ with corresponding eigenvector $w = S^{-1}v$.

(ii) The characteristic polynomial of the matrix $S^{-1}AS$ is

$$P_{S^{-1}AS}(t) \;=\; \det(tI - S^{-1}AS), \qquad (4.21)$$

where t is a real number. Note that

$$tI - S^{-1}AS \;=\; t \cdot S^{-1}S - S^{-1}AS \;=\; S^{-1}(tI - A)S, \qquad (4.22)$$

since $S^{-1}S = I$. Moreover,

$$\det(S^{-1})\det(S) \;=\; \det(S^{-1}S) \;=\; \det(I) \;=\; 1. \qquad (4.23)$$

From (4.21) and (4.22), and using (4.23), we obtain that

$$\begin{aligned} P_{S^{-1}AS}(t) &= \det(tI - S^{-1}AS) \\ &= \det(S^{-1}(tI - A)S) \\ &= \det(S^{-1})\det(tI - A)\det(S) \\ &= \det(tI - A)\,\det(S^{-1})\det(S) \\ &= \det(tI - A) \\ &= P_A(t). \quad \square \end{aligned}$$

Problem 13: Let A be a square matrix with real entries. If $\lambda = a + ib$ is a complex eigenvalue of A (i.e., with $b \neq 0$), show that $\overline{\lambda} = a - ib$, the complex conjugate of λ, is also an eigenvalue of A.

Solution: Let $v = (v_i)_{i=1:n}$ be an eigenvector corresponding to the eigenvalue $\lambda = a + ib$ of A. Then,

$$Av \;=\; \lambda v. \qquad (4.24)$$

By taking the complex conjugate of (4.24), we obtain that

$$\overline{Av} \;=\; \overline{\lambda v}, \qquad (4.25)$$

where $\overline{v} = (\overline{v_i})_{i=1:n}$. Note that

$$\overline{A} \;=\; A, \qquad (4.26)$$

since all the entries of the matrix A are real numbers and therefore $\overline{A}(j,k) = \overline{A(j,k)} = A(j,k)$ for all $1 \leq j,k \leq n$. From (4.25) and (4.26), we obtain that

$$A\overline{v} \;=\; \overline{\lambda}\,\overline{v}.$$

In other words, $\overline{\lambda} = a - ib$ is an eigenvalue of the matrix A with corresponding eigenvector $\overline{v}$. $\square$

Problem 14: The 2×2 matrix A has eigenvalues 1 and -2 with corresponding eigenvectors $\begin{pmatrix} -1 \\ 2 \end{pmatrix}$ and $\begin{pmatrix} 3 \\ 1 \end{pmatrix}$. If $v = \begin{pmatrix} 2 \\ -3 \end{pmatrix}$, find Av.

Solution 1: Let $\lambda_1 = 1$, $v_1 = \begin{pmatrix} -1 \\ 2 \end{pmatrix}$, and $\lambda_2 = -2$, $v_2 = \begin{pmatrix} 3 \\ 1 \end{pmatrix}$. Then,

$$Av_1 = \lambda_1 v_1 = v_1; \tag{4.27}$$

$$Av_2 = \lambda_2 v_2 = -2v_2. \tag{4.28}$$

We first find constants $c_1, c_2 \in \mathbb{R}$ such that

$$v = c_1 v_1 + c_2 v_2, \tag{4.29}$$

i.e., such that

$$\begin{pmatrix} 2 \\ -3 \end{pmatrix} = c_1 \begin{pmatrix} -1 \\ 2 \end{pmatrix} + c_2 \begin{pmatrix} 3 \\ 1 \end{pmatrix} = \begin{pmatrix} -c_1 + 3c_2 \\ 2c_1 + c_2 \end{pmatrix}$$

$$\Longleftrightarrow \quad \begin{cases} 2 = -c_1 + 3c_2 \\ -3 = 2c_1 + c_2 \end{cases}$$

The solution of this linear system is $c_1 = -\frac{11}{7}$ and $c_2 = \frac{1}{7}$, and therefore, we obtain from (4.29) that

$$v = -\frac{11}{7}v_1 + \frac{1}{7}v_2. \tag{4.30}$$

Using (4.27) and (4.28), we find from (4.30) that

$$Av = -\frac{11}{7}Av_1 + \frac{1}{7}Av_2 = -\frac{11}{7}v_1 + \frac{1}{7}(-2v_2) = -\frac{11}{7}\begin{pmatrix} -1 \\ 2 \end{pmatrix} - \frac{2}{7}\begin{pmatrix} 3 \\ 1 \end{pmatrix}$$

$$= \begin{pmatrix} \frac{5}{7} \\ -\frac{24}{7} \end{pmatrix}.$$

Solution 2: Since the 2×2 matrix A has two eigenvectors, it follows that A has the following diagonal form:

$$A = V\Lambda V^{-1}, \tag{4.31}$$

where

$$\Lambda = \begin{pmatrix} \lambda_1 & 0 \\ 0 & \lambda_2 \end{pmatrix} = \begin{pmatrix} 1 & 0 \\ 0 & -2 \end{pmatrix}; \tag{4.32}$$

$$V = (v_1 \quad v_2) = \begin{pmatrix} -1 & 3 \\ 2 & 1 \end{pmatrix}. \tag{4.33}$$

Recall that the inverse of a 2×2 matrix is given by

$$\begin{pmatrix} a & b \\ c & d \end{pmatrix}^{-1} = \frac{1}{ad - bc}\begin{pmatrix} d & -b \\ -c & a \end{pmatrix}. \tag{4.34}$$

Then,

$$V^{-1} = \begin{pmatrix} -1 & 3 \\ 2 & 1 \end{pmatrix}^{-1} = \frac{1}{(-1)\cdot 1 - 3\cdot 2}\begin{pmatrix} 1 & -3 \\ -2 & -1 \end{pmatrix}$$

$$= \frac{1}{7}\begin{pmatrix} -1 & 3 \\ 2 & 1 \end{pmatrix}. \tag{4.35}$$

From (4.31), (4.32), (4.33), and (4.35), it follows that

$$A = V\Lambda V^{-1} = \frac{1}{7}\begin{pmatrix} -1 & 3 \\ 2 & 1 \end{pmatrix}\begin{pmatrix} 1 & 0 \\ 0 & -2 \end{pmatrix}\begin{pmatrix} -1 & 3 \\ 2 & 1 \end{pmatrix}$$

$$= \frac{1}{7}\begin{pmatrix} -11 & -9 \\ -6 & 4 \end{pmatrix}. \tag{4.36}$$

From (4.36), we conclude that

$$Av = \frac{1}{7}\begin{pmatrix} -11 & -9 \\ -6 & 4 \end{pmatrix}\begin{pmatrix} 2 \\ -3 \end{pmatrix} = \begin{pmatrix} \frac{5}{7} \\ -\frac{24}{7} \end{pmatrix}. \quad \square$$

Problem 15: Let $A = \begin{pmatrix} -1 & 2 \\ 2 & 2 \end{pmatrix}$.

(i) Find the eigenvalues and the eigenvectors of the matrix A.

(ii) What is the diagonal form of A?

(iii) Compute A^{12}.

Solution: (i) The characteristic polynomial of the matrix A is[2]

$$\begin{aligned} P_A(t) &= \det(tI - A) \\ &= \det\begin{pmatrix} t+1 & -2 \\ -2 & t-2 \end{pmatrix} = (t+1)(t-2) - (-2)(-2) \\ &= t^2 - t - 6. \end{aligned}$$

By solving

$$P_A(t) = 0 \iff t^2 - t - 6 = 0 \iff (t-3)(t+2) = 0 \iff t_1 = 3; \ t_2 = -2,$$

and since the eigenvalues of the matrix A are the roots of its characteristic polynomial $P_A(t)$, we obtain that the matrix A has eigenvalues

$$\lambda_1 = 3; \ \lambda_2 = -2.$$

Let $v_1 = \begin{pmatrix} x \\ y \end{pmatrix}$ be an eigenvector corresponding to the eigenvalue $\lambda_1 = 3$. Then,

$$Av_1 = \lambda_1 v_1 \iff Av_1 = 3v_1 \iff \begin{pmatrix} -1 & 2 \\ 2 & 2 \end{pmatrix}\begin{pmatrix} x \\ y \end{pmatrix} = 3\begin{pmatrix} x \\ y \end{pmatrix}$$

---

[2]Since A is a 2×2 matrix, the characteristic polynomial of A can also be computed as

$$P_A(t) = t^2 - \text{tr}(A)t + \det(A) = t^2 - t - 6,$$

where the trace $\text{tr}(A)$ and the determinant $\det(A)$ of the matrix $A = \begin{pmatrix} -1 & 2 \\ 2 & 2 \end{pmatrix}$ are given by

$$\begin{aligned} \text{tr}(A) &= -1 + 2 = 1; \\ \det(A) &= (-1)\cdot 2 - 2\cdot 2 = -6. \end{aligned}$$

$$\Longleftrightarrow \quad \begin{cases} -x + 2y & = & 3x \\ 2x + 2y & = & 3y \end{cases} \quad \Longleftrightarrow \quad \begin{cases} 2y & = & 4x \\ 2x & = & y \end{cases}$$

$$\Longleftrightarrow \quad y = 2x.$$

Thus,

$$v_1 = \begin{pmatrix} x \\ y \end{pmatrix} = \begin{pmatrix} x \\ 2x \end{pmatrix} = x \begin{pmatrix} 1 \\ 2 \end{pmatrix}. \tag{4.37}$$

By choosing $x = 1$ in (4.37), we conclude that the vector $v_1 = \begin{pmatrix} 1 \\ 2 \end{pmatrix}$ is an eigenvector corresponding to the eigenvalue $\lambda_1 = 3$.

Let $v_2 = \begin{pmatrix} x \\ y \end{pmatrix}$ be an eigenvector corresponding to the eigenvalue $\lambda_2 = -2$.
Then,

$$Av_2 = \lambda_2 v_2 \quad \Longleftrightarrow \quad Av_2 = -2v_2 \quad \Longleftrightarrow \quad \begin{pmatrix} -1 & 2 \\ 2 & 2 \end{pmatrix} \begin{pmatrix} x \\ y \end{pmatrix} = -2 \begin{pmatrix} x \\ y \end{pmatrix}$$

$$\Longleftrightarrow \quad \begin{cases} -x + 2y & = & -2x \\ 2x + 2y & = & -2y \end{cases} \quad \Longleftrightarrow \quad \begin{cases} x & = & -2y \\ 2x & = & -4y \end{cases}$$

$$\Longleftrightarrow \quad x = -2y.$$

Thus,

$$v_2 = \begin{pmatrix} x \\ y \end{pmatrix} = \begin{pmatrix} -2y \\ y \end{pmatrix} = y \begin{pmatrix} -2 \\ 1 \end{pmatrix}. \tag{4.38}$$

By choosing $y = 1$ in (4.38), we conclude that the vector $v_2 = \begin{pmatrix} -2 \\ 1 \end{pmatrix}$ is an eigenvector corresponding to the eigenvalue $\lambda_2 = -2$.

(ii) The diagonal form of the matrix A is

$$A = V\Lambda V^{-1}, \tag{4.39}$$

where

$$\Lambda = \begin{pmatrix} \lambda_1 & 0 \\ 0 & \lambda_2 \end{pmatrix} = \begin{pmatrix} 3 & 0 \\ 0 & -2 \end{pmatrix}; \tag{4.40}$$

$$V = (v_1 \ v_2) = \begin{pmatrix} 1 & -2 \\ 2 & 1 \end{pmatrix}. \tag{4.41}$$

From (4.34), we obtain that the inverse matrix V^{-1} is given by

$$V^{-1} = \begin{pmatrix} 1 & -2 \\ 2 & 1 \end{pmatrix}^{-1} = \frac{1}{5} \begin{pmatrix} 1 & 2 \\ -2 & 1 \end{pmatrix}. \tag{4.42}$$

From (4.39–4.42), it follows that the diagonal form of the matrix A is

$$A = V\Lambda V^{-1} = \frac{1}{5} \begin{pmatrix} 1 & -2 \\ 2 & 1 \end{pmatrix} \begin{pmatrix} 3 & 0 \\ 0 & -2 \end{pmatrix} \begin{pmatrix} 1 & 2 \\ -2 & 1 \end{pmatrix}. \tag{4.43}$$

(iii) By using the diagonal form of the matrix A from (4.43), we find that

$$
\begin{aligned}
A^{12} &= (V\Lambda V^{-1})^{12} \\
&= (V\Lambda V^{-1}) \cdot (V\Lambda V^{-1}) \cdots (V\Lambda V^{-1}) \\
&= V \cdot \Lambda \cdot (V^{-1}V) \cdot \Lambda \cdot (V^{-1}V) \cdots (V^{-1}V) \cdot \Lambda V^{-1} \\
&= V \cdot \Lambda \cdot \Lambda \cdots \Lambda \cdot V^{-1} \\
&= V\Lambda^{12} V^{-1},
\end{aligned}
\tag{4.44}
$$

since $V^{-1}V = I$.

From (4.44), and using (4.40), (4.41), and (4.42), we conclude that

$$
\begin{aligned}
A^{12} &= V\Lambda^{12} V^{-1} \\
&= \begin{pmatrix} 1 & -2 \\ 2 & 1 \end{pmatrix} \begin{pmatrix} 3^{12} & 0 \\ 0 & 2^{12} \end{pmatrix} \frac{1}{5} \begin{pmatrix} 1 & 2 \\ -2 & 1 \end{pmatrix} \\
&= \begin{pmatrix} 109565 & 210938 \\ 210938 & 425972 \end{pmatrix}. \quad \square
\end{aligned}
$$

Problem 16: Let $A = \begin{pmatrix} a & b \\ c & d \end{pmatrix}$ be a 2×2 matrix, and let

$$
P_A(t) = t^2 - (a+d)t + (ad - bc)
$$

be the characteristic polynomial associated to A.

Show that $P_A(A) = 0$, i.e., show that

$$
A^2 - (a+d)A + (ad - bc)I = 0.
$$

Solution: By matrix multiplication, we obtain that

$$
\begin{aligned}
&A^2 - (a+d)A + (ad - bc)I \\
&= \begin{pmatrix} a^2 + bc & ab + bd \\ ac + cd & bc + d^2 \end{pmatrix} - \begin{pmatrix} a(a+d) & b(a+d) \\ c(a+d) & d(a+d) \end{pmatrix} + \begin{pmatrix} ad - bc & 0 \\ 0 & ad - bc \end{pmatrix} \\
&= \begin{pmatrix} a^2 + bc - a^2 - ad + ad - bc & ab + bd - ab - bd \\ ac + cd - ac - cd & bc + d^2 - ad - d^2 + ad - bc \end{pmatrix} \\
&= \begin{pmatrix} 0 & 0 \\ 0 & 0 \end{pmatrix}.
\end{aligned}
$$

Note that this problem is the 2×2 case of the Cayley-Hamilton theorem which states that $P_A(A) = 0$ for any square matrix A. $\quad \square$

Problem 17: Let A and B be square matrices of the same size.

(i) If B is nonsingular, show that the matrices AB and BA have the same characteristic polynomial, i.e., show that

$$
\det(tI - AB) = \det(tI - BA).
\tag{4.45}
$$

(ii) Show that the characteristic polynomial of AB is the same as the characteristic polynomial of BA even if B is a singular matrix.

(iii) Show that the matrices AB and BA have the same eigenvalues.

Solution: (i) If the matrix B is nonsingular, then

$$tI - AB = B^{-1}(tI - BA)B, \tag{4.46}$$

since $B^{-1}B = I$ and therefore

$$B^{-1}(tI - BA)B = t \cdot B^{-1}B - B^{-1}BAB = tI - AB.$$

From (4.46), we obtain that

$$
\begin{aligned}
\det(tI - AB) &= \det(B^{-1})\det(tI - BA)\det(B) \\
&= \det(tI - BA) \cdot \det(B^{-1})\det(B) \\
&= \det(tI - BA), \tag{4.47}
\end{aligned}
$$

since

$$\det(B^{-1})\det(B) = \det(B^{-1}B) = \det(I) = 1.$$

(ii) If the matrix B is singular and ϵ is a real number, note that the matrix $B - \epsilon I$ is singular if and only if ϵ is equal to an eigenvalue of B: if the matrix $B - \epsilon I$ is singular, then 0 is an eigenvalue of $B - \epsilon I$, and, if $v \neq 0$ is an eigenvector corresponding to the eigenvalue 0, then

$$(B - \epsilon I)v = 0 \iff Bv - \epsilon v = 0 \iff Bv = \epsilon v,$$

i.e., ϵ is an eigenvalue of B.

Since the $n \times n$ matrix B has at most n different eigenvalues, it follows that, except for a finite number of values of ϵ, the matrix $B - \epsilon I$ is nonsingular, in which case we obtain from (4.45) that

$$\det(tI - A(B - \epsilon I)) = \det(tI - (B - \epsilon I)A). \tag{4.48}$$

Since both sides of (4.48) are polynomials of degree n in ϵ, and therefore continuous functions of ϵ, we can let $\epsilon \to 0$ in (4.48) and obtain that

$$
\begin{aligned}
\lim_{\epsilon \to 0} \left(\det(tI - A(B - \epsilon I)) \right) &= \lim_{\epsilon \to 0} \left(\det(tI - (B - \epsilon I)A) \right) \\
\iff \det(tI - AB) &= \det(tI - BA). \tag{4.49}
\end{aligned}
$$

From (4.47) and (4.49), we conclude that $\det(tI - AB) = \det(tI - BA)$ regardless of whether the matrix B is nonsingular or singular.

(iii) Recall that the set of the eigenvalues of a matrix is the same as the set of the roots of the characteristic polynomial of the matrix.

Then, since the matrices AB and BA have the same characteristic polynomials, we conclude that AB and BA have the same eigenvalues. $\quad\square$

Problem 18: (i) Let A and B be square matrices of the same size. Show that the traces of the matrices AB and BA are equal, i.e., show that $\text{tr}(AB) = \text{tr}(BA)$.

(ii) Show that you cannot find two $n \times n$ matrices A and B such that

$$AB - BA = I,$$

where I is the $n \times n$ identity matrix.

Solution: (i) Recall that, for any two square matrices A and B of the same size, the matrices AB and BA have the same characteristic polynomial, i.e.,

$$P_{AB}(x) = \det(xI - AB) = \det(xI - BA) = P_{BA}(x), \quad \forall\, x \in \mathbb{R}; \qquad (4.50)$$

cf. (4.45).

Also, recall the following connection between the characteristic polynomial $P_M(x)$ of an $n \times n$ matrix M and the trace $\mathrm{tr}(M)$ of the matrix:

$$
\begin{aligned}
P_M(x) &= \det(xI - M) \\
&= x^n - \mathrm{tr}(M)x^{n-1} + \ldots + (-1)^n \det(M). \qquad (4.51)
\end{aligned}
$$

Since $P_{AB}(x) = P_{BA}(x)$ for all $x \in \mathbb{R}$, see (4.50), we obtain from (4.51) that

$$
\begin{aligned}
&x^n - \mathrm{tr}(AB)x^{n-1} + \ldots + (-1)^n \det(AB) \\
=\; &x^n - \mathrm{tr}(BA)x^{n-1} + \ldots + (-1)^n \det(BA), \quad \forall\, x \in \mathbb{R},
\end{aligned}
$$

and we conclude that

$$\mathrm{tr}(AB) = \mathrm{tr}(BA). \qquad (4.52)$$

(ii) We give a proof by contradiction. If it were possible to find $n \times n$ matrices A and B such that $AB - BA = I$, then

$$\mathrm{tr}(AB - BA) = \mathrm{tr}(I) = n. \qquad (4.53)$$

However,

$$\mathrm{tr}(AB - BA) = \mathrm{tr}(AB) - \mathrm{tr}(BA) = 0, \qquad (4.54)$$

since the trace of the sum of two matrices is the sum of the traces of the two matrices[3] and $\mathrm{tr}(AB) = \mathrm{tr}(BA)$; cf. (4.52).

Since (4.53) and (4.54) contradict each other, we conclude that there are no matrices A and B such that $AB - BA = I$. □

Problem 19: Recall that the matrix

$$
B_4 = \begin{pmatrix}
2 & -1 & 0 & 0 \\
-1 & 2 & -1 & 0 \\
0 & -1 & 2 & -1 \\
0 & 0 & -1 & 2
\end{pmatrix}
$$

---

[3] More precisely, for any two square matrices M_1 and M_2 of the same size ,

$$\mathrm{tr}(M_1 + M_2) = \mathrm{tr}(M_1) + \mathrm{tr}(M_2).$$

has four eigenvectors v_1, v_2, v_3, v_4 given by

$$v_j(i) = \sin\left(\frac{ij\pi}{5}\right), \quad \forall\, i = 1:4,$$

for $j = 1:4$.

Are the vectors v_1, v_2, v_3, and v_4 orthogonal and of norm 1?

Solution: Recall that

$$
\begin{aligned}
\sin(\pi - x) &= \sin(x), & \forall\, x \in \mathbb{R}; \\
\sin(\pi + x) &= -\sin(x), & \forall\, x \in \mathbb{R}; \\
\sin(2\pi - x) &= -\sin(x), & \forall\, x \in \mathbb{R}; \\
\sin(2\pi + x) &= \sin(x), & \forall\, x \in \mathbb{R},
\end{aligned}
$$

and therefore

$$\sin\left(\frac{3\pi}{5}\right) = \sin\left(\pi - \frac{2\pi}{5}\right) = \sin\left(\frac{2\pi}{5}\right);$$

$$\sin\left(\frac{4\pi}{5}\right) = \sin\left(\pi - \frac{\pi}{5}\right) = \sin\left(\frac{\pi}{5}\right);$$

$$\sin\left(\frac{6\pi}{5}\right) = \sin\left(\pi + \frac{\pi}{5}\right) = -\sin\left(\frac{\pi}{5}\right);$$

$$\sin\left(\frac{8\pi}{5}\right) = \sin\left(2\pi - \frac{2\pi}{5}\right) = -\sin\left(\frac{2\pi}{5}\right);$$

$$\sin\left(\frac{9\pi}{5}\right) = \sin\left(2\pi - \frac{\pi}{5}\right) = -\sin\left(\frac{\pi}{5}\right);$$

$$\sin\left(\frac{12\pi}{5}\right) = \sin\left(2\pi + \frac{2\pi}{5}\right) = \sin\left(\frac{2\pi}{5}\right);$$

$$\sin\left(\frac{16\pi}{5}\right) = \sin\left(2\pi + \frac{6\pi}{5}\right) = \sin\left(\frac{6\pi}{5}\right) = -\sin\left(\frac{\pi}{5}\right).$$

Then, the eigenvectors v_1, v_2, v_3, v_4 of the matrix B_4 are

$$v_1 = \left(\sin\left(\frac{i\pi}{5}\right)\right)_{i=1:4} = \begin{pmatrix} \sin\left(\frac{\pi}{5}\right) \\ \sin\left(\frac{2\pi}{5}\right) \\ \sin\left(\frac{3\pi}{5}\right) \\ \sin\left(\frac{4\pi}{5}\right) \end{pmatrix} = \begin{pmatrix} \sin\left(\frac{\pi}{5}\right) \\ \sin\left(\frac{2\pi}{5}\right) \\ \sin\left(\frac{2\pi}{5}\right) \\ \sin\left(\frac{\pi}{5}\right) \end{pmatrix}; \qquad (4.55)$$

$$v_2 = \left(\sin\left(\frac{2i\pi}{5}\right)\right)_{i=1:4} = \begin{pmatrix} \sin\left(\frac{2\pi}{5}\right) \\ \sin\left(\frac{4\pi}{5}\right) \\ \sin\left(\frac{6\pi}{5}\right) \\ \sin\left(\frac{8\pi}{5}\right) \end{pmatrix} = \begin{pmatrix} \sin\left(\frac{2\pi}{5}\right) \\ \sin\left(\frac{\pi}{5}\right) \\ -\sin\left(\frac{\pi}{5}\right) \\ -\sin\left(\frac{2\pi}{5}\right) \end{pmatrix}; \qquad (4.56)$$

$$v_3 = \left(\sin\left(\frac{3i\pi}{5}\right)\right)_{i=1:4} = \begin{pmatrix} \sin\left(\frac{3\pi}{5}\right) \\ \sin\left(\frac{6\pi}{5}\right) \\ \sin\left(\frac{9\pi}{5}\right) \\ \sin\left(\frac{12\pi}{5}\right) \end{pmatrix} = \begin{pmatrix} \sin\left(\frac{2\pi}{5}\right) \\ -\sin\left(\frac{\pi}{5}\right) \\ -\sin\left(\frac{\pi}{5}\right) \\ \sin\left(\frac{2\pi}{5}\right) \end{pmatrix} \qquad (4.57)$$

$$v_4 = \left(\sin\left(\frac{4i\pi}{5}\right)\right)_{i=1:4} = \begin{pmatrix} \sin\left(\frac{4\pi}{5}\right) \\ \sin\left(\frac{8\pi}{5}\right) \\ \sin\left(\frac{12\pi}{5}\right) \\ \sin\left(\frac{16\pi}{5}\right) \end{pmatrix} = \begin{pmatrix} \sin\left(\frac{\pi}{5}\right) \\ -\sin\left(\frac{2\pi}{5}\right) \\ \sin\left(\frac{2\pi}{5}\right) \\ -\sin\left(\frac{\pi}{5}\right) \end{pmatrix}. \qquad (4.58)$$

From (4.55–4.58), we obtain that

$$||v_1||^2 = ||v_2||^2 = ||v_3||^2 = ||v_4||^2 = 2\left(\sin\left(\frac{\pi}{5}\right)\right)^2 + 2\left(\sin\left(\frac{2\pi}{5}\right)\right)^2 = 2.5.$$

In other words, the eigenvectors v_1, v_2, v_3, v_4 do not have norm 1.

From (4.55–4.58), we find that

$$
\begin{aligned}
(v_1, v_2) &= \sin\left(\frac{\pi}{5}\right)\cdot\sin\left(\frac{2\pi}{5}\right) + \sin\left(\frac{2\pi}{5}\right)\cdot\sin\left(\frac{\pi}{5}\right) \\
&\quad + \sin\left(\frac{2\pi}{5}\right)\cdot\left(-\sin\left(\frac{\pi}{5}\right)\right) + \sin\left(\frac{\pi}{5}\right)\cdot\left(-\sin\left(\frac{2\pi}{5}\right)\right) \\
&= 2\sin\left(\frac{\pi}{5}\right)\cdot\sin\left(\frac{2\pi}{5}\right) - 2\sin\left(\frac{\pi}{5}\right)\cdot\sin\left(\frac{2\pi}{5}\right) \\
&= 0. \\[4pt]
(v_1, v_3) &= \sin\left(\frac{\pi}{5}\right)\cdot\sin\left(\frac{2\pi}{5}\right) + \sin\left(\frac{2\pi}{5}\right)\cdot\left(-\sin\left(\frac{\pi}{5}\right)\right) \\
&\quad + \sin\left(\frac{2\pi}{5}\right)\cdot\left(-\sin\left(\frac{\pi}{5}\right)\right) + \sin\left(\frac{\pi}{5}\right)\cdot\sin\left(\frac{2\pi}{5}\right) \\
&= 2\sin\left(\frac{\pi}{5}\right)\cdot\sin\left(\frac{2\pi}{5}\right) - 2\sin\left(\frac{\pi}{5}\right)\cdot\sin\left(\frac{2\pi}{5}\right) \\
&= 0. \\[4pt]
(v_1, v_4) &= \sin\left(\frac{\pi}{5}\right)\cdot\sin\left(\frac{\pi}{5}\right) + \sin\left(\frac{2\pi}{5}\right)\cdot\left(-\sin\left(\frac{2\pi}{5}\right)\right) \\
&\quad + \sin\left(\frac{2\pi}{5}\right)\cdot\sin\left(\frac{2\pi}{5}\right) + \sin\left(\frac{\pi}{5}\right)\cdot\left(-\sin\left(\frac{\pi}{5}\right)\right) \\
&= \sin^2\left(\frac{\pi}{5}\right) - \sin^2\left(\frac{2\pi}{5}\right) + \sin^2\left(\frac{2\pi}{5}\right) - \sin^2\left(\frac{\pi}{5}\right) \\
&= 0. \\[4pt]
(v_2, v_3) &= \sin\left(\frac{2\pi}{5}\right)\cdot\sin\left(\frac{2\pi}{5}\right) + \sin\left(\frac{\pi}{5}\right)\cdot\left(-\sin\left(\frac{\pi}{5}\right)\right) \\
&\quad + \left(-\sin\left(\frac{\pi}{5}\right)\right)\cdot\left(-\sin\left(\frac{\pi}{5}\right)\right) + \left(-\sin\left(\frac{2\pi}{5}\right)\right)\cdot\sin\left(\frac{2\pi}{5}\right) \\
&= \sin^2\left(\frac{2\pi}{5}\right) - \sin^2\left(\frac{\pi}{5}\right) + \sin^2\left(\frac{\pi}{5}\right) - \sin^2\left(\frac{2\pi}{5}\right) \\
&= 0. \\[4pt]
(v_2, v_4) &= \sin\left(\frac{2\pi}{5}\right)\cdot\sin\left(\frac{\pi}{5}\right) + \sin\left(\frac{\pi}{5}\right)\cdot\left(-\sin\left(\frac{2\pi}{5}\right)\right) \\
&\quad + \left(-\sin\left(\frac{\pi}{5}\right)\right)\cdot\sin\left(\frac{2\pi}{5}\right) + \left(-\sin\left(\frac{2\pi}{5}\right)\right)\cdot\left(-\sin\left(\frac{\pi}{5}\right)\right) \\
&= 2\sin\left(\frac{\pi}{5}\right)\cdot\sin\left(\frac{2\pi}{5}\right) - 2\sin\left(\frac{\pi}{5}\right)\cdot\sin\left(\frac{2\pi}{5}\right) \\
&= 0. \\[4pt]
(v_3, v_4) &= \sin\left(\frac{2\pi}{5}\right)\cdot\sin\left(\frac{\pi}{5}\right) + \left(-\sin\left(\frac{\pi}{5}\right)\right)\cdot\left(-\sin\left(\frac{2\pi}{5}\right)\right)
\end{aligned}
$$

$$+ \left(-\sin\left(\frac{\pi}{5}\right)\right) \cdot \sin\left(\frac{2\pi}{5}\right) + \left(\sin\left(\frac{2\pi}{5}\right)\right) \cdot \left(-\sin\left(\frac{\pi}{5}\right)\right)$$
$$= \ 2\sin\left(\frac{\pi}{5}\right) \cdot \sin\left(\frac{2\pi}{5}\right) - 2\sin\left(\frac{\pi}{5}\right) \cdot \sin\left(\frac{2\pi}{5}\right)$$
$$= \ 0.$$

In other words, the eigenvectors v_1, v_2, v_3, and v_4 are orthogonal.[4]   □

Problem 20: Show that the matrix $\begin{pmatrix} 3 & -1 & -1 \\ 0 & 2 & 2 \\ 0 & 1 & 1 \end{pmatrix}$ is a weakly diagonally dominant singular matrix.

Solution: Let

$$A = \begin{pmatrix} 3 & -1 & -1 \\ 0 & 2 & 2 \\ 0 & 1 & 1 \end{pmatrix}.$$

The matrix A is weakly diagonally dominant since, for every row of A, the absolute value of the main diagonal entry from the row is greater than or equal to the sum of the absolute values of all the other entries in that row:

$$3 > |-1| + |-1| = 2; \quad 2 = 2; \quad 1 = 1.$$

The matrix A is singular since its determinant is equal to 0:

$$\det(A) = 3 \cdot 2 \cdot 1 + (-1) \cdot 2 \cdot 0 + (-1) \cdot 0 \cdot 1 - (-1) \cdot 2 \cdot 0 - (-1) \cdot 0 \cdot 1 - 3 \cdot 2 \cdot 1$$
$$= 6 - 6 = 0. \quad \square$$

Problem 21: Show that the matrix

$$\begin{pmatrix} -5 & -1 & 0.25 & -1 \\ 2 & 4 & -1 & 3 \\ 0.5 & -2 & 3 & -1 \\ 1 & 0.5 & 0 & -6 \end{pmatrix}$$

is nonsingular.

Solution: Let

$$A = \begin{pmatrix} -5 & -1 & 0.25 & -1 \\ 2 & 4 & -1 & 3 \\ 0.5 & -2 & 3 & -1 \\ 1 & 0.5 & 0 & -6 \end{pmatrix}.$$

---

[4]Note that this was to be expected, since different eigenvectors of any symmetric matrix are orthogonal.

The matrix A is strictly column diagonally dominant[5] since, for every column of A, the absolute value of the main diagonal entry from the column is strictly greater than the sum of the absolute values of all the other entries in that column:

$$|-5| = 5 \quad > \quad 2 + 0.5 + 1 = 3.75;$$
$$4 \quad > \quad |-1| + |-2| + 0.5 = 3.5;$$
$$3 \quad > \quad 0.25 + |-1| = 1.25;$$
$$|-6| = 6 \quad > \quad |-1| + 3 + |-1| = 5.$$

Since any strictly column diagonally dominant matrix is nonsingular, we conclude that the matrix A is nonsingular. $\square$

Problem 22: Let A be an $n \times n$ matrix. For every $j = 1 : n$, let

$$R_j = \sum_{k=1:n, k \neq j} |A(j,k)|,$$

and denote by D_j the disc of center $A(j,j)$ and radius R_j, i.e.,

$$D_j = \{z \in \mathbb{C} \text{ such that } |z - A(j,j)| \leq R_j\};$$

note that D_j is called a Gershgorin disk corresponding to the matrix A.
A more general form of Gershgorin's theorem states that, if a Gershgorin disk D_i is disjoint from the union of the other $n - 1$ Gershgorin disks of A, then exactly one eigenvalue of the matrix A is in the disk D_i.
Use this result to show that all the eigenvalues of the matrix

$$\begin{pmatrix} -4 & 1 & 0 & -0.5 \\ 0 & 0.1 & 0 & -0.2 \\ 1 & 2 & 5 & -1 \\ 0.25 & -0.15 & 0.1 & -1 \end{pmatrix}$$

are real numbers.

Solution: Let

$$A = \begin{pmatrix} -4 & 1 & 0 & -0.5 \\ 0 & 0.1 & 0 & -0.2 \\ 1 & 2 & 5 & -1 \\ 0.25 & -0.15 & 0.1 & -1 \end{pmatrix}.$$

Note that

$$A(1,1) = -4; \quad R_1 = 1 + |-0.5| = 1.5;$$
$$A(2,2) = 0.1; \quad R_2 = |-0.2| = 0.2;$$
$$A(3,3) = 5; \quad R_3 = 1 + 2 + |-1| = 4;$$
$$A(4,4) = -1; \quad R_4 = 0.25 + |-0.15| + 0.1 = 0.5.$$

[5]Note that the matrix A is not strictly diagonally dominant, since
$$A(2,2) = 4 < |A(2,1)| + |A(2,3)| + |A(2,4)| = 6,$$
and therefore we cannot use the property that any strictly diagonally dominant matrix is nonsingular in order to show that the matrix A is nonsingular.

The four Gershgorin disks corresponding to the rows of the matrix A are:

$$
\begin{aligned}
D_1 &= D(-4, 1.5) = \{z \in \mathbb{C} \text{ such that } |z + 4| \leq 1.5\}; \\
D_2 &= D(0.1, 0.2) = \{z \in \mathbb{C} \text{ such that } |z - 0.1| \leq 0.2\}; \\
D_3 &= D(5, 4) = \{z \in \mathbb{C} \text{ such that } |z - 5| \leq 4\}; \\
D_4 &= D(-1, 0.5) = \{z \in \mathbb{C} \text{ such that } |z + 1| \leq 0.5\}.
\end{aligned}
$$

The Gershgorin disks above do not intersect. To see this, note that the centers of the disks are all on the real axis, and if the disks were to intersect, they would do so on the real axis as well. However, the disks cover disjoint intervals on the real axis:

$$
\begin{aligned}
\text{the interval} \quad [-5.5, -2.5] \quad &\text{corresponding to} \quad D_1 = D(-4, 1.5); \\
\text{the interval} \quad [-1.5, -0.5] \quad &\text{corresponding to} \quad D_4 = D(-1, 0.5); \\
\text{the interval} \quad [-0.1, 0.3] \quad &\text{corresponding to} \quad D_2 = D(0.1, 0.2); \\
\text{the interval} \quad [1, 9] \quad &\text{corresponding to} \quad D_3 = D(5, 4).
\end{aligned}
$$

The 4×4 matrix A has four eigenvalues, counted with their multiplicities. Since the matrix A has four disjoint Gershgorin disks, it follows from the more general form of Gershgorin's theorem stated above that every Gershgorin disk contains exactly one eigenvalue of the matrix A.[6]

To see how this implies that all the eigenvalues of the matrix A are real numbers, assume that $\lambda_1 = a + ib$ is a complex eigenvalue of the matrix A with $b \neq 0$. Then, $\lambda_1 = a + ib$ is a root of the characteristic polynomial $P_A(t)$ of the matrix A. The polynomial $P_A(t)$ has real coefficients. Thus, the complex conjugate of the root λ_1, i.e., $\lambda_2 = \overline{\lambda_1} = a - ib$, is also a root of $P_A(t)$, and therefore an eigenvalue of the matrix A.

However, $\lambda_1 = a + ib$ and $\lambda_2 = a - ib$ have the same real part a, and therefore both λ_1 and λ_2 belong to the Gershgorin disk whose real axis interval contains the point a. (Recall that the four Gershgorin disks of this matrix cover the disjoint intervals $[-5.5, -2.5]$, $[-1.5, -0.5]$, $[-0.1, 0.3]$, and $[1, 9]$.)

This contradicts the fact that exactly one eigenvalue of the matrix A belongs to one Gershgorin disk. The contradiction comes from assuming that the matrix A has a complex eigenvalue.

We conclude that all the eigenvalues of the matrix A are real numbers. $\square$

Problem 23: Show that all the eigenvalues of the matrix

$$
\begin{pmatrix}
2 & 0.0012 & -0.0003 & 0.0015 \\
-0.0002 & -1.25 & 0.0010 & -0.0001 \\
0 & 0.0016 & 3 & 0.0009 \\
-0.0011 & -0.0008 & -0.0002 & -2.5
\end{pmatrix}
$$

are real numbers, and find estimates for the values of the eigenvalues of the matrix with 0.005 accuracy.

---

[6]The eigenvalues of the matrix A are -3.9637, -1.0412, 0.1239, and 4.9809.

Solution: Let

$$A = \begin{pmatrix} 2 & 0.0012 & -0.0003 & 0.0015 \\ -0.0002 & -1.25 & 0.0010 & -0.0001 \\ 0 & 0.0016 & 3 & 0.0009 \\ -0.0011 & -0.0008 & -0.0002 & -2.5 \end{pmatrix}.$$

We will show that all the eigenvalues of the matrix A are real numbers by using the more general form of Gershgorin's theorem stating that, if a Gershgorin disk D_i is disjoint from the union of the other $n-1$ Gershgorin disks of A, then exactly one eigenvalue of the matrix A is in the disk D_i.

To compute the Gershgorin disks of the matrix A, note that

$$\begin{aligned} A(1,1) &= 2; & R_1 &= 0.0012 + |-0.0003| + 0.0015 = 0.003; \\ A(2,2) &= -1.25; & R_2 &= |-0.0002| + 0.0010 + |-0.0001| = 0.0013; \\ A(3,3) &= 3; & R_3 &= 0.0016 + 0.0009 = 0.0025; \\ A(4,4) &= -2.5; & R_4 &= |-0.0011| + |-0.0008| + |-0.0002| = 0.0021. \end{aligned}$$

The four Gershgorin disks corresponding to the rows of the matrix A are:

$$\begin{aligned} D_1 &= D(2, 0.003) = \{z \in \mathbb{C} \text{ such that } |z-2| \leq 0.003\}; \\ D_2 &= D(-1.25, 0.0013) = \{z \in \mathbb{C} \text{ such that } |z+1.25| \leq 0.0013\}; \\ D_3 &= D(3, 0.0025) = \{z \in \mathbb{C} \text{ such that } |z-3| \leq 0.0025\}; \\ D_4 &= D(-2.5, 0.0021) = \{z \in \mathbb{C} \text{ such that } |z+2.5| \leq 0.0021\}. \end{aligned}$$

The Gershgorin disks above do not intersect. To see this, note that the centers of the disks are all on the real axis, and if the disks were to intersect, they would do so on the real axis as well. However, the disks cover disjoint intervals on the real axis:

the interval	$[-2.5021, -2.4979]$	corresponding to	$D_4 = D(-2.5, 0.0021)$;
the interval	$[-1.2513, -1.2487]$	corresponding to	$D_2 = D(-1.25, 0.0013)$;
the interval	$[1.997, 2.003]$	corresponding to	$D_1 = D(2, 0.003)$;
the interval	$[2.9975, 3.0025]$	corresponding to	$D_3 = D(3, 0.0025)$.

The 4×4 matrix A has four eigenvalues, counted with their multiplicities. Since the matrix A has four disjoint Gershgorin disks, it follows from the more general form of Gershgorin's theorem stated above that every Gershgorin disk contains exactly one eigenvalue of the matrix A. Since the radius of each of the disks $D_1 = D(2, 0.003)$, $D_2 = D(-1.25, 0.0013)$, $D_3 = D(3, 0.0025)$, and $D_4 = D(-2.5, 0.0021)$ is smaller than 0.005, the center of each disk approximates the eigenvalue from that disk with 0.005 accuracy. In other words, -2, -1.25, 3 and -2.5 approximate the eigenvalues of A with 0.005 accuracy.[7]

To show that all the eigenvalues of the matrix A are real numbers, assume that $\lambda_1 = a + ib$ is a complex eigenvalue of the matrix A with $b \neq 0$. Then, $\lambda_1 = a + ib$ is a root of the characteristic polynomial $P_A(t)$ of the matrix A. The polynomial $P_A(t)$

---

[7]The eigenvalues of the matrix A are -2.4999996646, -1.2500002386, 1.9999995594, and 3.0000003438. Note that the diagonal elements -2.5, -1.25, 2, and 3 of the matrix A are, in fact, within 10^{-6} of the eigenvalues of A. This is a consequence of the fact that all the off-diagonal elements of the matrix A are small, a property which is used in the convergence of the QR algorithm for computing matrix eigenvalues.

has real coefficients. Thus, the complex conjugate of the root λ_1, i.e., $\lambda_2 = \overline{\lambda_1} = a - ib$, is also a root of $P_A(t)$, and therefore an eigenvalue of the matrix A.

However, $\lambda_1 = a + ib$ and $\lambda_2 = a - ib$ have the same real part a, and therefore both λ_1 and λ_2 belong to the Gershgorin disk whose real axis interval contains the point a. (Recall that the four Gershgorin disks of this matrix cover the disjoint intervals $[-2.5021, -2.4979]$, $[-1.2513, -1.2487]$, $[1.997, 2.003]$, and $[2.9975, 3.0025]$.)

This contradicts the fact that exactly one eigenvalue of the matrix A belongs to one Gershgorin disk. The contradiction comes from assuming that the matrix A has a complex eigenvalue.

We conclude that all the eigenvalues of the matrix A are real numbers. $\qquad\square$

Problem 24: Let A be an $n \times n$ matrix given by

$$\begin{aligned}
A(i, i) &= 2, \quad \forall\, i = 1 : n; \\
A(i, i - 1) &= 1, \quad \forall\, i = 2 : n; \\
A(j, k) &= 0, \quad \text{otherwise.}
\end{aligned}$$

Find the eigenvalues and the eigenvectors of A.

Solution: Note that the matrix A is bidiagonal lower triangular:

$$A = \begin{pmatrix} 2 & 0 & \cdots & 0 \\ 1 & \ddots & \ddots & \vdots \\ \vdots & \ddots & \ddots & 0 \\ 0 & \cdots & 1 & 2 \end{pmatrix}.$$

The characteristic polynomial of the matrix A is

$$\begin{aligned}
P_A(t) &= \det(tI - A) = \det \begin{pmatrix} t-2 & 0 & \cdots & 0 \\ -1 & \ddots & \ddots & \vdots \\ \vdots & \ddots & \ddots & 0 \\ 0 & \cdots & -1 & t-2 \end{pmatrix} \\
&= (t-2)^n.
\end{aligned}$$

By solving

$$P_A(t) = 0 \iff (t-2)^n = 0 \iff t = 2,$$

we find that the only root of the characteristic polynomial $P_A(t)$ is 2 (with multiplicity n) and therefore $\lambda = 2$ is the only eigenvalue of the matrix A and has multiplicity n.

Let $v = (v_i)_{i=1:n}$ be an eigenvector corresponding to the eigenvalue $\lambda = 2$. Then,

$$\begin{aligned}
Av = \lambda v &\iff Av = 2v \\
&\iff \begin{pmatrix} 2 & 0 & \cdots & 0 \\ 1 & \ddots & \ddots & \vdots \\ \vdots & \ddots & \ddots & 0 \\ 0 & \cdots & 1 & 2 \end{pmatrix} \begin{pmatrix} v_1 \\ v_2 \\ \vdots \\ v_n \end{pmatrix} = 2 \begin{pmatrix} v_1 \\ v_2 \\ \vdots \\ v_n \end{pmatrix}
\end{aligned}$$

$$\Longleftrightarrow \begin{cases} 2v_1 & = & 2v_1 \\ 2v_2 + v_1 & = & 2v_2 \\ & \vdots & \\ 2v_n + v_{n-1} & = & 2v_n \end{cases} \Longleftrightarrow \begin{cases} v_1 & = & 0 \\ & \vdots & \\ v_{n-1} & = & 0 \end{cases}$$

Thus,

$$v = \begin{pmatrix} v_1 \\ v_2 \\ \vdots \\ v_{n-1} \\ v_n \end{pmatrix} = \begin{pmatrix} 0 \\ 0 \\ \vdots \\ 0 \\ v_n \end{pmatrix} = v_n \begin{pmatrix} 0 \\ 0 \\ \vdots \\ 0 \\ 1 \end{pmatrix}.$$

In other words, the eigenvalue $\lambda = 2$ has only one linearly independent eigenvector. We conclude that the matrix A has one eigenvalue, $\lambda = 2$, with multiplicity n, and only one eigenvector, e.g., $\begin{pmatrix} 0 \\ \vdots \\ 0 \\ 1 \end{pmatrix}$, corresponding to the eigenvalue $\lambda = 2$. □

Problem 25: Let J be an $n \times n$ matrix given by

$$\begin{aligned} A(i,i) & = & a, \quad \forall\, i = 1:n; \\ A(i,i+1) & = & b, \quad \forall\, i = 1:(n-1); \\ A(j,k) & = & 0, \quad \text{otherwise}, \end{aligned}$$

where $a, b \in \mathbb{R}$ are constants. Find the eigenvalues and the eigenvectors of J.

Note: The matrix J is called a Jordan block if $b = 1$.

Solution: Let

$$A = \begin{pmatrix} a & b & \cdots & 0 \\ 0 & \ddots & \ddots & \vdots \\ \vdots & \ddots & \ddots & b \\ 0 & \cdots & 0 & a \end{pmatrix}.$$

The characteristic polynomial of the matrix A is

$$P_A(t) = \det(tI - A) = (t-a)^n.$$

By solving

$$P_A(t) = 0 \iff (t-a)^n = 0 \iff t = a,$$

we find that the only root of the characteristic polynomial $P_A(t)$ is a (with multiplicity n) and therefore $\lambda = a$ is the only eigenvalue of the matrix A and has multiplicity n.

• If $b = 0$, then the matrix A is the identity matrix multiplied by the constant a, i.e., $A = aI$. Thus, A has n eigenvectors $e_1, e_2, \ldots e_n$, where e_k is the k-th column of the identity matrix given by

$$e_k(i) = 0, \text{ for } 1 \le i \ne k \le n \quad \text{and} \quad e_k(k) = 1.$$

• If $b \neq 0$, let $v = (v_i)_{i=1:n}$ be an eigenvector of the matrix A corresponding to the eigenvalue $\lambda = a$. Then,

$$Av = av \iff \begin{pmatrix} a & b & \cdots & 0 \\ 0 & \ddots & \ddots & \vdots \\ \vdots & \ddots & \ddots & b \\ 0 & \cdots & 0 & a \end{pmatrix} \begin{pmatrix} v_1 \\ v_2 \\ \vdots \\ v_n \end{pmatrix} = a \begin{pmatrix} v_1 \\ v_2 \\ \vdots \\ v_n \end{pmatrix}$$

$$\iff \begin{cases} av_1 + bv_2 &= av_1 \\ av_2 + bv_3 &= av_2 \\ &\vdots & \vdots \\ av_{n-1} + bv_n &= av_{n-1} \\ av_n &= av_n \end{cases} \iff \begin{cases} bv_2 &= 0 \\ bv_3 &= 0 \\ \vdots & \vdots \\ bv_n &= 0 \end{cases}$$

$$\iff v_2 = v_3 = \ldots = v_n = 0,$$

since $b \neq 0$. Thus, $v = \begin{pmatrix} v_1 \\ 0 \\ \vdots \\ 0 \end{pmatrix} = v_1 \begin{pmatrix} 1 \\ 0 \\ \vdots \\ 0 \end{pmatrix}$, and we conclude that the eigenvalue

$\lambda = a$ of A has one linearly independent eigenvector, e.g., $\begin{pmatrix} 1 \\ 0 \\ \vdots \\ 0 \end{pmatrix}$. $\quad\square$

Problem 26: The 2–norm of a symmetric matrix A is given by

$$||A||_2 = \max_{\lambda \text{ eigenvalue of } A} |\lambda|,$$

i.e., $||A||_2$ is equal to the largest absolute value of all the eigenvalues of A.

The Forward Euler finite difference discretization of the heat PDE is convergent if and only if $||A_N||_2 \leq 1$ for all $N \geq 2$, where A_N is the $N \times N$ matrix given by

$$A_N = \begin{pmatrix} 1 - 2\alpha & -\alpha & \cdots & 0 \\ -\alpha & \ddots & \ddots & \vdots \\ \vdots & \ddots & \ddots & -\alpha \\ 0 & \cdots & -\alpha & 1 - 2\alpha \end{pmatrix}, \tag{4.59}$$

with $\alpha > 0$ a positive constant.

Show that $||A_N||_2 \leq 1$ for all $N \geq 2$ if and only if $0 < \alpha \leq \frac{1}{2}$.

Solution: Recall that the eigenvalues of the matrix

$$T_N = \begin{pmatrix} d & -a & \cdots & 0 \\ -a & \ddots & \ddots & \vdots \\ \vdots & \ddots & \ddots & -a \\ 0 & \cdots & -a & d \end{pmatrix}$$

are

$$\lambda_j \;=\; d - 2a\cos\left(\frac{\pi j}{N+1}\right), \quad \text{for}\quad j = 1:N. \tag{4.60}$$

By letting $d = 1 - 2\alpha$ and $a = \alpha$ in (4.60), we obtain that the eigenvalues of the matrix A_N given by (4.59) are

$$\lambda_j \;=\; 1 - 2\alpha - 2\alpha\cos\left(\frac{\pi j}{N+1}\right), \quad \text{for}\quad j = 1:N. \tag{4.61}$$

Note that

$$1 \;>\; \cos\left(\frac{\pi}{N+1}\right) \;>\; \cos\left(\frac{2\pi}{N+1}\right) \;>\; \dots \;>\; \cos\left(\frac{N\pi}{N+1}\right) \;>\; -1,$$

and, since $\alpha > 0$,

$$-2\alpha \;<\; -2\alpha\cos\left(\frac{\pi}{N+1}\right) \;<\; -2\alpha\cos\left(\frac{2\pi}{N+1}\right) \;<\dots<\; -2\alpha\cos\left(\frac{N\pi}{N+1}\right) \;<\; 2\alpha. \tag{4.62}$$

By adding $1 - 2\alpha$ to (4.62) and using (4.61), we obtain that

$$1 - 4\alpha \;<\; \lambda_1 \;<\; \lambda_2 \;<\; \dots \;<\; \lambda_N \;<\; 1. \tag{4.63}$$

Note that

$$\|A_N\|_2 \;=\; \max_{i=1:N}|\lambda_i| \;\leq\; 1 \quad\Longleftrightarrow\quad -1 \leq \lambda_i \leq 1, \;\; \forall\, i = 1:N, \tag{4.64}$$

i.e., all the eigenvalues of the matrix A_N are between -1 and 1. Since the eigenvalues of A_N are between $1 - 4\alpha$ and 1, see (4.63), it follows that all the eigenvalues of A_N are between -1 and 1 if and only if

$$-1 \leq 1 - 4\alpha \quad\Longleftrightarrow\quad 4\alpha \leq 2 \quad\Longleftrightarrow\quad \alpha \leq \frac{1}{2}.$$

In other words,

$$-1 \leq \lambda_i \leq 1, \;\; \forall\, i = 1:N \quad\Longleftrightarrow\quad \alpha \leq \frac{1}{2}. \tag{4.65}$$

From (4.64) and (4.65), and since $\alpha > 0$, we conclude that

$$\|A_N\|_2 \leq 1 \quad\Longleftrightarrow\quad 0 < \alpha \leq \frac{1}{2}. \quad\square$$

Problem 27: The Backward Euler finite difference discretization of the heat PDE is convergent if and only if $\|A_N^{-1}\|_2 \leq 1$ for all $N \geq 2$, where

$$A_N \;=\; \begin{pmatrix} 1+2\alpha & \alpha & \dots & 0 \\ \alpha & \ddots & \ddots & \vdots \\ \vdots & \ddots & \ddots & \alpha \\ 0 & \dots & \alpha & 1+2\alpha \end{pmatrix}, \tag{4.66}$$

with $\alpha > 0$ a positive constant.

Show that $||A_N^{-1}||_2 \le 1$ for all $N \ge 2$ for any $\alpha > 0$.

Solution: Since the matrix

$$
T_N = \begin{pmatrix}
d & -a & \cdots & 0 \\
-a & \ddots & \ddots & \vdots \\
\vdots & \ddots & \ddots & -a \\
0 & \cdots & -a & d
\end{pmatrix}
$$

has eigenvalues

$$\lambda_j = d - 2a \cos\left(\frac{\pi j}{N+1}\right), \quad \text{for} \quad j = 1 : N. \tag{4.67}$$

By letting $d = 1 + 2\alpha$ and $a = -\alpha$ in (4.67), it follows that the eigenvalues of the matrix A_N given by (4.66) are

$$\lambda_j = 1 + 2\alpha + 2\alpha \cos\left(\frac{\pi j}{N+1}\right), \quad \text{for} \quad j = 1 : N. \tag{4.68}$$

Note that

$$1 > \cos\left(\frac{\pi}{N+1}\right) > \cos\left(\frac{2\pi}{N+1}\right) > \cdots > \cos\left(\frac{N\pi}{N+1}\right) > -1. \tag{4.69}$$

Since $\alpha > 0$, we obtain from (4.68) and (4.69) that $\lambda_1 > \lambda_2 > \ldots > \lambda_N$. Moreover, we find that

$$\lambda_1 = 1 + 2\alpha + 2\alpha \cos\left(\frac{\pi}{N+1}\right) < 1 + 2\alpha + 2\alpha = 1 + 4\alpha;$$

$$\lambda_N = 1 + 2\alpha + 2\alpha \cos\left(\frac{N\pi}{N+1}\right) > 1 + 2\alpha - 2\alpha = 1.$$

Thus,

$$1 + 4\alpha > \lambda_1 > \lambda_2 > \ldots > \lambda_N > 1. \tag{4.70}$$

Recall that, if λ_j, $j = 1 : N$, are the eigenvalues of the matrix A_N, then $\mu_j = \frac{1}{\lambda_j}$, $j = 1 : N$, are the eigenvalues of the matrix A_N^{-1}. Then, from (4.70), we obtain that

$$1 > \mu_N > \ldots > \mu_2 > \mu_1 = \frac{1}{1+4\alpha},$$

and therefore

$$||A_N^{-1}||_2 = \max_{i=1:N} |\mu_i| = \mu_N < 1$$

for all $\alpha > 0$. $\square$

Chapter 5

Symmetric matrices and symmetric positive definite matrices

5.1 Exercises

1. Show that the matrix $\begin{pmatrix} 1 & 1 \\ 1 & 1 \end{pmatrix}$ is symmetric positive semidefinite but is not symmetric positive definite.

2. Show that a 2×2 matrix A is symmetric positive definite if and only if A is symmetric, $\text{tr}(A) > 0$, and $\det(A) > 0$.

3. Let A be a symmetric positive definite matrix. Use the diagonal form of the matrix A to find a matrix B such that $B^2 = A$.

4. Show that a matrix A is symmetric positive semidefinite if and only if there exists a symmetric matrix B such that $B^2 = A$.

5. Show that the inner product of two vectors is bounded from above by the product of the norms of the vectors, i.e., show that

$$(u, v) \leq ||u|| \, ||v||, \quad \forall \, u, v \in \mathbb{R}^n. \tag{5.1}$$

Hint: Use the facts that

$$||tu + v||^2 = t^2 ||u||^2 + 2t(u, v) + ||v||^2 \geq 0, \quad \forall t \in \mathbb{R},$$

and that a quadratic polynomial is nonnegative if and only if its discriminant is less than ot equal to 0 , i.e.,

$$at^2 + bt + c \geq 0, \forall \, t \in \mathbb{R} \quad \Longleftrightarrow \quad b^2 - 4ac \leq 0,$$

where a, b, and c are fixed real numbers.

Note: The inequality (5.1) is the Cauchy–Schwarz inequality for inner products. For the Euclidean inner product and Euclidean vector norm, the inequality (5.1) becomes

$$\left(\sum_{i=1}^{n} u_i v_i\right)^2 \le \left(\sum_{i=1}^{n} u_i^2\right)\left(\sum_{i=1}^{n} v_i^2\right), \quad \forall\; u_i, v_i \in \mathbb{R},\; i = 1:n,$$

which is the classical version of the Cauchy–Schwarz inequality for real numbers.

6. Let A be a symmetric positive semidefinite matrix, and let B be a symmetric matrix such that $B^2 = A$.

(i) Show that
$$(Ax, y) \;=\; (Bx, By), \quad \forall\; x, y \in \mathbb{R}^n, \tag{5.2}$$
where $(\cdot, \cdot)$ denotes the Euclidean inner product.

(ii) Show that
$$(Ax, y)^2 \;\le\; (Ax, x)\,(Ay, y), \quad \forall\; x, y \in \mathbb{R}^n. \tag{5.3}$$

Hint: Note that, from the Cauchy–Schwartz inequality (5.1), it follows that $(Bx, By) \le \|Bx\|\, \|By\|$.

7. Let λ and v be an eigenvalue and a corresponding eigenvector of the square matrix A of size n. Let Q be an orthogonal matrix of size n. Show that λ is also an eigenvalue of the matrix $Q^t AQ$. What is a corresponding eigenvector?

8. Let Q be an orthogonal matrix of size n and let A be a square matrix of size n. Show that the matrix $Q^t AQ$ has the same characteristic polynomial as A, i.e., show that
$$P_{Q^t AQ}(t) \;=\; P_A(t), \quad \forall\; t \in \mathbb{R}.$$

9. Let A be a symmetric positive definite matrix with diagonal form $A = Q\Lambda Q^t$, where Q is an orthogonal matrix and $\Lambda = \mathrm{diag}(\lambda_k)_{k=1:n}$ is a diagonal matrix. Recall that λ_k, $k = 1:n$, are the eigenvalues of A, and that $\lambda_k > 0$ for all $k = 1:n$, since A is symmetric positive definite.
Let
$$\Lambda^{1/2} \;=\; \mathrm{diag}\left(\sqrt{\lambda_k}\right)_{k=1:n} \quad \text{and} \quad \Lambda^{-1/2} = \mathrm{diag}\left(\frac{1}{\sqrt{\lambda_k}}\right)_{k=1:n},$$
and let $A^{1/2}$ and $A^{-1/2}$ be the matrices given by
$$A^{1/2} \;=\; Q\Lambda^{1/2}Q^t \quad \text{and} \quad A^{-1/2} \;=\; Q\Lambda^{-1/2}Q^t.$$

Show that
$$\begin{aligned}
(A^{1/2})^2 &= A;\\
(A^{-1/2})^2 &= A^{-1};\\
(A^{1/2})^{-1} &= A^{-1/2}.
\end{aligned}$$

10. Show that an $n \times n$ matrix has $2^n - 1$ principal minors.

11. Show that the matrix

$$
A = \begin{pmatrix}
1 & 0.2 & -0.2 & 0.1 \\
0.2 & 1 & -0.25 & 0.05 \\
-0.2 & -0.25 & 1 & -0.15 \\
0.1 & 0.05 & -0.15 & 1
\end{pmatrix}
$$

is symmetric positive definite.

12. Let

$$
A = \begin{pmatrix}
1 & 0.05 & 0.25 & 0.55 \\
0.05 & 2 & 1 & 1.25 \\
0.25 & 1 & 4 & 2.5 \\
0.55 & 1.25 & 2.5 & 6
\end{pmatrix}.
$$

(i) Show that the matrix A is not strictly diagonally dominant.

(ii) Show that the matrix A is symmetric positive definite.

13. Let A be a symmetric positive semidefinite matrix. Show that the matrix A^k is also symmetric positive semidefinite for any positive integer k.

14. Recall that any symmetric positive definite matrix is nonsingular. Let A be a symmetric positive definite matrix. Show that A^{-1} is a symmetric positive definite matrix.

15. Let A be a square matrix such that $A^2 = A$. Show that the matrix A cannot be strictly diagonally dominant unless A is the identity matrix.

16. Let A be an $n \times n$ symmetric positive definite matrix, and let M be an $n \times m$ matrix.

(i) Show that the matrix $M^t A M$ is symmetric positive semidefinite.

(ii) Show that the matrix $M^t A M$ is symmetric positive definite if and only if the columns of the matrix M are linearly independent.

5.2 Solutions to Chapter 5 Exercises

Problem 1: Show that the matrix $\begin{pmatrix} 1 & 1 \\ 1 & 1 \end{pmatrix}$ is symmetric positive semidefinite but is not symmetric positive definite.

Solution: Let $A = \begin{pmatrix} 1 & 1 \\ 1 & 1 \end{pmatrix}$. Note that $A^t = A$, i.e., the matrix A is symmetric.

Let $x = \begin{pmatrix} x_1 \\ x_2 \end{pmatrix} \in \mathbb{R}^2$. Then,

$$
\begin{aligned}
x^t A x &= (x_1 \; x_2) \begin{pmatrix} 1 & 1 \\ 1 & 1 \end{pmatrix} \begin{pmatrix} x_1 \\ x_2 \end{pmatrix} \\
&= (x_1 \; x_2) \begin{pmatrix} x_1 + x_2 \\ x_1 + x_2 \end{pmatrix} \\
&= x_1(x_1 + x_2) + x_2(x_1 + x_2) \\
&= x_1^2 + 2x_1 x_2 + x_2^2 \\
&= (x_1 + x_2)^2 \\
&\geq 0, \quad \forall \, x_1, x_2 \in \mathbb{R}.
\end{aligned}
$$

In other words,

$$x^t A x \geq 0, \quad \forall \, x \in \mathbb{R}^2.$$

By definition, we conclude that the matrix A is symmetric positive semidefinite.

To see that the matrix A is not symmetric positive definite, note that

$$A \begin{pmatrix} 1 \\ -1 \end{pmatrix} = \begin{pmatrix} 1 & 1 \\ 1 & 1 \end{pmatrix} \begin{pmatrix} 1 \\ -1 \end{pmatrix} = \begin{pmatrix} 0 \\ 0 \end{pmatrix}.$$

Then,

$$(1 \; -1) \, A \begin{pmatrix} 1 \\ -1 \end{pmatrix} = 0,$$

and therefore A is not a symmetric positive definite matrix, since, by definition, the matrix A is symmetric positive definite if and only if $x^t A x > 0$ for all $x \neq 0$. $\quad \square$

Problem 2: Show that a 2×2 matrix A is symmetric positive definite if and only if A is symmetric, $\text{tr}(A) > 0$, and $\det(A) > 0$.

Solution: Let $A = \begin{pmatrix} a & b \\ b & d \end{pmatrix}$ The characteristic polynomial of the matrix A is

$$
\begin{aligned}
P_A(t) &= \det(tI - A) = \det \begin{pmatrix} t - a & -b \\ -b & t - d \end{pmatrix} \\
&= (t - a)(t - d) - (-b)^2 \\
&= t^2 - (a + d)t + ad - b^2 \\
&= t^2 - \text{tr}(A)x + \det(A), \quad\quad\quad\quad\quad\quad (5.4)
\end{aligned}
$$

where $\text{tr}(A) = a + d$ and $\det(A) = ad - b^2$ are the trace and the determinant of A, respectively.

Let λ_1 and λ_2 be the eigenvalues of A, and recall that λ_1 and λ_2 are the roots of the characteristic polynomial $P_A(t)$. Thus,

$$P_A(t) = (t - \lambda_1)(t - \lambda_2) = t^2 - (\lambda_1 + \lambda_2)t + \lambda_1\lambda_2. \tag{5.5}$$

Since A is a symmetric matrix with real entries, it follows that λ_1 and λ_2 are real numbers. From (5.4) and (5.5), we obtain that

$$\lambda_1 + \lambda_2 = \text{tr}(A); \quad \lambda_1\lambda_2 = \det(A). \tag{5.6}$$

Recall that a symmetric matrix is symmetric positive definite if and only if all the eigenvalues of the matrix are positive. Then,

$$A \text{ symmetric positive definite} \iff (\lambda_1 > 0 \text{ and } \lambda_2 > 0). \tag{5.7}$$

Note that two real numbers are positive if and only if both their product and their sum are positive. From (5.6), we find that

$$(\lambda_1 > 0 \text{ and } \lambda_2 > 0) \iff (\lambda_1 + \lambda_2 > 0 \text{ and } \lambda_1\lambda_2 > 0)$$
$$\iff (\text{tr}(A) > 0 \text{ and } \det(A) > 0). \tag{5.8}$$

From (5.7) and (5.8), we conclude that

$$A \text{ symmetric positive definite} \iff (\text{tr}(A) > 0 \text{ and } \det(A) > 0),$$

i.e., the matrix A is symmetric positive definite if and only if the matrix A is symmetric and $\text{tr}(A) > 0$ and $\det(A) > 0$. $\square$

Problem 3: Let A be a symmetric positive definite matrix. Use the diagonal form of the matrix A to find a matrix B such that $B^2 = A$.

Solution: Let A be an $n \times n$ symmetric positive definite matrix and let $A = Q\Lambda Q^t$ be the diagonal form of A, where Q is an $n \times n$ orthogonal matrix and $\Lambda = \text{diag}(\lambda_k)_{k=1:n}$ is a diagonal matrix whose entries $\lambda_1, \lambda_2, \ldots, \lambda_n$ are the eigenvalues of A.

Note that $\lambda_k > 0$ for all $k = 1 : n$, since A is symmetric positive definite. Let

$$\Lambda^{1/2} = \text{diag}\left(\sqrt{\lambda_k}\right)_{k=1:n}.$$

Then,

$$\Lambda^{1/2} \cdot \Lambda^{1/2} = \text{diag}\left(\sqrt{\lambda_k}\right)_{k=1:n} \cdot \text{diag}\left(\sqrt{\lambda_k}\right)_{k=1:n}$$
$$= \text{diag}(\lambda_k)_{k=1:n}$$
$$= \Lambda. \tag{5.9}$$

Using (5.9) and the fact that $Q^t Q = I$ since Q is an orthogonal matrix, we obtain that

$$A = Q\Lambda Q^t$$

$$
\begin{aligned}
&= \; Q \, \Lambda^{1/2} \cdot \Lambda^{1/2} \, Q^t \\
&= \; Q\Lambda^{1/2}(Q^t Q)\Lambda^{1/2}Q^t \\
&= \; (Q\Lambda^{1/2}Q^t)\,(Q\Lambda^{1/2}Q^t) \\
&= \; B^2,
\end{aligned}
$$

where $B = Q\Lambda^{1/2}Q^t$. $\square$

Problem 4: Show that a matrix A is symmetric positive semidefinite if and only if there exists a symmetric matrix B such that $B^2 = A$.

Solution: We give a proof by double implication.

"If A is symmetric positive semidefinite, then there exists a symmetric matrix B such that $B^2 = A$."

Let A be an $n \times n$ symmetric positive semidefinite matrix and let $A = Q\Lambda Q^t$ be the diagonal form of A, where Q is an $n \times n$ orthogonal matrix and $\Lambda = \mathrm{diag}(\lambda_k)_{k=1:n}$ is a diagonal matrix whose entries $\lambda_1, \lambda_2, \ldots, \lambda_n$ are the eigenvalues of A.

Note that $\lambda_k \geq 0$ for all $k = 1 : n$, since A is symmetric positive semidefinite. Let

$$
\Lambda^{1/2} \; = \; \mathrm{diag}\left(\sqrt{\lambda_k}\right)_{k=1:n}.
$$

Then,

$$
\Lambda^{1/2} \cdot \Lambda^{1/2} \; = \; \mathrm{diag}(\lambda_k)_{k=1:n} \; = \; \Lambda. \tag{5.10}
$$

Using (5.10) and the fact that $Q^t Q = I$ since Q is an orthogonal matrix, it follows that

$$
\begin{aligned}
A \; &= \; Q\Lambda Q^t \; = \; Q \, \Lambda^{1/2} \cdot \Lambda^{1/2} \, Q^t \\
&= \; Q\Lambda^{1/2}(Q^t Q)\Lambda^{1/2}Q^t \; = \; (Q\Lambda^{1/2}Q^t)\,(Q\Lambda^{1/2}Q^t) \\
&= \; B^2, \tag{5.11}
\end{aligned}
$$

where

$$
B \; = \; Q\Lambda^{1/2}Q^t.
$$

Note that B is a symmetric matrix:

$$
B^t \; = \; \left(Q\Lambda^{1/2}Q^t\right)^t \; = \; (Q^t)^t(\Lambda^{1/2})^t Q^t \; = \; Q\Lambda^{1/2}Q^t,
$$

since $\Lambda^{1/2}$ is a diagonal matrix and therefore $(\Lambda^{1/2})^t = \Lambda^{1/2}$.

From (5.11), it follows that we found a symmetric matrix B such that $B^2 = A$.

"If there exists a symmetric matrix B such that $B^2 = A$, then A is symmetric positive semidefinite."

Recall that,[1] for any $n \times n$ symmetric matrix M,

$$
(Mv, w) \; = \; (v, Mw), \quad \forall \; v, w \in \mathbb{R}^n. \tag{5.14}
$$

---

[1] For completeness, we include a proof of (5.14) here. Note that

$$
(Mv, w) \; = \; w^t M v \tag{5.12}
$$

$$
(v, Mw) \; = \; (Mw)^t v \; = \; w^t M^t v \; = \; w^t M v, \tag{5.13}
$$

where for the last equality we used the fact that the matrix M is symmetric and therefore $M^t = M$. From (5.12) and (5.12), we conclude that $(Mv, w) = (v, Mw)$.

From (5.14) for $M = B$, $v = Bx$, and $w = x$, we find that

$$(B \cdot Bx, x) = (Bx, Bx), \quad \forall \, x \in \mathbb{R}^n.$$

Thus,

$$(B^2 x, x) = (Bx, Bx) = ||Bx||^2, \quad \forall \, x \in \mathbb{R}^n. \tag{5.15}$$

Since $A = B^2$, we obtain from (5.15) that

$$(Ax, x) = ||Bx||^2 \geq 0, \quad \forall \, x \in \mathbb{R}^n.$$

Thus,

$$x^t A x = (Ax, x) \geq 0, \quad \forall \, x \in \mathbb{R}^n,$$

and we conclude by definition that the matrix A is symmetric positive semidefinite.
$\square$

Problem 5: Show that the inner product of two vectors is bounded from above by the product of the norms of the vectors, i.e., show that

$$(u, v) \leq ||u|| \, ||v||, \quad \forall \, u, v \in \mathbb{R}^n. \tag{5.16}$$

Note that (5.16) is the Cauchy–Schwartz inequality for inner products.

Solution: Let $t \in \mathbb{R}$, and let $u, v \in \mathbb{R}^n$. Since the inner product is bilinear and using the fact that $(u, v) = (v, u)$, it follows that

$$\begin{aligned} ||u + tv||^2 &= (u + tv, u + tv) = (u, u) + t(v, u) + t(u, v) + t^2(v, v) \\ &= t^2 ||v||^2 + 2t(u, v) + ||u||^2. \end{aligned}$$

Since $||u + tv||^2 \geq 0$, we conclude that

$$t^2 ||v||^2 + 2t(u, v) + ||u||^2 \geq 0, \quad \forall \, t \in \mathbb{R}. \tag{5.17}$$

Note that the left hand side of (5.17) is a quadratic polynomial of t. The inequality (5.17) holds true for any real number t if and only if this polynomial either has no real roots or has one real double root. This happens if and only if its discriminant is less than or equal to 0, i.e., if and only if

$$\begin{aligned} & (2(u, v))^2 - 4||u||^2 ||v||^2 \leq 0 \\ \iff & 4(u, v)^2 \leq 4||u||^2 ||v||^2 \\ \iff & (u, v)^2 \leq ||u||^2 ||v||^2 \\ \iff & |(u, v)| \leq ||u|| \, ||v||. \end{aligned} \tag{5.18}$$

Since $(u, v) \leq |(u, v)|$ for any $u, v \in \mathbb{R}^n$, we conclude from (5.18) that

$$|(u, v)| \leq ||u|| \, ||v||, \quad \forall \, u, v \in \mathbb{R}^n. \quad \square$$

Problem 6: Let A be a symmetric positive semidefinite matrix, and let B be a symmetric matrix such that $B^2 = A$.

(i) Show that
$$(Ax, y) = (Bx, By), \quad \forall\, x, y \in \mathbb{R}^n,$$
where $(\cdot, \cdot)$ denotes the Euclidean inner product.

(ii) Show that
$$(Ax, y)^2 \le (Ax, x)\,(Ay, y), \quad \forall\, x, y \in \mathbb{R}^n. \tag{5.19}$$

Solution: For any $n \times n$ symmetric matrix M,
$$(Mv, w) = (v, Mw), \quad \forall\, v, w \in \mathbb{R}^n; \tag{5.20}$$
see also (5.14).

From (5.20) for $M = B$, $v = Bx$, and $w = y$, we find that
$$(B \cdot Bx, y) = (Bx, By), \quad \forall\, x, y \in \mathbb{R}^n.$$

Thus,
$$(B^2 x, y) = (Bx, By), \quad \forall\, x, y \in \mathbb{R}^n. \tag{5.21}$$

Since $A = B^2$, we obtain from (5.21) that
$$(Ax, y) = (Bx, By), \quad \forall\, x, y \in \mathbb{R}^n. \tag{5.22}$$

(ii) From the Cauchy–Schwartz inequality (5.1), it follows that
$$(Bx, By) \le ||Bx||\,||By||, \quad \forall\, x, y \in \mathbb{R}^n,$$
and therefore
$$(Bx, By)^2 \le ||Bx||^2\,||By||^2, \quad \forall\, x, y \in \mathbb{R}^n. \tag{5.23}$$

Then, from (5.22) and (5.23), we obtain that
$$(Ax, y)^2 = (Bx, By)^2 \le ||Bx||^2\,||By||^2, \quad \forall\, x, y \in \mathbb{R}^n. \tag{5.24}$$

Moreover, from (5.22), we find that
$$(Ax, x) = (Bx, Bx) = ||Bx||^2, \quad \forall\, x \in \mathbb{R}^n; \tag{5.25}$$
$$(Ay, y) = (By, By) = ||By||^2, \quad \forall\, y \in \mathbb{R}^n. \tag{5.26}$$

From (5.24–5.26), we conclude that
$$(Ax, y)^2 \le ||Bx||^2\,||By||^2 = (Ax, x)\,(Ay, y), \quad \forall\, x, y \in \mathbb{R}^n,$$
which is the same as (5.19). □

Problem 7: Let λ and v be an eigenvalue and a corresponding eigenvector of the square matrix A of size n. Let Q be an orthogonal matrix of size n. Show that λ is also an eigenvalue of the matrix $Q^t A Q$. What is a corresponding eigenvector?

Solution: If λ and $v \neq 0$ are an eigenvalue and a corresponding eigenvector of A, then $Av = \lambda v$. Since Q is an orthogonal matrix, it follows that Q is nonsingular and $QQ^t = I$. Then,

$$
\begin{aligned}
Av = \lambda v \quad &\Longleftrightarrow \quad Q^t Av = Q^t(\lambda v) &\text{(5.27)} \\
&\Longleftrightarrow \quad Q^t Av = \lambda Q^t v &\text{(5.28)} \\
&\Longleftrightarrow \quad Q^t A(QQ^t)v = \lambda Q^t v \\
&\Longleftrightarrow \quad Q^t AQ \cdot Q^t v = \lambda Q^t v \\
&\Longleftrightarrow \quad Q^t AQw = \lambda w, &\text{(5.29)}
\end{aligned}
$$

where $w = Q^t v \neq 0$.

From (5.29), we conclude that λ is an eigenvalue of $Q^t AQ$ with corresponding eigenvector $w = Q^t v$. $\square$

Problem 8: Let Q be an orthogonal matrix of size n and let A be a square matrix of size n. Show that the matrix $Q^t AQ$ has the same characteristic polynomial as A, i.e., show that

$$
P_{Q^t AQ}(t) = P_A(t), \quad \forall \, t \in \mathbb{R}.
$$

Solution: The characteristic polynomial of a matrix $Q^t AQ$ is

$$
P_{Q^t AQ}(t) = \det(tI - Q^t AQ), \tag{5.30}
$$

where t is a real number.

Note that $QQ^t = Q^t Q = I$ since Q is an orthogonal matrix, and therefore

$$
tI - Q^t AQ = t\,Q^t Q - Q^t AQ = Q^t(tI - A)Q. \tag{5.31}
$$

Moreover,

$$
\det(Q^t)\det(Q) = \det(Q^t Q) = \det(I) = 1. \tag{5.32}
$$

From (5.30) and (5.31), and using (5.32), we obtain that

$$
\begin{aligned}
P_{Q^t AQ}(t) &= \det(tI - Q^t AQ) \\
&= \det(Q^t(tI - A)Q) \\
&= \det(Q^t)\,\det(tI - A)\,\det(Q) \\
&= \det(tI - A) \cdot \det(Q^t)\det(Q) \\
&= \det(tI - A) \\
&= P_A(t). \quad \square
\end{aligned}
$$

Problem 9: Let A be a symmetric positive definite matrix with diagonal form $A = Q\Lambda Q^t$, where Q is an orthogonal matrix and $\Lambda = \mathrm{diag}(\lambda_k)_{k=1:n}$ is a diagonal matrix. Recall that λ_k, $k = 1 : n$, are the eigenvalues of A, and that $\lambda_k > 0$ for all $k = 1 : n$, since A is symmetric positive definite.

Let

$$\Lambda^{1/2} = \text{diag} \left(\sqrt{\lambda_k} \right)_{k=1:n} \quad \text{and} \quad \Lambda^{-1/2} = \text{diag} \left(\frac{1}{\sqrt{\lambda_k}} \right)_{k=1:n}, \tag{5.33}$$

and let $A^{1/2}$ and $A^{-1/2}$ be the matrices given by

$$A^{1/2} = Q\Lambda^{1/2}Q^t \quad \text{and} \quad A^{-1/2} = Q\Lambda^{-1/2}Q^t. \tag{5.34}$$

Show that

$$(A^{1/2})^2 = A; \tag{5.35}$$
$$(A^{-1/2})^2 = A^{-1}; \tag{5.36}$$
$$(A^{1/2})^{-1} = A^{-1/2}. \tag{5.37}$$

Solution: To establish (5.35), we find from (5.34) that

$$\begin{aligned}
(A^{1/2})^2 &= (Q\Lambda^{1/2}Q^t)(Q\Lambda^{1/2}Q^t) = Q\Lambda^{1/2}(Q^tQ)\Lambda^{1/2}Q^t \\
&= Q\Lambda^{1/2}\Lambda^{1/2}Q^t = Q\Lambda Q^t \\
&= A,
\end{aligned}$$

where we used the facts that $Q^tQ = I$ since Q is an orthogonal matrix and that

$$\begin{aligned}
\Lambda^{1/2} \cdot \Lambda^{1/2} &= \text{diag} \left(\sqrt{\lambda_k} \right)_{k=1:n} \cdot \text{diag} \left(\sqrt{\lambda_k} \right)_{k=1:n} \\
&= \text{diag} \left(\sqrt{\lambda_k} \cdot \sqrt{\lambda_k} \right)_{k=1:n} \\
&= \text{diag}(\lambda_k)_{k=1:n} \\
&= \Lambda.
\end{aligned}$$

Furthermore,

$$\begin{aligned}
\Lambda^{-1/2} \cdot \Lambda^{-1/2} &= \text{diag} \left(\frac{1}{\sqrt{\lambda_k}} \right)_{k=1:n} \cdot \text{diag} \left(\frac{1}{\sqrt{\lambda_k}} \right)_{k=1:n} \\
&= \text{diag} \left(\frac{1}{\sqrt{\lambda_k}} \cdot \frac{1}{\sqrt{\lambda_k}} \right)_{k=1:n} \\
&= \text{diag} \left(\frac{1}{\lambda_k} \right)_{k=1:n} \\
&= \Lambda^{-1}; \tag{5.38}
\end{aligned}$$

for the last equality we used the fact that the inverse matrix of a nonsingular diagonal matrix $D = \text{diag}(d_k)_{k=1:n}$ with $d_k \neq 0$ for $k = 1 : n$ is

$$D^{-1} = \text{diag} \left(\frac{1}{d_k} \right)_{k=1:n}. \tag{5.39}$$

Using (5.38) and the fact that $Q^tQ = I$ since Q is an orthogonal matrix, we find from (5.34) that

$$\begin{aligned}
(A^{-1/2})^2 &= (Q\Lambda^{-1/2}Q^t)(Q\Lambda^{-1/2}Q^t) = Q\Lambda^{-1/2}(Q^tQ)\Lambda^{-1/2}Q^t \\
&= Q\Lambda^{-1/2}\Lambda^{-1/2}Q^t \\
&= Q\Lambda^{-1}Q^t. \tag{5.40}
\end{aligned}$$

Also, note that $(Q^t)^{-1} = Q$ since $Q^t Q = I$, and therefore

$$\begin{aligned} A^{-1} &= (Q\Lambda Q^t)^{-1} = (Q^t)^{-1} \cdot \Lambda^{-1} \cdot Q^{-1} \\ &= Q\Lambda^{-1}Q^t. \end{aligned} \tag{5.41}$$

From (5.40) and (5.41), we find that $(A^{-1/2})^2 = A^{-1}$ and therefore (5.36) is proved.

To establish (5.37), recall that $Q^t Q = QQ^t = I$ since Q is an orthogonal matrix and therefore

$$(Q^t)^{-1} = Q; \quad Q^{-1} = Q^t. \tag{5.42}$$

Moreover, from (5.39) and using (5.33), it follows that the inverse matrix of the diagonal matrix $\Lambda^{1/2} = \text{diag}\left(\sqrt{\lambda_k}\right)_{k=1:n}$ is

$$\left(\Lambda^{1/2}\right)^{-1} = \text{diag}\left(\frac{1}{\sqrt{\lambda_k}}\right)_{k=1:n} = \Lambda^{-1/2}. \tag{5.43}$$

From (5.34) and using (5.42) and (5.43), we obtain that

$$\begin{aligned} (A^{1/2})^{-1} &= (Q\Lambda^{1/2}Q^t)^{-1} \\ &= (Q^t)^{-1}(\Lambda^{1/2})^{-1}Q^{-1} \\ &= Q\Lambda^{-1/2}Q^t \\ &= A^{-1/2}, \end{aligned}$$

and therefore (5.37) is proved. □

Problem 10: Show that an $n \times n$ matrix has $2^n - 1$ principal minors.

Solution: The principal minors of an $n \times n$ matrix A are the determinants of all the square matrices obtained by eliminating the same rows and columns from the matrix A. Recall that there are $\binom{n}{k}$ ways[2] to choose k of the n columns of the matrix A. Thus, there are $\binom{n}{k}$ principal minors of the matrix A that correspond to determinants of $(n-k) \times (n-k)$ matrices obtained by eliminating the same k rows and k columns from the matrix A and therefore the total number of principal minors of the matrix A is

$$\sum_{k=1}^{n} \binom{n}{k}. \tag{5.45}$$

Using the binomial formula

$$(x+y)^n - \sum_{k=0}^{n} \binom{n}{k} x^k y^{n-k}, \quad \forall\, x, y \subset \mathbb{R}, \tag{5.46}$$

---

[2] The binomial coefficient $\binom{n}{k}$ is given by

$$\binom{n}{k} = \frac{n!}{k!\,(n-k)!}, \tag{5.44}$$

with $k! = 1 \cdot 2 \cdot \ldots \cdot k$ for all positive integers k and $0! = 1$, by definition.

for $x = y = 1$, we obtain from (5.46) that

$$2^n = \sum_{k=0}^{n} \binom{n}{k}. \tag{5.47}$$

Then, it follows from (5.47) that

$$\sum_{k=1}^{n} \binom{n}{k} = \sum_{k=0}^{n} \binom{n}{k} - \binom{n}{0}$$
$$= 2^n - 1, \tag{5.48}$$

since, from (5.44) and using the fact that $0! = 1$, we find that $\binom{n}{0} = \frac{n!}{0!\,n!} = 1$.

From (5.45) and (5.48), we conclude that the number of principal minors of the $n \times n$ matrix A is

$$\sum_{k=1}^{n} \binom{n}{k} = 2^n - 1. \quad \square$$

Problem 11: Show that the matrix

$$A = \begin{pmatrix} 1 & 0.2 & -0.2 & 0.1 \\ 0.2 & 1 & -0.25 & 0.05 \\ -0.2 & -0.25 & 1 & -0.15 \\ 0.1 & 0.05 & -0.15 & 1 \end{pmatrix} \tag{5.49}$$

is symmetric positive definite.

Solution: Note that $A^t = A$ and therefore the matrix A is symmetric.

Solution 1 (using Sylvester's Criterion): Recall that a symmetric matrix is symmetric positive definite if and only if all the leading principal minors of the matrix are positive.

The matrix A from (5.49) has the following four leading principal minors:

$$\det(1) = 1; \quad \det \begin{pmatrix} 1 & 0.2 \\ 0.2 & 1 \end{pmatrix} = 0.96;$$

$$\det \begin{pmatrix} 1 & 0.2 & -0.2 \\ 0.2 & 1 & -0.25 \\ -0.2 & -0.25 & 1 \end{pmatrix} = 0.8775; \quad \det(A) = 0.853275,$$

which are all positive. From Sylvester's Criterion, we conclude that the matrix A is symmetric positive definite.

Solution 2 (using Gershgorin's Theorem): Recall that a strictly diagonally dominant symmetric matrix with positive entries on the main diagonal is symmetric positive definite.

Note that the matrix A is strictly diagonally dominant, since, for every row of A, the absolute value of the main diagonal entry from the row is greater than the sum of the absolute values of all the other entries in that row:

$$1 \quad > \quad 0.2 + |-0.2| + 0.1 = 0.5;$$
$$1 \quad > \quad 0.2 + |-0.25| + 0.05 = 0.5;$$
$$1 \quad > \quad |-0.2| + |-0.25| + |-0.15| = 0.6;$$
$$1 \quad > \quad 0.1 + 0.05 + |-0.15| = 0.3.$$

Since the matrix A is symmetric and its main diagonal entries are positive, we conclude that A is a symmetric positive definite matrix. □

Problem 12: Let

$$A = \begin{pmatrix} 1 & 0.05 & 0.25 & 0.55 \\ 0.05 & 2 & 1 & 1.25 \\ 0.25 & 1 & 4 & 2.5 \\ 0.55 & 1.25 & 2.5 & 6 \end{pmatrix}.$$

(i) Show that the matrix A is not strictly diagonally dominant.

(ii) Show that the matrix A is symmetric positive definite.

Solution: (i) Note that $A(2,2) < A(2,1) + A(2,3) + A(2,4)$, since

$$2 \quad < \quad 0.05 + 1 + 1.25 = 2.30.$$

Thus, the matrix A is not strictly diagonally dominant.

(ii) The leading principal minors of the matrix A are

$$\det(1) = 1; \quad \det \begin{pmatrix} 1 & 0.05 \\ 0.05 & 2 \end{pmatrix} = 1.9975;$$

$$\det \begin{pmatrix} 1 & 0.05 & 0.25 \\ 0.05 & 2 & 1 \\ 0.25 & 1 & 4 \end{pmatrix} = 6.89; \quad \det(A) = 27.9264.$$

Since all the leading principal minors of the matrix A are positive, it follows from Sylvester's criterion that the matrix A is symmetric positive definite. □

Problem 13: Let A be a symmetric positive semidefinite matrix. Show that the matrix A^k is also symmetric positive semidefinite for any positive integer k.

Solution: Let A be an $n \times n$ symmetric positive semidefinite matrix, and let λ_1, λ_2, ..., λ_n be the eigenvalues of A. Since the matrix A is symmetric, there exist an orthogonal matrix Q and a diagonal matrix Λ whose diagonal entries are the eigenvalues of A, i.e., $\Lambda = \text{diag}(\lambda_i)_{i=1:n}$, such that $A = Q\Lambda Q^t$.

If k is a positive integer, then

$$\begin{aligned} A^k &= \left(Q\Lambda Q^t \right)^k = \left(Q\Lambda Q^t \right) \left(Q\Lambda Q^t \right) \dots \left(Q\Lambda Q^t \right) \left(Q\Lambda Q^t \right) \\ &= Q\Lambda (Q^t Q)\Lambda (Q^t Q) \dots (Q^t Q)\Lambda Q^t \\ &= Q\Lambda^k Q^t, \end{aligned}$$

since $Q^t Q = I$. Thus, the diagonal form of the matrix A^k is

$$A^k = Q \Lambda^k Q^t.$$

Since the eigenvalues of A^k are the same as the diagonal entries of the matrix $\Lambda^k = \mathrm{diag}(\lambda_i^k)_{i=1:n}$, it follows that the eigenvalues of the matrix A^k are λ_1^k, λ_2^k, ..., λ_n^k.

Recall that A is a symmetric positive semidefinite matrix and therefore $\lambda_i \geq 0$ for all $i = 1 : n$. Then, $\lambda_i^k \geq 0$ for all $i = 1 : n$. In other words, all the eigenvalues of the matrix A^k are nonnegative, and we conclude that the matrix A^k is symmetric positive semidefinite matrix. □

Problem 14: Recall that any symmetric positive definite matrix is nonsingular. Let A be a symmetric positive definite matrix. Show that A^{-1} is a symmetric positive definite matrix.

Solution: Denote by λ_1, λ_2, ..., λ_n the eigenvalues of the $n \times n$ matrix A. Since A is symmetric positive definite, if follows that $\lambda_k > 0$ for all $k = 1 : n$.

Recall that λ is an eigenvalue of A if and only if $\frac{1}{\lambda}$ is an eigenvalues of the inverse matrix A^{-1}. Then, the eigenvalues of the matrix A^{-1} are $\frac{1}{\lambda_1}$, $\frac{1}{\lambda_2}$, ..., $\frac{1}{\lambda_n}$. Since $\lambda_k > 0$ for all $k = 1 : n$, it follows that $\frac{1}{\lambda_k} > 0$ for all $k = 1 : n$. Thus, all the eigenvalues of A^{-1} are positive and we conclude that the matrix A^{-1} is symmetric positive definite. □

Problem 15: Let A be a square matrix such that $A^2 = A$. Show that the matrix A cannot be strictly diagonally dominant unless A is the identity matrix.

Solution: We will show that, if the matrix A is strictly diagonally dominant, then A is the identity matrix. Recall that any strictly diagonally dominant matrix is nonsingular. Then, if A is strictly diagonally dominant, we can multiply $A^2 = A$ by the inverse matrix A^{-1} of A to the left to obtain that

$$
\begin{aligned}
A^2 = A \quad &\Longleftrightarrow \quad A^{-1} A^2 = A^{-1} A \\
&\Longleftrightarrow \quad (A^{-1} A) \cdot A = I \\
&\Longleftrightarrow \quad A = I,
\end{aligned}
$$

since $A^{-1} A = I$.

We conclude that, if A is a strictly diagonally dominant matrix such that $A^2 = A$, then A is the identity matrix. □

Problem 16: Let A be an $n \times n$ symmetric positive definite matrix, and let M be an $n \times m$ matrix.

(i) Show that the matrix $M^t A M$ is symmetric positive semidefinite.

(ii) Show that the matrix $M^t A M$ is symmetric positive definite if and only if the columns of the matrix M are linearly independent.

Solution: The matrix $M^t A M$ is symmetric since

$$(M^t A M)^t \;=\; M^t A^t (M^t)^t \;=\; M^t A M;$$

here, we used the facts that $(M^t)^t = M$ and $A^t = A$, since A is symmetric. Also, note that $M^t A M$ is an $m \times m$ matrix since it is the product of the $m \times n$ matrix M^t, the $n \times n$ matrix A, and the $n \times m$ matrix M.

(i) Let x be an $m \times 1$ column vector. Then, $(Mx)^t = x^t M^t$ and

$$x^t M^t A M x \;=\; (Mx)^t A M x \;=\; (AMx, Mx). \qquad (5.50)$$

Note that
$$(Av, v) \;\geq\; 0, \quad \forall\, v \in \mathbb{R}^m, \qquad (5.51)$$

since A is a symmetric positive definite matrix.

From (5.50) and using (5.51) with $v = Mx$, we obtain that

$$x^t M^t A M x \;=\; (AMx, Mx) \;\geq\; 0, \quad \forall\, x \in \mathbb{R}^m, \qquad (5.52)$$

and we conclude that the matrix $M^t A M$ is symmetric positive semidefinite.

(ii) From (5.50), it follows that

$$x^t M^t A M x = 0 \;\Longleftrightarrow\; (AMx, Mx) = 0 \;\Longleftrightarrow\; Mx = 0, \qquad (5.53)$$

since A is a symmetric positive definite matrix and therefore

$$(Av, v) = 0 \;\Longleftrightarrow\; v = 0.$$

Let $M = \mathrm{col}\,(c_k)_{k=1:m}$ be the column form of the matrix M, and let $x = (x_k)_{k=1:m}$. Then,

$$Mx \;=\; \sum_{k=1}^{m} x_k c_k = 0. \qquad (5.54)$$

From (5.53) and (5.54), we obtain that

$$x^t M^t A M x = 0 \;\Longleftrightarrow\; Mx = 0 \;\Longleftrightarrow\; \sum_{k=1}^{m} x_k c_k = 0.$$

Thus, there exists a vector $x \neq 0$ such that $x^t M^t A M x = 0$ if and only if $\sum_{k=1}^{m} x_k c_k = 0$, i.e., if and only if the columns of the matrix M are not linearly independent. Since $x^t M^t A M x \geq 0$ for all $x \in \mathbb{R}^m$, see (5.52), we conclude that the matrix $M^t M$ is symmetric positive definite if and only if the columns of M are linearly independent. □

Chapter 6

Cholesky decomposition. Efficient cubic spline interpolation.

6.1 Exercises

1. Let

$$A = \begin{pmatrix} 2 & -2 \\ -2 & 5 \end{pmatrix}.$$

(i) Find a 2×2 matrix M such that $A = M^2$;

(ii) Find a 2×2 matrix M such that $A = MM^t$.

2. Let $A = \begin{pmatrix} 41 & 12 \\ 12 & 34 \end{pmatrix}$.

(i) Show that the matrix A is symmetric positive definite.

(ii) Find a 2×2 matrix B such that $A = B^2$.

(iii) Find a 2×2 upper triangular matrix U such that $A = U^tU$.

3. Let $A = \begin{pmatrix} 1 & -0.2 & 0.8 \\ -0.2 & 2 & 0.3 \\ 0.8 & 0.3 & 1.1 \end{pmatrix}$.

(i) Show that the matrix A is weakly diagonally dominated, and therefore symmetric positive semidefinite.

(ii) Use Sylvester's criterion to show that the matrix A is symmetric positive definite.

4. Let A be a symmetric positive definite matrix, and let U be the Cholesky factor of A. If V is an upper triangular matrix such that $A = V^tV$, show that there exists a diagonal matrix D whose entries on the main diagonal are either -1 or 1 such that $V = DU$.

5. Let B_N be the following $N \times N$ symmetric positive definite matrix:

$$
B_N = \begin{pmatrix} 2 & -1 & \cdots & 0 \\ -1 & \ddots & \ddots & \vdots \\ \vdots & \ddots & \ddots & -1 \\ 0 & \cdots & -1 & 2 \end{pmatrix}.
\tag{6.1}
$$

Let C_N be an $N \times (N+1)$ matrix with

$$
\begin{aligned}
C_N(i,i) &= 1, \quad \forall\, i = 1 : N; \\
C_N(i,i+1) &= -1, \quad \forall\, i = 1 : N,
\end{aligned}
$$

and with all the other entries equal to 0.

Show that $B_N = C_N C_N^t$. Why is this not the Cholesky decomposition of the matrix B_N?

6. Let B_N be the $N \times N$ tridiagonal symmetric positive definite matrix given by (6.1). Show that the Cholesky factor U_N of the matrix B_N is the upper triangular bidiagonal matrix given by

$$
U_N(i,i) = \sqrt{\frac{i+1}{i}}, \quad \forall\, i = 1 : N;
\tag{6.2}
$$

$$
U_N(i,i+1) = -\sqrt{\frac{i}{i+1}}, \quad \forall\, i = 1 : (N-1).
\tag{6.3}
$$

7. Let B_N be the $N \times N$ tridiagonal symmetric positive definite matrix given by (6.1), and let U_N be the Cholesky factor of B_N given by (6.2–6.3).

(i) Show that the solution to a linear system $B_N x = b$, where b and x are $N \times 1$ column vectors can be obtained by using the explicit pseudocode from Table 6.1.

(ii) What is the operation count for the pseudocode above, and how does it compare to $8N + O(1)$, the operation count for the optimal linear solver for tridiagonal symmetric positive definite matrices?

8. Let B_N be the $N \times N$ tridiagonal symmetric positive definite matrix given by (6.1). Show that the LU factors L and U of the matrix B_N are the lower triangular bidiagonal matrix and the upper triangular bidiagonal matrix given by

$$
L(i,i) = 1, \quad \forall\, i = 1 : N; \quad L(i+1,i) = -\frac{i}{i+1}, \quad \forall\, i = 1 : (N-1);
\tag{6.4}
$$

$$
U(i,i) = \frac{i+1}{i}, \quad \forall\, i = 1 : N; \quad U(i,i+1) = -1, \quad \forall\, i = 1 : (N-1).
\tag{6.5}
$$

Table 6.1: Linear system solution for a special tridiagonal system

Function Call:
$x =$ linear_solve_cholesky_B_N(b)

Input:
$b = N \times 1$ column vector

Output:
$x =$ solution to $B_N x = b$

$y(1) = \frac{b(1)}{\sqrt{2}}$
for $i = 2 : N$
$\qquad y(i) = \frac{b(i) + y(i-1)\sqrt{(i-1)/i}}{\sqrt{(i+1)/i}}$
end
$x(N) = \frac{y(N)\sqrt{N}}{\sqrt{N+1}}$
for $i = (N-1) : 1$
$\qquad x(i) = \frac{y(i) + x(i+1)\sqrt{i/(i+1)}}{\sqrt{(i+1)/i}}$
end

9. Let B_N be the $N \times N$ tridiagonal symmetric positive definite matrix given by (6.1), and let L and U be the LU factors of B_N given by (6.4–6.5).

(i) Show that the solution to a linear system $B_N x = b$, where b and x are $N \times 1$ column vectors can be obtained by using the explicit pseudocode from Table 6.2.

(ii) What is the operation count for the pseudocode above, and how does it compare to $8N + O(1)$, the operation count for the optimal linear solver for tridiagonal symmetric positive definite matrices?

10. Write an explicit optimal pseudocode for solving linear systems corresponding to the same tridiagonal symmetric positive definite matrix. In other words, write a pseudocode for solving p linear systems $Ax_i = b_i$, for $i = 1 : p$, where A is an $n \times n$ tridiagonal symmetric positive definite matrix, by using the explicit solver linear_solve_LU_tridiag_spd from Table 6.7 from [3] and the method from Table 6.3 from [3]. What is the operation count for solving the p linear systems?

11. Write the pseudocode for the Cholesky decomposition of symmetric positive definite banded matrices of band m. What is the corresponding operation count?

Hint: Use (without proving) the fact that the Cholesky factor of a symmetric positive definite banded matrix of band m is a banded upper triangular matrix of band m.

Table 6.2: Linear system solution for a special tridiagonal system

Function Call:
$x = \text{linear\_solve\_lu\_B\_N}(b)$

Input:
$b = N \times 1$ column vector

Output:
$x =$ solution to $B_N x = b$

$y(1) = b(1)$
for $i = 2 : N$
$\quad y(i) = b(i) + \frac{(i-1)y(i-1)}{i}$
end
$x(N) = \frac{N y(N)}{N+1}$
for $i = (N-1) : 1$
$\quad x(i) = \frac{i(y(i)+x(i+1))}{i+1}$
end

12. What is the operation count for solving a linear system corresponding to a symmetric positive definite banded matrix of band m using a Cholesky linear solver?

13. The following discount factors were obtained from market data:

Date	Discount Factor
2 months	0.9980
5 months	0.9935
11 months	0.9820
15 months	0.9775

The overnight rate is 0.75%.

(i) What are the corresponding 2 months, 5 months, 11 months and 15 months zero rates?

(ii) What is the tridiagonal system that must be solved in the efficient implementation of the natural cubic spline interpolation for finding the zero rate curve for all times less than 15 months?

(iii) Use the efficient implementation of the natural cubic spline interpolation to find a zero rate curve for all times less than 15 months matching the discount factors above.

(iv) Find the value of a 14 months quarterly coupon bond with 2.5% coupon rate.

Note: A quarterly coupon bond with face value $100, coupon rate C, and maturity T pays the holder of the bond a coupon payment equal to $\frac{C}{4} \cdot 100$

every three months, except at maturity. The final payment at maturity T is equal to the face value of the bond plus one coupon payment, i.e., $100 + \frac{C}{4}100$.

6.2   Solutions to Chapter 6 Exercises

Problem 1: Let

$$A = \begin{pmatrix} 2 & -2 \\ -2 & 5 \end{pmatrix}.$$

(i) Find a 2×2 matrix M such that $A = M^2$;

(ii) Find a 2×2 matrix M such that $A = MM^t$.

Solution: (i) Recall that any symmetric matrix has diagonal form

$$A = Q\Lambda Q^t, \tag{6.6}$$

where Λ is the diagonal matrix whose main diagonal entries are the eigenvalues of A and Q is the orthogonal matrix whose columns are the corresponding eigenvectors of A of norm 1, i.e.,

$$\Lambda = \begin{pmatrix} \lambda_1 & 0 \\ 0 & \lambda_2 \end{pmatrix}; \quad Q = (v_1 \ v_2), \tag{6.7}$$

where $Av_1 = \lambda_1 v_1$ and $Av_2 = \lambda_2 v_2$, with $||v_1|| = ||v_2|| = 1$.

If the matrix A has nonnegative eigenvalues, i.e., if $\lambda_1 \geq 0$ and $\lambda_2 \geq 0$, then the matrix

$$M = Q\Lambda^{1/2}Q^t \tag{6.8}$$

with

$$\Lambda^{1/2} = \begin{pmatrix} \sqrt{\lambda_1} & 0 \\ 0 & \sqrt{\lambda_2} \end{pmatrix} \tag{6.9}$$

has the property that $M^2 = A$:

$$
\begin{aligned}
M^2 &= \left(Q\Lambda^{1/2}Q^t\right)\left(Q\Lambda^{1/2}Q^t\right) \\
&= Q\Lambda^{1/2}(Q^tQ)\Lambda^{1/2}Q^t \\
&= Q\Lambda^{1/2}\Lambda^{1/2}Q^t \\
&= Q\Lambda Q^t \\
&= A, \tag{6.10}
\end{aligned}
$$

since Q is an orthogonal matrix and therefore $Q^tQ = I$, and since, from (6.9), it follows that $\Lambda^{1/2}\Lambda^{1/2} = \begin{pmatrix} \lambda_1 & 0 \\ 0 & \lambda_2 \end{pmatrix} = \Lambda$.

We now proceed to compute the eigenvalues and the eigenvectors of the matrix A. The eigenvalues of the matrix A are the roots of the characteristic polynomial $P_A(x)$ of the matrix A given by[1]

$$
\begin{aligned}
P_A(x) &= \det(xI - A) = \det\begin{pmatrix} x-2 & 2 \\ 2 & x-5 \end{pmatrix} \\
&= (x-2)(x-5) - 4 = x^2 - 7x + 6 \\
&= (x-1)(x-6).
\end{aligned}
$$

---

[1] The characteristic polynomial of the matrix A can also be obtained as follows:

$$P_A(x) = x^2 - \mathrm{tr}(A)x + \det(A) = x^2 - 7x + 6,$$

where $\mathrm{tr}(A) = 2 + 5 = 7$ and $\det(A) = 2 \cdot 5 - (-2) \cdot (-2) = 6$.

By solving $P_A(x) = 0$, we obtain that the roots of $P_A(x)$ are 1 and 6, and therefore the eigenvalues of A are $\lambda_1 = 1$ and $\lambda_2 = 6$. The corresponding eigenvectors of norm 1 are[2]

$$v_1 = \begin{pmatrix} \frac{2}{\sqrt{5}} \\ \frac{1}{\sqrt{5}} \end{pmatrix} \quad \text{and} \quad v_2 = \begin{pmatrix} \frac{1}{\sqrt{5}} \\ -\frac{2}{\sqrt{5}} \end{pmatrix}. \tag{6.11}$$

Then, it follows from (6.8), (6.9), and (6.11) that the matrix M given by

$$\begin{aligned} M &= \begin{pmatrix} \frac{2}{\sqrt{5}} & \frac{1}{\sqrt{5}} \\ \frac{1}{\sqrt{5}} & -\frac{2}{\sqrt{5}} \end{pmatrix} \begin{pmatrix} \sqrt{1} & 0 \\ 0 & \sqrt{6} \end{pmatrix} \begin{pmatrix} \frac{2}{\sqrt{5}} & \frac{1}{\sqrt{5}} \\ \frac{1}{\sqrt{5}} & -\frac{2}{\sqrt{5}} \end{pmatrix} \\ &= \frac{1}{5} \begin{pmatrix} 2 & 1 \\ 1 & -2 \end{pmatrix} \begin{pmatrix} 1 & 0 \\ 0 & \sqrt{6} \end{pmatrix} \begin{pmatrix} 2 & 1 \\ 1 & -2 \end{pmatrix} \\ &= \frac{1}{5} \begin{pmatrix} 4 + \sqrt{6} & 2 - 2\sqrt{6} \\ 2 - 2\sqrt{6} & 1 + 4\sqrt{6} \end{pmatrix} \end{aligned}$$

has the property that $M^2 = A$.

(ii) Since the eigenvalues of A are positive, the matrix A is symmetric positive definite and therefore has a Cholesky decomposition. Recall that, if U is the Cholesky factor of the matrix A, then $A = U^t U$. Thus, in order to find a matrix M such that $A = MM^t$, it is enough to compute the Cholesky factor U of the matrix A and let $M = U^t$.

The Cholesky factor of a 2×2 symmetric positive definite matrix $\begin{pmatrix} a & b \\ b & d \end{pmatrix}$ is

$\begin{pmatrix} \sqrt{a} & \frac{b}{\sqrt{a}} \\ 0 & \sqrt{\frac{ad-b^2}{a}} \end{pmatrix}$. Then, the Cholesky factor of the matrix $A = \begin{pmatrix} 2 & -2 \\ -2 & 5 \end{pmatrix}$ is

$U = \begin{pmatrix} \sqrt{2} & -\sqrt{2} \\ 0 & \sqrt{3} \end{pmatrix}$, and therefore the matrix $M = U^t$ given by

$$M = \begin{pmatrix} \sqrt{2} & 0 \\ -\sqrt{2} & \sqrt{3} \end{pmatrix}$$

has the property that $A = MM^t$. $\quad\Box$

Problem 2: Let $A = \begin{pmatrix} 41 & 12 \\ 12 & 34 \end{pmatrix}$.

[2] For example, if $\lambda_2 = 6$, any corresponding eigenvector $v_2 = \begin{pmatrix} a \\ b \end{pmatrix} \neq 0$ is a solution to $Av = 6v$, which can be written as

$$\begin{cases} 2a - 2b &= 6a \\ -2a + 5b &= 6b \end{cases} \iff \begin{cases} -4a &= 2b \\ -2a &= b \end{cases} \iff b = -2a.$$

Thus, any eigenvector corresponding to the eigenvalue $\lambda_2 = 6$ is of the form $v_2 = \begin{pmatrix} a \\ -2a \end{pmatrix} = a \begin{pmatrix} 1 \\ -2 \end{pmatrix}$. By choosing $a = \frac{1}{\sqrt{5}}$, we obtain that an eigenvector of norm 1 corresponding to the eigenvalue $\lambda_2 = 6$ is $v_2 = \begin{pmatrix} \frac{1}{\sqrt{5}} \\ -\frac{2}{\sqrt{5}} \end{pmatrix}$.

(i) Show that the matrix A is symmetric positive definite.

(ii) Find a 2×2 matrix B such that $A = B^2$.

(iii) Find a 2×2 upper triangular matrix U such that $A = U^t U$.

Solution: (i) The eigenvalues of the matrix A are the roots of the characteristic polynomial $P_A(x)$ of the matrix A given by

$$
\begin{aligned}
P_A(x) &= x^2 - \operatorname{tr}(A)x + \det(A) \\
&= x^2 - 75x + 1250 \\
&= (x - 25)(x - 50),
\end{aligned}
$$

where the trace and the determinant of the matrix A are

$$
\begin{aligned}
\operatorname{tr}(A) &= A(1,1) + A(2,2) = 41 + 34 = 75; \\
\det(A) &= A(1,1)A(2,2) - A(1,2)A(2,1) \\
&= 41 \cdot 34 - 12 \cdot 12 = 1250.
\end{aligned}
$$

By solving $P_A(x) = 0$, we obtain that the roots of the characteristic polynomial $P_A(x)$, which are the same as the eigenvalues of A, are 25 and 50.

Thus, the eigenvalues of the symmetric matrix A are positive, and therefore A is a symmetric positive definite matrix.

(ii) The symmetric matrix A has diagonal form

$$
A = Q \Lambda Q^t, \tag{6.12}
$$

where Λ is the diagonal matrix whose main diagonal entries are the eigenvalues of A and Q is the orthogonal matrix whose columns are the corresponding eigenvectors of A of norm 1, i.e.,

$$
\Lambda = \begin{pmatrix} \lambda_1 & 0 \\ 0 & \lambda_2 \end{pmatrix}; \quad Q = (v_1 \ v_2), \tag{6.13}
$$

where $Av_1 = \lambda_1 v_1$ and $Av_2 = \lambda_2 v_2$, with $||v_1|| = ||v_2|| = 1$.

In part (i), we found that the eigenvalues of the matrix A are $\lambda_1 = 25$ and $\lambda_2 = 50$. Since the eigenvalues of A are positive, it follows that the matrix

$$
M = Q \Lambda^{1/2} Q^t \tag{6.14}
$$

with

$$
\Lambda^{1/2} = \begin{pmatrix} \sqrt{\lambda_1} & 0 \\ 0 & \sqrt{\lambda_2} \end{pmatrix} \tag{6.15}
$$

has the property that

$$
M^2 = A; \tag{6.16}
$$

see (6.8–6.10)

The eigenvectors of norm 1 corresponding to the eigenvalues $\lambda_1 = 25$ and $\lambda_2 = 50$ are $v_1 = \begin{pmatrix} 0.6 \\ -0.8 \end{pmatrix}$ and $v_2 = \begin{pmatrix} -0.8 \\ -0.6 \end{pmatrix}$, respectively.

Then, it follows from (6.14), (6.15), and (6.16) that the matrix M given by

$$M = \begin{pmatrix} 0.6 & -0.8 \\ -0.8 & -0.6 \end{pmatrix} \begin{pmatrix} \sqrt{25} & 0 \\ 0 & \sqrt{50} \end{pmatrix} \begin{pmatrix} 0.6 & -0.8 \\ -0.8 & -0.6 \end{pmatrix}$$

$$= \frac{1}{5} \begin{pmatrix} 3 & -4 \\ -4 & -3 \end{pmatrix} \cdot \begin{pmatrix} 5 & 0 \\ 0 & 5\sqrt{2} \end{pmatrix} \frac{1}{5} \begin{pmatrix} 3 & -4 \\ -4 & -3 \end{pmatrix}$$

$$= \frac{1}{5} \begin{pmatrix} 3 & -4 \\ -4 & -3 \end{pmatrix} \cdot \begin{pmatrix} 1 & 0 \\ 0 & \sqrt{2} \end{pmatrix} \begin{pmatrix} 3 & -4 \\ -4 & -3 \end{pmatrix}$$

$$= \frac{1}{5} \begin{pmatrix} 9 + 16\sqrt{2} & -12 + 12\sqrt{2} \\ -12 + 12\sqrt{2} & 16 + 9\sqrt{2} \end{pmatrix}$$

has the property that $M^2 = A$.

(iii) The Cholesky factor U of the symmetric positive definite matrix A is a 2×2 upper triangular matrix U such that $A = U^t U$. Since the Cholesky factor of a 2×2 symmetric positive definite matrix $\begin{pmatrix} a & b \\ b & d \end{pmatrix}$ is $\begin{pmatrix} \sqrt{a} & \frac{b}{\sqrt{a}} \\ 0 & \sqrt{\frac{ad-b^2}{a}} \end{pmatrix}$, we obtain that

the Cholesky factor of the matrix $A = \begin{pmatrix} 41 & 12 \\ 12 & 34 \end{pmatrix}$ is

$$U = \begin{pmatrix} \sqrt{41} & \frac{12}{\sqrt{41}} \\ 0 & \frac{25\sqrt{2}}{\sqrt{41}} \end{pmatrix}. \quad \square$$

Problem'3: Let $A = \begin{pmatrix} 1 & -0.2 & 0.8 \\ -0.2 & 2 & 0.3 \\ 0.8 & 0.3 & 1.1 \end{pmatrix}$.

(i) Show that the matrix A is weakly diagonally dominated, and therefore symmetric positive semidefinite.

(ii) Use Sylvester's criterion to show that the matrix A is symmetric positive definite.

Solution: (i) The matrix A is weakly diagonally dominant since, for every row of A, the absolute value of the main diagonal entry from the row is greater than or equal to the sum of the absolute values of all the other entries in that row:

$$1 = |-0.2| + 0.8 = 1;$$
$$2 > |-0.2| + 0.3 = 0.5;$$
$$1.1 = 0.8 + 0.3 = 1.1.$$

Recall that any weakly diagonally dominant matrix with positive entries of the main diagonal is symmetric positive semidefinite. Then, since the main diagonal entries of the matrix A are 1, 2, and 1.1, we conclude that the matrix A is symmetric positive semidefinite.

(ii) The leading principal minors of the matrix A are

$$\det(1) = 1; \quad \det \begin{pmatrix} 1 & -0.2 \\ -0.2 & 2 \end{pmatrix} = 1.96; \quad \det(A) = 0.69$$

Since all the leading principal minors of the matrix A are positive, it follows from Sylvester's criterion that the matrix A is symmetric positive definite.[3]   □

Problem 4: Let A be a symmetric positive definite matrix, and let U be the Cholesky factor of A. If V is an upper triangular matrix such that $A = V^t V$, show that there exists a diagonal matrix D whose entries on the main diagonal are either -1 or 1 such that $V = DU$.

Solution: If U is the Cholesky factor of the symmetric positive definite matrix A, then U is an upper triangular matrix with all diagonal entries positive such that $A = U^t U$. Since $A = V^t V$, it follows that

$$U^t U \; = \; V^t V. \tag{6.17}$$

Note that the matrix U is a Cholesky factor, and therefore, by definition, U is nonsingular. Moreover, since A is nonsingular and $A = V^t V$, it follows that V is a nonsingular matrix as well. Otherwise, $\det(V) = 0$ and therefore $\det(A) = \det(V^t)\det(V) = 0$, which contradicts the fact that A is a nonsingular matrix.

Thus, U and V are nonsingular matrices, and we can multiply (6.17) by $(V^t)^{-1}$ on the left and by U^{-1} on the right and obtain

$$
\begin{aligned}
(V^t)^{-1} \left(U^t U \right) U^{-1} &= (V^t)^{-1} \left(V^t V \right) U^{-1} \\
\Longleftrightarrow \quad \left((V^t)^{-1} U^t \right) \cdot \left(U U^{-1} \right) &= \left((V^t)^{-1} V^t \right) \cdot \left(V U^{-1} \right) \\
\Longleftrightarrow \quad (V^t)^{-1} U^t &= V U^{-1},
\end{aligned} \tag{6.18}
$$

since $U U^{-1} = I$ and $(V^t)^{-1} V^t = I$.

Recall that the inverse of a lower triangular matrix is lower triangular and the inverse of an upper triangular matrix is upper triangular. Also, recall that the product of two lower triangular matrices is lower triangular, and the product of two upper triangular matrices is upper triangular.

Thus, since the matrices U and V are upper triangular, the matrix U^{-1} is upper triangular and therefore the matrix $V U^{-1}$ is also upper triangular. Moreover, the matrices U^t and V^t are lower triangular. Then, the matrix $(V^t)^{-1}$ is lower triangular and therefore the matrix $(V^t)^{-1} U^t$ is also lower triangular. Since the lower triangular matrix $(V^t)^{-1} U^t$ and the upper triangular matrix $V U^{-1}$ are equal, see (6.18), it follows that they must be diagonal matrices.

Let $D = \text{diag}(d_j)_{j=1:n}$ be a diagonal matrix such that

$$(V^t)^{-1} U^t \; = \; V U^{-1} \; = \; D.$$

Then, by multiplying $V U^{-1} = D$ to the right by U, we find that

$$V \; = \; DU \tag{6.19}$$

---

[3]Note that the result of part (ii) does not contradict the result of part (i). For this matrix, diagonal dominance can only be used to show that the matrix is symmetric positive semidefinite and cannot be used to show that the matrix is symmetric positive definite. Sylvester's Criterion gives a necessary and sufficient condition for a matrix to be symmetric positive definite, and therefore can be used successfully to establish whether the matrix is positive definite or not.

and therefore

$$A = V^t V = (DU)^t DU = U^t D^t DU = U^t D^2 U, \tag{6.20}$$

where the last equality follows from the fact that $D^t = D$ since D is a diagonal matrix. Since $A = U^t U$ and U is a nonsingular matrix, it follows from (6.20) that

$$U^t U = U^t D^2 U \iff (U^t)^{-1} U^t U (U)^{-1} = (U^t)^{-1} U^t D^2 U (U)^{-1}$$
$$\iff I = D^2.$$

Note that $D^2 = \operatorname{diag}(d_j^2)_{j=1:n}$, and therefore $I = D^2$ if and only if $d_j^2 = 1$ for all $j = 1 : n$. Thus,

$$d_j = 1 \quad \text{or} \quad d_j = -1, \quad \forall\, j = 1 : n. \tag{6.21}$$

From (6.19) and (6.21), we conclude that the matrix V can be written as $V = DU$, where the diagonal matrix $D = \operatorname{diag}(d_j)_{j=1:n}$ has main diagonal entries equal either to 1 or to -1. $\quad\square$

Problem 5: Let B_N be the following $N \times N$ symmetric positive definite matrix:

$$B_N = \begin{pmatrix} 2 & -1 & \cdots & 0 \\ -1 & \ddots & \ddots & \vdots \\ \vdots & \ddots & \ddots & -1 \\ 0 & \cdots & -1 & 2 \end{pmatrix}. \tag{6.22}$$

Let C_N be an $N \times (N+1)$ matrix with

$$C_N(i, i) = 1, \quad \forall\, i = 1 : N;$$
$$C_N(i, i+1) = -1, \quad \forall\, i = 1 : N,$$

and with all the other entries equal to 0.
Show that $B_N = C_N C_N^t$. Why is this not the Cholesky decomposition of the matrix B_N?

Solution: The row form of the $N \times (N+1)$ matrix C_N is $C_N = \operatorname{row}(r_j)_{j=1:N}$, where the $1 \times (N+1)$ vector r_j has all entries equal to 0 except for the j-th and $(j+1)$-th entries which are 1 and -1, respectively, i.e., $r_j(j) = 1$ and $r_j(j+1) = -1$. Then, the column form of the $(N+1) \times N$ matrix C_N^t is $C_N^t = \operatorname{col}(r_k^t)_{k=1:N}$. By direct computation, we find that the entries of the $N \times N$ matrix $C_N C_N^t$ are

$$C_N C_N^t(j, k) = r_j \cdot r_k^t = \begin{cases} 2, & \text{if} \quad j = k; \\ -1, & \text{if} \quad k = j - 1 \text{ or } k = j + 1; \\ 0, & \text{else.} \end{cases}$$

Thus, $C_N C_N^t(j, k) = B_N(j, k)$ for all $1 \le j, k \le n$, see (6.22), and therefore $B_N = C_N C_N^t$.

The matrix C_N is not the Cholesky factor of the matrix B_N since C_N is not a square matrix. $\quad\square$

Problem 6: Let B_N be the $N \times N$ tridiagonal symmetric positive definite matrix given by (6.22). Show that the Cholesky factor U_N of the matrix B_N is the upper triangular bidiagonal matrix given by

$$U_N(i,i) = \sqrt{\frac{i+1}{i}}, \quad \forall\, i = 1:N; \tag{6.23}$$

$$U_N(i,i+1) = -\sqrt{\frac{i}{i+1}}, \quad \forall\, i = 1:(N-1). \tag{6.24}$$

Solution: Recall that the Cholesky factor U of a tridiagonal symmetric positive definite matrix A is an upper triangular bidiagonal matrix, and can be computed by using the pseudocode from Table 6.3.

Table 6.3: Cholesky decomposition for tridiagonal spd matrices

```
Function Call:
U = cholesky_tridiag_spd(A)

Input:
A = tridiagonal symmetric positive definite matrix of size n

Output:
U = upper triangular bidiagonal matrix such that UᵗU = A

for i = 1 : (n − 1)
    U(i,i) = √(A(i,i));
    U(i,i+1) = A(i,i+1)/U(i,i);
    A(i+1,i+1) = A(i+1,i+1) − U(i,i+1)²;
end
U(n,n) = √(A(n,n))
```

For

$$A = B_N = \begin{pmatrix} 2 & -1 & \cdots & 0 \\ -1 & \ddots & \ddots & \vdots \\ \vdots & \ddots & \ddots & -1 \\ 0 & \cdots & -1 & 2 \end{pmatrix},$$

the "for" loop from Table 6.3 used to compute the Cholesky factor $U = U_N$ of B_N becomes

$$
\begin{aligned}
&\text{for } i = 1:(N-1) \\
&\quad U_N(i,i) = \sqrt{B_N(i,i)}; &&\text{(BN1)}\\
&\quad U_N(i,i+1) = -\tfrac{1}{U_N(i,i)}; &&\text{(BN2)}\\
&\quad B_N(i+1,i+1) = 2 - U_N(i,i+1)^2; &&\text{(BN3)}\\
&\text{end}
\end{aligned}
$$

Using (BN1) and (BN2), the formula (BN3) for $B_N(i+1,i+1)$ can be written as follows:

$$B_N(i+1,i+1) = 2 - \frac{1}{U_N(i,i)^2} = 2 - \frac{1}{B_N(i,i)}. \tag{6.25}$$

Recall that both $B_N(i, i)$ and $B_N(i+1, i+1)$ from (6.25) are the updated entries of the matrix B_N after $i-1$ steps of the Cholesky decomposition, and after i steps of the Cholesky decomposition, respectively. For simplicity, denote by x_i the value of the updated entry $B_N(i, i)$, i.e., let $x_i = B_N(i, i)$ for $1 \le i \le N$. Then, from (6.25), it follows that the values of x_i satisfy the following recursion:

$$x_{i+1} = 2 - \frac{1}{x_i}, \quad \forall\, i = 1 : (N - 1). \tag{6.26}$$

Note that x_1 is equal to the initial value of $B_N(1, 1)$, i.e., $x_1 = 2$.

By letting $i = 1$, $i = 2$, and $i = 3$ in (6.26), we find that $x_2 = \frac{3}{2}$, $x_3 = \frac{4}{3}$, and $x_4 = \frac{5}{4}$. This suggests the following general formula for x_i:

$$x_i = \frac{i+1}{i}, \quad \forall\, i = 1 : N. \tag{6.27}$$

We prove formula (6.27) by induction.

For $i = 1$ in (6.27), $x_1 = 2$, which is the same as the initial value of x_1.
Assume that $x_i = \frac{i+1}{i}$. Then, from (6.27), we obtain that

$$x_{i+1} = 2 - \frac{1}{x_i} = 2 - \frac{i}{i+1} = \frac{i+2}{i+1},$$

which is the same as the formula (6.27) for x_{i+1}. We conclude by induction that $x_i = \frac{i+1}{i}$ for all $i = 1 : N$ and therefore that

$$B_N(i, i) = \frac{i+1}{i}, \quad \forall\, i = 1 : N. \tag{6.28}$$

From (BN1) and (BN2), and using (6.28), we obtain that the diagonal and upper diagonal entries of the Cholesky factor U_N of the matrix B_N are

$$U_N(i, i) = \sqrt{B_N(i, i)} = \sqrt{\frac{i+1}{i}}, \quad \forall\, i = 1 : N; \tag{6.29}$$

$$U_N(i, i+1) = -\frac{1}{U_N(i, i)} = -\sqrt{\frac{i}{i+1}}, \quad \forall\, i = 1 : (N - 1). \quad \square \tag{6.30}$$

Problem 7: Let B_N be the $N \times N$ tridiagonal symmetric positive definite matrix given by (6.22), and let U_N be the Cholesky factor of B_N given by (6.23–6.24).

(i) Show that the solution to a linear system $B_N x = b$, where b and x are $N \times 1$ column vectors can be obtained by using the explicit pseudocode from Table 6.4.

(ii) What is the operation count for the pseudocode above, and how does it compare to $8N + O(1)$, the operation count for the optimal linear solver for tridiagonal symmetric positive definite matrices?

Solution: (i) A linear solver using Cholesky decomposition for an $n \times n$ tridiagonal symmetric positive matrix A requires three steps: the Cholesky decomposition of

Table 6.4: Linear system solution for the tridiagonal B_N matrix

---

Function Call:
$x = \text{linear\_solve\_cholesky\_B\_N}(b)$

Input:
$b = N \times 1$ column vector

Output:
$x = $ solution to $B_N x = b$

$y(1) = \frac{b(1)}{\sqrt{2}}$
for $i = 2 : N$
$\qquad y(i) = \frac{b(i) + y(i-1)\sqrt{(i-1)/i}}{\sqrt{(i+1)/i}}$
end
$x(N) = \frac{y(N)\sqrt{N}}{\sqrt{N+1}}$
for $i = (N-1) : 1$
$\qquad x(i) = \frac{y(i) + x(i+1)\sqrt{i/(i+1)}}{\sqrt{(i+1)/i}}$
end

---

the matrix A, followed by a banded forward substitution and a banded backward substitution.

Since the Cholesky factor of B_N is given by (6.4–6.5), i.e.,

$$U_N(i,i) \quad = \quad \sqrt{\frac{i+1}{i}}, \quad \forall\, i = 1 : N; \tag{6.31}$$

$$U_N(i,i+1) \quad = \quad -\sqrt{\frac{i}{i+1}}, \quad \forall\, i = 1 : (N-1), \tag{6.32}$$

there is no need to do the Cholesky decomposition of the matrix B_N within the linear solver using Cholesky decomposition for B_N. A simplified explicit interpretation of a Cholesky linear solver corresponding to the matrix B_N can be found in Table 6.5.

Since $U_N(i-1,i) = -\sqrt{\frac{i-1}{i}}$, see (6.32), and $U_N(i,i) = \sqrt{\frac{i+1}{i}}$, see (6.31), it follows that

$$y(i) \quad = \quad \frac{b(i) - U_N(i-1,i)y(i-1)}{U_N(i,i)} \quad \text{is the same as}$$

$$y(i) \quad = \quad \frac{b(i) + y(i-1)\sqrt{(i-1)/i}}{\sqrt{(i+1)/i}}. \tag{6.33}$$

Since $U_N(N,N) = \sqrt{\frac{N+1}{N}}$, see (6.31), we find that

$$x(N) \quad = \quad \frac{y(N)}{U_N(N,N)} \quad \text{is the same as}$$

$$x(N) \quad = \quad \frac{y(N)\sqrt{N}}{\sqrt{N+1}}; \tag{6.34}$$

Table 6.5: Explicit tridiagonal solver with Cholesky decomposition for B_N

Function Call:
$x = $ linear_solve_Cholesky_B_n(b)

Input:
$U_N = $ Cholesky factor of the matrix U_n
$b = $ column vector of size n

Output:
$x = $ solution to $B_N x = b$

$y(1) = \frac{b(1)}{U_N(1,1)}$;
for $i = 2 : N$
$\quad y(i) = \frac{b(i) - U_N(i-1,i)y(i-1)}{U_N(i,i)}$;
end $\qquad\qquad$ // forward substitution for $U_N^t y = b$
$x(N) = \frac{y(N)}{U_N(N,N)}$;
for $i = (N-1) : 1$
$\quad x(i) = \frac{y(i) - U_N(i,i+1)x(i+1)}{U_N(i,i)}$;
end $\qquad\qquad$ // backward substitution for $U_N x = y$

Since $U_N(i, i+1) = -\sqrt{\frac{i}{i+1}}$, see (6.32), and $U_N(i,i) = \sqrt{\frac{i+1}{i}}$, see (6.31), it follows that

$$x(i) = \frac{y(i) - U_N(i,i+1)x(i+1)}{U_N(i,i)} \quad \text{is the same as}$$

$$x(i) = \frac{y(i) + x(i+1)\sqrt{i/(i+1)}}{\sqrt{(i+1)/i}}. \tag{6.35}$$

From (6.33), (6.34), and (6.35), and since $U_N(1,1) = \sqrt{2}$, see (6.31), we obtain that the code from Table 6.5 can be simplified to the code from Table 6.6.

(ii) Computing $y(1) = \frac{b(1)}{\sqrt{2}}$ requires 2 operations, and the first "for" loop from Table 6.6, i.e.,

---

for $i = 2 : N$
$\quad y(i) = \frac{b(i) + y(i-1)\sqrt{(i-1)/i}}{\sqrt{(i+1)/i}}$
end

---

requires $9(N-1)$ operations. The second "for" loop from Table 6.6, i.e.,

---

for $i = (N-1) : 1$
$\quad x(i) = \frac{y(i) + x(i+1)\sqrt{i/(i+1)}}{\sqrt{(i+1)/i}}$
end

---

also requires $9(N-1)$ operations, while computing $x(N) = \frac{y(N)\sqrt{N}}{\sqrt{N+1}}$ requires 5 operations.

Table 6.6: Linear system solution for the tridiagonal B_N matrix

Function Call:
$x = \text{linear\_solve\_cholesky\_B\_N}(b)$

Input:
$b = N \times 1$ column vector

Output:
$x = $ solution to $B_N x = b$

$y(1) = \frac{b(1)}{\sqrt{2}}$
for $i = 2 : N$
$\qquad y(i) = \dfrac{b(i) + y(i-1)\sqrt{(i-1)/i}}{\sqrt{(i+1)/i}}$
end
$x(N) = \dfrac{y(N)\sqrt{N}}{\sqrt{N+1}}$
for $i = (N-1) : 1$
$\qquad x(i) = \dfrac{y(i) + x(i+1)\sqrt{i/(i+1)}}{\sqrt{(i+1)/i}}$
end

Thus, the operation count for the pseudocode from Table 6.6 for solving a linear system corresponding to the matrix B_N using Cholesky decomposition requires

$$2 + 9(N-1) + 9(N-1) + 5 \;=\; 18N - 11 \;=\; 18N + O(1)$$

operations, which is more than twice more than $8N + O(1)$, the operation count for the optimal linear solver for tridiagonal symmetric positive definite matrices. $\square$

Problem 8: Let B_N be the $N \times N$ tridiagonal symmetric positive definite matrix given by (6.22). Show that the LU factors L and U of the matrix B_N are the lower triangular bidiagonal matrix and the upper triangular bidiagonal matrix given by

$$L(i,i) = 1, \;\; \forall\, i = 1 : N; \quad L(i+1,i) = -\frac{i}{i+1}, \;\; \forall\, i = 1 : (N-1); \qquad (6.36)$$

$$U(i,i) = \frac{i+1}{i}, \;\; \forall\, i = 1 : N; \quad U(i,i+1) = -1, \;\; \forall\, i = 1 : (N-1). \qquad (6.37)$$

Solution: Note that B_N is a symmetric positive definite matrix, and therefore it has an LU decomposition without pivoting. Moreover, since B_N is tridiagonal, the L factor of B_N is a lower triangular bidiagonal matrix and the U factor of B_N is an upper triangular bidiagonal matrix.

Thus, it is enough to show that $B_N = LU$, which can be done by direct computation as follows:

$$(LU)(i,i) \;=\; \sum_{l=1}^{N} L(i,l)U(l,i)$$

$$= L(i,i-1)\,U(i-1,i) \;+\; L(i,i)\,U(i,i)$$
$$= \left(-\frac{i-1}{i}\right)\cdot(-1)+1\cdot\frac{i+1}{i} \;=\; \frac{i-1+i+1}{i}$$
$$= 2,\;\; \forall\, i=1:N; \tag{6.38}$$

$$(LU)(i-1,i) = \sum_{l=1}^{N} L(i-1,l)U(l,i)$$
$$= L(i-1,i)\,U(i,i)\;+\;L(i-1,i-1)\,U(i-1,i)$$
$$= 0\cdot\frac{i+1}{i}+1\cdot(-1)$$
$$= -1,\;\; \forall\, i=2:N; \tag{6.39}$$

$$(LU)(i+1,i) = \sum_{l=1}^{N} L(i+1,l)U(l,i)$$
$$= L(i+1,i)\,U(i,i)\;+\;L(i+1,i+1)\,U(i+1,1)$$
$$= \left(-\frac{i}{i+1}\right)\cdot\frac{i+1}{i}+1\cdot 0$$
$$= -1,\;\; \forall\, i=1:(N-1). \tag{6.40}$$

The matrix LU is bidiagonal, since it is the product of a lower triangular bidiagonal matrix and an upper triangular bidiagonal matrix. More precisely, if $|j-k|\geq 2$, i.e., if either $1\leq j\leq k-2$ or $k+2\leq j\leq n$, then $L(j,k)=0$, $L(j,k+1)=0$, $U(j,k)=0$, $U(j-1,k)=0$, and therefore

$$(LU)(j,k) = \sum_{l=1}^{N} L(j,l)U(l,k)$$
$$= L(j,j-1)\,U(j-1,k)\;+\;L(j,j)\,U(j,k)$$
$$\quad +\;L(j,k)\,U(k,k)\;+\;L(j,k+1)\,U(k+1,k)$$
$$= -\frac{j-1}{j}U(j-1,k)+U(j,k)+L(j,k)\frac{k+1}{k}-L(j,k+1)$$
$$= 0,\;\; \forall\, 1\leq j,k\leq n \;\text{ with } |j-k|\geq 2. \tag{6.41}$$

From (6.38–6.41), we obtain that

$$LU = \begin{pmatrix} 2 & -1 & \cdots & 0 \\ -1 & \ddots & \ddots & \vdots \\ \vdots & \ddots & \ddots & -1 \\ 0 & \cdots & -1 & 2 \end{pmatrix} = B_N;$$

see also (6.22). □

Problem 9: Let B_N be the $N\times N$ tridiagonal symmetric positive definite matrix given by (6.22), and let L and U be the LU factors of B_N given by (6.36–6.37).

(i) Show that the solution to a linear system $B_N x = b$, where b and x are $N\times 1$ column vectors can be obtained by using the explicit pseudocode from Table 6.7.

Table 6.7: Linear system solution corresponding to matrix B_N

Function Call:
$x = $ linear_solve_lu_B_N(b)
Input:
$b = N \times 1$ column vector
Output:
$x = $ solution to $B_N x = b$
$y(1) = b(1)$
for $i = 2 : N$
$\quad y(i) = b(i) + \frac{(i-1)y(i-1)}{i}$
end
$x(N) = \frac{Ny(N)}{N+1}$
for $i = (N-1) : 1$
$\quad x(i) = \frac{i(y(i)+x(i+1))}{i+1}$
end

(ii) What is the operation count for the pseudocode above, and how does it compare to $8N+O(1)$, the operation count for the optimal linear solver for tridiagonal symmetric positive definite matrices?

Solution: (i) A linear solver using LU decomposition for an $n \times n$ tridiagonal symmetric positive matrix A requires three steps: the LU decomposition of the matrix A, followed by a banded forward substitution and a banded backward substitution.

Since the LU factors L and U of B_N are given by (6.36–6.37), i.e.,

$$L(i,i) = 1, \quad \forall\, i = 1 : N; \quad L(i+1,i) = -\frac{i}{i+1}, \quad \forall\, i = 1 : (N-1); \tag{6.42}$$

$$U(i,i) = \frac{i+1}{i}, \quad \forall\, i = 1 : N; \quad U(i,i+1) = -1, \quad \forall\, i = 1 : (N-1), \tag{6.43}$$

there is no need to do the LU decomposition of the matrix B_N within the linear solver using LU decomposition for B_N. A simplified explicit interpretation of the linear solver using the LU decomposition without pivoting corresponding to the matrix B_N can be found in Table 6.5.

Since $L(i, i-1) = -\frac{i-1}{i}$, see (6.42), it follows that

$$y(i) \;=\; b(i) - L(i,i-1)y(i-1) \quad \text{is the same as}$$

$$y(i) \;=\; b(i) + \frac{(i-1)y(i-1)}{i}. \tag{6.44}$$

Since $U(N,N) = \frac{N+1}{N}$, $U(i,i+1) = -1$, and $U(i,i) = \frac{i+1}{i}$, see (6.43), we find that

$$x(N) \;=\; \frac{y(N)}{U(N,N)} \quad \text{is the same as} \tag{6.45}$$

$$x(N) \;=\; \frac{Ny(N)}{N+1}; \tag{6.46}$$

Table 6.8: Explicit tridiagonal solver with LU decomposition corresponding to B_N

Function Call:
$x = \text{linear\_solve\_lu\_B\_N}(b)$

Input:
$b = N \times 1$ column vector
$L, U = $ the L and U factors from the LU decomposition of the matrix B_N

Output:
$x = $ solution to $B_N x = b$

$y(1) = b(1);$
for $i = 2 : N$
$\quad y(i) = b(i) - L(i, i - 1)y(i - 1);$
end   // forward substitution for $Ly = b$
$x(N) = \frac{y(N)}{U(N,N)};$
for $i = (N - 1) : 1$
$\quad x(i) = \frac{y(i) - U(i,i+1)x(i+1)}{U(i,i)};$
end   // backward substitution for $Ux = y$

$$x(i) = \frac{y(i) - U(i, i + 1)x(i + 1)}{U(i, i)} \quad \text{is the same as} \quad (6.47)$$

$$x(i) = \frac{i(y(i) + x(i + 1))}{i + 1}. \quad\quad (6.48)$$

From (6.44), (6.46), and (6.48), we obtain that the code from Table 6.8 can be simplified to the code from Table 6.9.

(ii) The first "for" loop from Table 6.9, i.e.,

---
for $i = 2 : N$
$\quad y(i) = b(i) + \frac{(i-1)y(i-1)}{i}$
end

---

requires $4(N - 1)$ operations. The second 'for" loop from Table 6.9, i.e.,

---
for $i = (N - 1) : 1$
$\quad x(i) = \frac{i(y(i) + x(i+1))}{i+1}$
end

---

also requires $4(N - 1)$ operations, while computing $x(N) = \frac{Ny(N)}{N+1}$ requires 3 operations.

Thus, the operation count for the pseudocode from Table 6.9 for solving a linear system corresponding to the matrix B_N requires

$$4(N - 1) + 4(N - 1) + 3 = 8N - 5 = 8N + O(1)$$

operations, which is the same as the operation count for the optimal linear solver for tridiagonal symmetric positive definite matrices. $\square$

Table 6.9: Linear system solution corresponding to matrix B_N

Function Call:
$x = $ linear_solve_lu_B_N(b)

Input:
$b = N \times 1$ column vector

Output:
$x = $ solution to $B_N x = b$

$y(1) = b(1)$
for $i = 2 : N$
$\quad y(i) = b(i) + \frac{(i-1)y(i-1)}{i}$
end
$x(N) = \frac{N y(N)}{N+1}$
for $i = (N-1) : 1$
$\quad x(i) = \frac{i(y(i)+x(i+1))}{i+1}$
end

Problem 10: Write an explicit optimal pseudocode for solving linear systems corresponding to the same tridiagonal symmetric positive definite matrix. In other words, write a pseudocode for solving p linear systems $Ax_i = b_i$, for $i = 1 : p$, where A is an $n \times n$ tridiagonal symmetric positive definite matrix, by using the explicit solver linear_solve_LU_tridiag_spd from Table 6.7 from [3] and the method from Table 6.3 from [3]. What is the operation count for solving the p linear systems?

Solution: Recall from Table 6.3 from [3] that solving p linear systems corresponding to the same symmetric positive definite matrix is done efficiently by first computing the Cholesky decomposition of the matrix, and then implementing a "for" loop to solve each linear system using first a forward substitution and then a backward substitution. However, a linear system corresponding to a symmetric positive definite matrix is solved efficiently by using the LU decomposition without pivoting of the matrix.

We conclude that an efficient way to solve p linear systems $Ax_i = b_i$, for $i = 1 : p$, corresponding to the $n \times n$ tridiagonal symmetric positive definite matrix A is to compute the LU decomposition without pivoting of A given by the first part of the routine linear_solve_LU_tridiag_spd from Table 6.7 from [3], and then use forward substitution and backward substitution in a for loop to solve each linear system, i.e.,

$[L, U] = $ LU_tridiag_spd(A)
for $i = 1 : p$
$\quad y = $ forward_subst(U^t, b_i)
$\quad x_i = $ backward_subst(U, y)
end

The routine for computing the LU decomposition without pivoting of the matrix A was included as the first part of the routine linear_solve_LU_tridiag_spd from Table 6.7 from [3]. Based on that, a detailed pseudocode for this method can be found in Table 6.10.

Table 6.10: Solution of multiple linear systems corresponding to the same tridiagonal spd matrix

Input:
A = tridiagonal symmetric positive definite matrix of size n
b_i = column vectors of size n, $i = 1 : p$

Output:
x_i = solution to $Ax_i = b_i$, $i = 1 : p$

for $i = 1 : (n-1)$
   $L(i,i) = 1; L(i+1,i) = A(i+1,i)/A(i,i);$
   $U(i,i) = A(i,i); U(i,i+1) = A(i,i+1);$
   $A(i+1,i+1) = A(i+1,i+1) - L(i+1,i)U(i,i+1);$
end
$L(n,n) = 1; U(n,n) = A(n,n);$ // LU decomposition of A
for $i = 1 : p$
   $y(1) = b_i(1);$
   for $j = 2 : n$
      $y(j) = b_i(j) - L(j,j-1)y(j-1);$
   end   // forward substitution for $Ly = b_i$
   $x_i(n) = \frac{y(n)}{U(n,n)};$
   for $j = (n-1) : 1$
      $x_i(j) = \frac{y(j) - U(j,j+1)x_i(j+1)}{U(j,j)};$
   end   // backward substitution for $Ux_i = y$
end

The operation count for the pseudocode from Table 6.10 is as follows:
- $3n - 3$ operations for the LU decomposition of A;
- $2n - 2$ operations for the forward substitution for $Ly = b_i$, which is run in the for loop p times, for a total of $2np - 2p$ operations;
- $3n - 2$ operations for the backward substitution for $Ux_i = y$, which is run in the for loop p times, for a total of $3np - 2p$ operations.

Summarizing, the operation count for the pseudocode from Table 6.10 is

$$5np + 3n - 4p - 3. \quad \square$$

Problem 11: Write the pseudocode for the Cholesky decomposition of symmetric positive definite banded matrices of band m. What is the corresponding operation count?

Solution: Note that the Cholesky factor of a symmetric positive definite banded matrix of band m is a banded upper triangular matrix of band m; see, e.g., [1], [4], [5]. Then, for a banded matrix of band m, computing the updated form of the matrix A after the i–th row of U was computed only requires updating the upper diagonal part and the main diagonal of the $m \times m$ matrix $A(i+1 : i+m, i+1 : i+m)$, if $i + m \leq n$.

The corresponding pseudocode for the banded Cholesky decomposition can be found in Table 6.11.

Table 6.11: Pseudocode for banded Cholesky decomposition

```
Function Call:
U = cholesky_banded(A)

Input:
A = banded symmetric positive definite matrix of size n and band m

Output:
U = banded upper triangular matrix of band m such that U^t U = A

for i = 1 : (n - 1)
    U(i, i) = √A(i, i);
    for k = (i + 1) : min(i + m, n)        // compute up to entry i + m of U
        U(i, k) = A(i, k)/U(i, i);
    end
    for j = (i + 1) : min(i + m, n)        // update up to row i + m of A
        for k = j : min(i + m, n)          // update up to column i + m of A
            A(j, k) = A(j, k) - U(i, j)U(i, k);
        end
    end
end
U(n, n) = √A(n, n)
```

We now proceed to finding the operation count for the pseudocode from Table 6.11.

If $i + m \leq n$, i.e., for $i \leq n - m$, it follows that $\min(i + m, n) = i + m$ and the outside "for" loop from the pseudocode from Table 6.11 becomes

```
for i = 1 : (n - m)
    U(i, i) = √A(i, i);
    for k = (i + 1) : (i + m)
        U(i, k) = A(i, k)/U(i, i);
    end
    for j = (i + 1) : (i + m)
        for k = j : (i + m)
            A(j, k) = A(j, k) - U(i, j)U(i, k);
        end
    end
end
```

In this case, at step i, the i-th row of U is computed by executing

```
U(i, i) = √A(i, i);
for k = (i + 1) : (i + m)
    U(i, k)  =  A(i, k)/U(i, i)
end
```

which requires

$$m + 1 \tag{6.49}$$

operations. Also at step i, the double "for" loop

```
for j = (i + 1) : (i + m)
    for k = j : (i + m)
        A(j, k) = A(j, k) − U(i, j)U(i, k);
    end
end
```

to update the banded upper diagonal part of the symmetric matrix $A(i+1 : n, i+1 : n)$ requires

$$\sum_{j=i+1}^{i+m} \sum_{k=j}^{i+m} 2 = \sum_{j=i+1}^{i+m} 2(i + m - j + 1) = 2 \sum_{j=i+1}^{i+m} (i + m - j + 1) \quad (6.50)$$

operations. By letting $l = i + m - j + 1$ in (6.50), we find that

$$2 \sum_{j=i+1}^{i+m} (i + m - j + 1) = 2 \sum_{l=1}^{m} l = 2 \cdot \frac{m(m + 1)}{2} = m(m + 1), \quad (6.51)$$

since $\sum_{l=1}^{m} l = \frac{m(m+1)}{2}$.

From (6.49) and (6.51), we obtain that the number of operations required to go through the outside "for" loop for $i = 1 : (n - m)$ is

$$\sum_{i=1}^{n-m} (m + 1 + m(m + 1)) = \sum_{i=1}^{n-m} (m + 1)^2 = (n - m)(m + 1)^2. \quad (6.52)$$

If $i + m > n$, i.e., for $i \geq n - m + 1$, it follows that $\min(i + m, n) = n$ and the outside "for" loop from the pseudocode from Table 6.11 becomes

```
for i = (n − m + 1) : (n − 1)
    U(i, i) = √A(i, i);
    for k = (i + 1) : n
        U(i, k) = A(i, k)/U(i, i);
    end
    for j = (i + 1) : n
        for k = j : n
            A(j, k) = A(j, k) − U(i, j)U(i, k);
        end
    end
end
```

In this case, at step i, the i–th row of U is computed by executing

```
U(i, i) = √A(i, i);
for k − (i + 1) : n
    U(i, k)  =  A(i, k)/U(i, i)
end
```

which requires

$$n - i + 1 \quad (6.53)$$

operations. Also at step i, the double "for" loop

```
for j = (i + 1) : n
  for k = j : n
    A(j, k) = A(j, k) − U(i, j)U(i, k);
  end
end
```

to update the banded upper diagonal part of the symmetric matrix $A(i+1 : n, i+1 : n)$ requires

$$\sum_{j=i+1}^{n} \sum_{k=j}^{n} 2 \; = \; \sum_{j=i+1}^{n} 2(n - j + 1) \; = \; 2 \sum_{j=i+1}^{n} (n - j + 1) \qquad (6.54)$$

operations. By letting $l = n - j + 1$ in (6.54), we find that

$$2 \sum_{j=i+1}^{n} (n - j + 1) \; = \; 2 \sum_{l=1}^{n-i} l \; = \; 2 \cdot \frac{(n - i)(n - i + 1)}{2} \; = \; (n - i)(n - i + 1), \quad (6.55)$$

where we used the fact that $\sum_{l=1}^{p} l = \frac{p(p+1)}{2}$ for $p = n - i$.

From (6.53) and (6.55), we obtain that the number of operations required to go through the outside "for" loop for $i = (n − m + 1) : n$ is

$$\sum_{i=n-m+1}^{n} (n - i + 1 + (n - i)(n - i + 1)) \; = \; \sum_{i=n-m+1}^{n} (n - i + 1)^2. \qquad (6.56)$$

By letting $l = n - i + 1$ in (6.56), we find that

$$\sum_{i=m-n+1}^{n} (n - i + 1)^2 \; = \; \sum_{l=1}^{m} l^2 \; = \; \frac{m(m + 1)(2m + 1)}{6}. \qquad (6.57)$$

Thus, when accounting for the outside "for" loop "for $i = 1 : (n − 1)$" from Table 6.11 separately for $i = 1 : (n − m)$ and for $i = (n − m + 1) : n$, and for the one extra operation required for $U(n, n) = \sqrt{A(n, n)}$, we obtain using (6.52) and (6.57) that the operation count for the banded Cholesky decomposition is

$$(n - m)(m + 1)^2 + \frac{m(m + 1)(2m + 1)}{6} + 1$$

$$= \; n(m + 1)^2 - \frac{2m^3}{3} - \frac{3m^2}{2} - \frac{5m}{6} + 1. \quad \square \qquad (6.58)$$

Problem 12: What is the operation count for solving a linear system corresponding to a symmetric positive definite banded matrix of band m using a Cholesky linear solver?

Solution: Let A be a banded symmetric positive definite matrix of size n and band m, and let b be a column vector of size n. If $A = U^t U$ is the Cholesky decomposition of A, then solving a linear system $Ax = b$ is equivalent to solving $U^t U x = b$. This can be done by solving $U^t y = b$ for y using the banded forward substitution routine from Table 2.2, and then solving $Ux = y$ for x using the banded backward substitution routine from Table 2.3.

Table 6.12: Linear solver using Cholesky decomposition

Function Call:
$x = $ linear_solve_cholesky_banded(A,b)

Input:
$A = $ banded symmetric positive definite matrix of size n and band m
$b = $ column vector of size n

Output:
$x = $ solution to $Ax = b$

$U = $ cholesky_banded(A);
$y = $ forward_subst_banded(U^t, b);
$x = $ backward_subst_banded(U, y);

The pseudocode for solving a linear system corresponding to a symmetric positive definite banded matrix of band m using a Cholesky linear solver can be found in Table 6.12.

The operation count for the linear solver from Table 6.12 is as follows:

- $n(m+1)^2 - \frac{2m^3}{3} - \frac{3m^2}{2} - \frac{5m}{6} + 1$ for $U = $ cholesky_banded(A), the banded Cholesky decomposition of A; cf. (6.58);
- $2mn + 2n - m^2 - m - 1$ for $y = $ forward_subst_banded(U^t, b), the banded forward substitution for solving $U^t y = b$; cf. (2.34);
- $2mn + 2n - m^2 - m - 1$ for $x = $ backward_subst_banded(U, y), the banded backward substitution for solving $Ux = y$; cf. (2.44),

for a total operation count of

$$\left(n(m+1)^2 - \frac{2m^3}{3} - \frac{3m^2}{2} - \frac{5m}{6} + 1 \right) + 2\left(2mn + 2n - m^2 - m - 1 \right)$$
$$= n(m^2 + 6m + 5) - \frac{2m^3}{3} - \frac{7m^2}{2} - \frac{17m}{6} - 1. \quad \square$$

Problem 13: The following discount factors were obtained from market data:

Date	Discount Factor
2 months	0.9980
5 months	0.9935
11 months	0.9820
15 months	0.9775

The overnight rate is 0.75%.

(i) What are the corresponding 2 months, 5 months, 11 months and 15 months zero rates?

(ii) What is the tridiagonal system that must be solved in the efficient implementation of the natural cubic spline interpolation for finding the zero rate curve for all times less than 15 months?

(iii) Use the efficient implementation of the natural cubic spline interpolation to find a zero rate curve for all times less than 15 months matching the discount factors above.

(iv) Find the value of a 14 months quarterly coupon bond with 2.5% coupon rate.

Solution: (i) For continuously compounded interest, the discount factor $\text{Disc}(t)$ at time t can be written in terms of the risk–free zero rate $r(0,t)$ at time t as follows:

$$\text{Disc}(t) \;=\; e^{-tr(0,t)}, \quad \forall\, t > 0,$$

and therefore

$$r(0,t) \;=\; -\frac{1}{t}\ln(\text{Disc}(t)), \quad \forall\, t > 0. \tag{6.59}$$

Note that

$$\text{Disc}\left(\frac{2}{12}\right) = 0.9980; \qquad \text{Disc}\left(\frac{5}{12}\right) = 0.9935;$$

$$\text{Disc}\left(\frac{11}{12}\right) = 0.9820; \qquad \text{Disc}\left(\frac{15}{12}\right) = 0.9775.$$

Formula (6.59) can be used to find the 2 months, 5 months, 11 months, and 15 months continuously compounded zero rates:

$$r\left(0,\frac{2}{12}\right) = -\frac{12}{2}\ln\left(\text{Disc}\left(\frac{2}{12}\right)\right) = 0.012012;$$

$$r\left(0,\frac{5}{12}\right) = -\frac{12}{5}\ln\left(\text{Disc}\left(\frac{5}{12}\right)\right) = 0.015651;$$

$$r\left(0,\frac{11}{12}\right) = -\frac{12}{11}\ln\left(\text{Disc}\left(\frac{11}{12}\right)\right) = 0.019815;$$

$$r\left(0,\frac{15}{12}\right) = -\frac{12}{15}\ln\left(\text{Disc}\left(\frac{15}{12}\right)\right) = 0.018206.$$

Also, $r(0,0) = 0.0075$, since the overnight rate if 0.75%.

(ii) Assume that the zero rate $r(0,t)$ is a cubic polynomial on each of the intervals $\left[0,\frac{2}{12}\right]$, $\left[\frac{2}{12},\frac{5}{12}\right]$, $\left[\frac{5}{12},\frac{11}{12}\right]$, and $\left[\frac{11}{12},\frac{15}{12}\right]$. The tridiagonal system that must be solved in the efficient implementation of the natural cubic spline interpolation is

$$Mw \;=\; z, \tag{6.60}$$

where M is the 3×3 symmetric positive matrix given by

$$M \;=\; \begin{pmatrix} 0.8333 & 0.25 & 0 \\ 0.25 & 1.5 & 0.5 \\ 0 & 0.5 & 1.6667 \end{pmatrix}$$

and

$$z \;=\; \begin{pmatrix} -0.0751 \\ -0.0374 \\ -0.0789 \end{pmatrix}.$$

(iii) The solution to the linear system (6.60) is $w = \begin{pmatrix} -0.0922 \\ 0.0069 \\ -0.0495 \end{pmatrix}$. These represent

the values of the second order derivatives of $r(0,t)$ with respect to t at the times $\frac{2}{12}$, $\frac{5}{12}$, and $\frac{11}{12}$. Using these values, we compute the coefficients of the cubic polynomials that are equal to the zero rate curve $r(0,t)$ on the four intervals above and obtain that $r(0,t)$ is given by

$$r(0,t) = \begin{cases} 0.0075 + 0.0296t - 0.0922t^3, & \text{if } 0 \le t \le \frac{2}{12}; \\ 0.0068 + 0.0428t - 0.0791t^2 + 0.0661t^3, & \text{if } \frac{2}{12} \le t \le \frac{5}{12}; \\ 0.0129 - 0.0014t + 0.0270t^2 - 0.0188t^3, & \text{if } \frac{5}{12} \le t \le \frac{11}{12}; \\ -0.0206 + 0.1083t - 0.0927t^2 + 0.0247t^3, & \text{if } \frac{11}{12} \le t \le \frac{15}{12}, \end{cases} \quad (6.61)$$

(iv) A quarterly coupon bond with face value \$100, coupon rate C, and maturity T pays the holder of the bond a coupon payment equal to $\frac{C}{4} \cdot 100$ every three months, except at maturity. The final payment at maturity T is equal to the face value of the bond plus one coupon payment, i.e., $100 + \frac{C}{4} \cdot 100$.

Then, a 14 months quarterly coupon bond with 2.5% coupon rate and face value \$100 has the following cash flows:

Date	Cash flow
2 months	\$0.625
5 months	\$0.625
8 months	\$0.625
11 months	\$0.625
14 months	\$100.625

Using the zero rate $r(0,t)$ given by (6.61) and continuous compounding, we obtain that the value of the 14 months quarterly coupon bond with 2.5% coupon rate and face value \$100 is

$$\begin{aligned} B = &\ 0.625 \exp\left(-\frac{2}{12} r\left(0, \frac{2}{12}\right)\right) + 0.625 \exp\left(-\frac{5}{12} r\left(0, \frac{5}{12}\right)\right) \\ &+ 0.625 \exp\left(-\frac{8}{12} r\left(0, \frac{8}{12}\right)\right) + 0.625 \exp\left(-\frac{11}{12} r\left(0, \frac{11}{12}\right)\right) \\ &+ 100.625 \exp\left(-\frac{14}{12} r\left(0, \frac{14}{12}\right)\right). \end{aligned}$$

From (6.61), we find that

$$r\left(0, \frac{2}{12}\right) = 0.012012; \ r\left(0, \frac{5}{12}\right) = 0.015651; \ r\left(0, \frac{8}{12}\right) = 0.018397;$$

$$r\left(0, \frac{11}{12}\right) = 0.019815; \ r\left(0, \frac{14}{12}\right) = 0.018823.$$

Thus, $B = 100.9152$, i.e., the value of the bond is \$100.92. □

Chapter 7

Covariance and correlation matrices. Linear Transformation Property. Multivariate normal random variables.

7.1 Exercises

1. Let X_1 and X_2 be random variables with correlation matrix $\begin{pmatrix} 1 & \rho \\ \rho & 1 \end{pmatrix}$, where $-1 \le \rho \le 1$. Let $Y_1 = X_1$ and $Y_2 = -X_2$. Show that the correlation matrix of Y_1 and Y_2 is $\begin{pmatrix} 1 & -\rho \\ -\rho & 1 \end{pmatrix}$.

2. What is the correlation matrix of three random variables whose covariance matrix is $\begin{pmatrix} 1 & 0.36 & -1.44 \\ 0.36 & 4 & 0.80 \\ -1.44 & 0.80 & 9 \end{pmatrix}$?

3. Find all the values of ρ such that the matrix $\begin{pmatrix} 1 & 0.6 & -0.3 \\ 0.6 & 1 & \rho \\ -0.3 & \rho & 1 \end{pmatrix}$ is a correlation matrix.

4. Let X_1, X_2, X_3 be random variables such that
$$\text{corr}(X_1, X_3) = 0.3; \quad \text{corr}(X_2, X_3) = 0.1.$$
Find upper and lower bounds for $\rho = \text{corr}(X_1, X_2)$.

5. Consider three random variables with pairwise correlation ρ, i.e., such that the correlation between any two of the random variables is ρ. What is the smallest possible value for ρ?

6. Let X_1, X_2, X_3, X_4 be random variables such that
$$\text{corr}(X_1, X_3) = 0.3; \quad \text{corr}(X_1, X_4) = 0.2; \quad \text{corr}(X_2, X_3) = 0.1;$$

$$\text{corr}(X_2, X_4) = -0.1; \quad \text{corr}(X_3, X_4) = -0.2.$$

Find upper and lower bounds for $\rho = \text{corr}(X_1, X_2)$.

7. Assume that all the entries of an $n \times n$ correlation matrix which are not on the main diagonal are equal to q. Find upper and lower bounds on the possible values of q.

8. Show that the matrix $\begin{pmatrix} 1 & 0.1 & 0.2 \\ 0.1 & 1 & -0.3 \\ 0.2 & -0.3 & 1 \end{pmatrix}$ is a correlation matrix, and find its Cholesky factor.

9. Show that it is not possible to find three random variables on the same probability space with correlations 0.8, 0.7, and -0.5.

10. Consider three random variables given by the following time series data at five data points:

$$\begin{pmatrix} 0.25 & -0.50 & 1.50 \\ 1 & -1 & 1.25 \\ -0.50 & -0.25 & 2 \\ 0 & 0.50 & 0.75 \\ -1 & 0.75 & 1.50 \end{pmatrix}.$$

(i) Show that the time series vectors $\begin{pmatrix} 0.25 \\ 1 \\ -0.50 \\ 0 \\ -1 \end{pmatrix}$, $\begin{pmatrix} -0.50 \\ -1 \\ -0.25 \\ 0.50 \\ 0.75 \end{pmatrix}$, $\begin{pmatrix} 1.50 \\ 1.25 \\ 2 \\ 0.75 \\ 1.50 \end{pmatrix}$ of the three random variables are linearly independent.

(ii) Show that the vectors $\begin{pmatrix} 0.25 \\ 1 \\ -0.50 \\ 0 \\ -1 \end{pmatrix}$, $\begin{pmatrix} -0.50 \\ -1 \\ -0.25 \\ 0.50 \\ 0.75 \end{pmatrix}$, $\begin{pmatrix} 1.50 \\ 1.25 \\ 2 \\ 0.75 \\ 1.50 \end{pmatrix}$, $\begin{pmatrix} 1 \\ 1 \\ 1 \\ 1 \\ 1 \end{pmatrix}$ are not linearly independent.

(iii) Compute the sample covariance matrix of the three random variables and show that it is singular.

Note: Recall that the necessary and sufficient condition for the sample covariance matrix to be nonsingular is that the time series data column vectors and the column vector of the same size with all entries equal to 1 are linearly independent. This exercise shows that random variables with linearly independent time series data do not necessarily have a nonsingular sample covariance matrix.

11. The file *data-DJ30-july2011-june2013.xlsx* from fepress.org/nla-primer contains the end of week and end of month closing prices for eight financial and technology Dow Jones components (AXP; BAC; JPM; CSCO; HPQ; IBM; INTC; MSFT) between July 1, 2011, and June 30, 2013.

(i) Compute the weekly and monthly log returns of these eight stocks. Recall that the log return between t_1 and t_2 of an asset with price $S(t)$ is $\ln\left(\frac{S(t_2)}{S(t_1)}\right)$.

(ii) Compute the sample covariance matrix and the sample correlation matrix of the weekly and monthly log returns of the stocks, and compare them with the sample covariance and correlation matrices of the weekly and monthly percentage returns computed in an example in the corresponding chapter of Stefanica [3].

12. Let X_1, X_2, ..., X_n be nonconstant random variables, and let Y_1, Y_2, ..., Y_n be the random variables given by $Y_i = d_i X_i$, $i = 1:n$, where $d_i \neq 0$, $i = 1:n$, are constants. Denote by $\Sigma_\mathbf{X}$, $\Omega_\mathbf{X}$ and $\Sigma_\mathbf{Y}$, $\Omega_\mathbf{Y}$ the covariance and correlation matrices of X_1, X_2, ..., X_n and of Y_1, Y_2, ..., Y_n, respectively, and let $D = \mathrm{diag}(d_i)_{i=1:n}$.

(i) Show that $\Sigma_\mathbf{Y} = D\Sigma_\mathbf{X}D$.

(ii) If $d_i > 0$ for all $i = 1:n$, show that $\Omega_\mathbf{Y} = \Omega_\mathbf{X}$.

13. Let
$$\begin{pmatrix} X_1 \\ X_2 \\ X_3 \end{pmatrix} \sim N\left(\begin{pmatrix} 1 \\ -2 \\ 1 \end{pmatrix}, \begin{pmatrix} 1 & -1 & 0 \\ -1 & 3 & -2 \\ 0 & -2 & 3 \end{pmatrix} \right)$$
be a 3–dimensional multivariate normal random variable.

Find the probability that $X_1 + 2X_2 + 2X_3$ is positive.

14. Let
$$\begin{pmatrix} X_1 \\ X_2 \\ X_3 \end{pmatrix} \sim N\left(\begin{pmatrix} 2 \\ -1 \\ -2 \end{pmatrix}, \begin{pmatrix} 2 & 0 & 0.5 \\ 0 & 1 & -0.25 \\ 0.5 & -0.25 & 1 \end{pmatrix} \right)$$
be a 3–dimensional multivariate normal random variable, and let
$$Y_1 = X_1; \quad Y_2 = X_1 + X_2; \quad Y_3 = X_1 + X_2 + X_3.$$

(i) What are the mean vector and the covariance matrix of $\begin{pmatrix} Y_1 \\ Y_2 \\ Y_3 \end{pmatrix}$?

(ii) Find the probability that $Y_1 + Y_2 + Y_3$ is negative.

15. Let
$$\begin{pmatrix} X_1 \\ X_2 \\ X_3 \end{pmatrix} \sim N\left(\begin{pmatrix} 1 \\ 0 \\ -2 \end{pmatrix}, \begin{pmatrix} 2 & 1 & 0.5 \\ 1 & 3 & -1.5 \\ 0.5 & -1.5 & 4 \end{pmatrix} \right)$$

be a 3–dimensional multivariate normal random variable.

(i) What are the expected value and the covariance matrix of $\begin{pmatrix} X_1 - X_2 \\ X_2 - X_3 \\ X_3 - X_1 \end{pmatrix}$?

(ii) What is the probability density function of $\begin{pmatrix} X_1 - X_2 \\ X_2 - X_3 \\ X_3 - X_1 \end{pmatrix}$?

16. Let Z_1, Z_2, Z_3 be three independent standard normal variables, and let X_1, X_2, X_3 and Y_1, Y_2, Y_3 be normal random variables given by

$$X_1 = Z_3; \quad X_2 = Z_1 + 2Z_2 - 2Z_3; \quad X_3 = -3Z_1 + Z_2 + Z_3;$$

$$Y_1 = Z_1; \quad Y_2 = -2Z_1 + \sqrt{5}Z_2; \quad Y_3 = Z_1 - \frac{1}{\sqrt{5}}Z_2 + \sqrt{9.8}Z_3.$$

Show that the covariance matrix of X_1, X_2, X_3 and the covariance matrix of Y_1, Y_2, Y_3 are equal.

17. Given Z_1 and Z_2 independent standard normal variables, find two normal random variables X_1 and X_2 with covariance matrix $\begin{pmatrix} 4 & -1 \\ -1 & 9 \end{pmatrix}$.

18. Given Z_1 and Z_2 independent standard normal variables, find two normal random variables X_1 and X_2 with correlation matrix $\begin{pmatrix} 1 & 0.25 \\ 0.25 & 1 \end{pmatrix}$.

19. Given Z_1, Z_2, Z_3 independent standard normal variables, find three normal random variables X_1, X_2, X_3 with covariance matrix $\begin{pmatrix} 1 & 1 & 0.5 \\ 1 & 4 & -2 \\ 0.5 & -2 & 9 \end{pmatrix}$.

20. Given Z_1, Z_2, Z_3 independent standard normal variables, find three normal random variables X_1, X_2, X_3 with correlation matrix $\begin{pmatrix} 1 & 0.3 & 0.4 \\ 0.3 & 1 & 0.5 \\ 0.4 & 0.5 & 1 \end{pmatrix}$.

7.2 Solutions to Chapter 7 Exercises

Problem 1: Let X_1 and X_2 be random variables with correlation matrix $\begin{pmatrix} 1 & \rho \\ \rho & 1 \end{pmatrix}$, where $-1 \le \rho \le 1$. Let $Y_1 = X_1$ and $Y_2 = -X_2$. Show that the correlation matrix of Y_1 and Y_2 is $\begin{pmatrix} 1 & -\rho \\ -\rho & 1 \end{pmatrix}$.

Solution: If the correlation matrix of X_1 and X_2 is $\begin{pmatrix} 1 & \rho \\ \rho & 1 \end{pmatrix}$, then $\mathrm{corr}(X_1, X_2) = \rho$ and therefore

$$\mathrm{corr}(Y_1, Y_2) = \mathrm{corr}(X_1, -X_2) = -\mathrm{corr}(X_1, X_2) = -\rho.$$

Thus, the correlation matrix of Y_1 and Y_2 is

$$\begin{pmatrix} 1 & \mathrm{corr}(Y_1, Y_2) \\ \mathrm{corr}(Y_1, Y_2) & 1 \end{pmatrix} = \begin{pmatrix} 1 & -\rho \\ -\rho & 1 \end{pmatrix}. \quad \square$$

Problem 2: What is the correlation matrix of three random variables whose covariance matrix is $\begin{pmatrix} 1 & 0.36 & -1.44 \\ 0.36 & 4 & 0.80 \\ -1.44 & 0.80 & 9 \end{pmatrix}$?

Solution: Let $\Sigma_{\mathbf{X}} = \begin{pmatrix} 1 & 0.36 & -1.44 \\ 0.36 & 4 & 0.80 \\ -1.44 & 0.80 & 9 \end{pmatrix}$. The standard deviations of the random variables with covariance matrix $\Sigma_{\mathbf{X}}$ are

$$\sigma_1 = \sqrt{\Sigma_{\mathbf{X}}(1,1)} = 1; \quad \sigma_2 = \sqrt{\Sigma_{\mathbf{X}}(2,2)} = 2; \quad \sigma_3 = \sqrt{\Sigma_{\mathbf{X}}(2,2)} = 3.$$

The correlation matrix $\Omega_{\mathbf{X}}$ of the three random variables is

$$\Omega_{\mathbf{X}} = \begin{pmatrix} \frac{1}{\sigma_1} & 0 & 0 \\ 0 & \frac{1}{\sigma_2} & 0 \\ 0 & 0 & \frac{1}{\sigma_3} \end{pmatrix} \Sigma_{\mathbf{X}} \begin{pmatrix} \frac{1}{\sigma_1} & 0 & 0 \\ 0 & \frac{1}{\sigma_2} & 0 \\ 0 & 0 & \frac{1}{\sigma_3} \end{pmatrix}$$

$$= \begin{pmatrix} 1 & 0 & 0 \\ 0 & \frac{1}{2} & 0 \\ 0 & 0 & \frac{1}{3} \end{pmatrix} \begin{pmatrix} 1 & 0.36 & -1.44 \\ 0.36 & 4 & 0.80 \\ -1.44 & 0.80 & 9 \end{pmatrix} \begin{pmatrix} 1 & 0 & 0 \\ 0 & \frac{1}{2} & 0 \\ 0 & 0 & \frac{1}{3} \end{pmatrix}$$

$$= \begin{pmatrix} 1 & 0.18 & -0.48 \\ 0.18 & 1 & 0.1333 \\ -0.48 & 0.1333 & 1 \end{pmatrix}. \quad \square$$

Problem 3: Find all the values of ρ such that the matrix $\begin{pmatrix} 1 & 0.6 & -0.3 \\ 0.6 & 1 & \rho \\ -0.3 & \rho & 1 \end{pmatrix}$ is a correlation matrix.

Solution: A symmetric matrix with diagonal entries equal to 1 is a correlation matrix if and only if the matrix is symmetric positive semidefinite. Thus, we need to find all the values of ρ such that the matrix

$$\Omega = \begin{pmatrix} 1 & 0.6 & -0.3 \\ 0.6 & 1 & \rho \\ -0.3 & \rho & 1 \end{pmatrix} \tag{7.1}$$

is symmetric positive semidefinite.

Recall from Sylvester's criterion that a matrix is symmetric positive semidefinite if and only if all its principal minors are greater than or equal to 0.

Also, recall that the principal minors of a matrix are the determinants of all the square matrices obtained by eliminating the same rows and columns from the matrix. In particular, the matrix Ω from (7.1) has the following principal minors:

$$\det(1) = 1; \quad \det(1) = 1; \quad \det(1) = 1;$$

$$\det\begin{pmatrix} 1 & 0.6 \\ 0.6 & 1 \end{pmatrix} = 0.64; \quad \det\begin{pmatrix} 1 & -0.3 \\ -0.3 & 1 \end{pmatrix} = 0.91; \quad \det\begin{pmatrix} 1 & \rho \\ \rho & 1 \end{pmatrix} = 1 - \rho^2;$$

$$\begin{aligned} \det(\Omega) &= 1 - 0.18\rho - 0.18\rho - 0.09 - 0.36 - \rho^2 \\ &= 0.55 - 0.36\rho - \rho^2. \end{aligned}$$

Thus, it follows from Sylvester's criterion that Ω is a symmetric positive semidefinite matrix if and only if

$$\begin{aligned} 1 - \rho^2 &\geq 0; \\ 0.55 - 0.36\rho - \rho^2 &\geq 0, \end{aligned}$$

which is equivalent to $-1 \leq \rho \leq 1$ and

$$\rho^2 + 0.36\rho - 0.55 \leq 0. \tag{7.2}$$

Since the roots of the quadratic equation corresponding to (7.2) are -0.9432 and 0.5832, we conclude that the matrix Ω is symmetric positive semidefinite, and therefore a correlation matrix, if and only if

$$-0.9432 \leq \rho \leq 0.5832. \quad \square \tag{7.3}$$

Problem 4: Let X_1, X_2, X_3 be random variables such that

$$\operatorname{corr}(X_1, X_3) = 0.3; \quad \operatorname{corr}(X_2, X_3) = 0.1.$$

Find upper and lower bounds for $\rho = \operatorname{corr}(X_1, X_2)$.

Solution: Denote by

$$\Omega = \begin{pmatrix} 1 & \rho & 0.3 \\ \rho & 1 & 0.1 \\ 0.3 & 0.1 & 1 \end{pmatrix} \tag{7.4}$$

the correlation matrix of the random variables X_1, X_2, X_3.

A symmetric matrix with diagonal entries equal to 1 is a correlation matrix if and only if the matrix is symmetric positive semidefinite, i.e., if and only if all the principal minors of the matrix are greater than or equal to 0 (Sylvester's criterion). Thus, we need to find all the values of ρ such that all the principal minors of the matrix Ω from (7.4) are greater than or equal to 0.

The principal minors of a matrix are the determinants of all the square matrices obtained by eliminating the same rows and columns from the matrix. In particular, the matrix Ω from (7.4) has the following principal minors:

$$\det(1) = 1; \quad \det(1) = 1; \quad \det(1) = 1;$$

$$\det \begin{pmatrix} 1 & \rho \\ \rho & 1 \end{pmatrix} = 1 - \rho^2; \quad \det \begin{pmatrix} 1 & 0.3 \\ 0.3 & 1 \end{pmatrix} = 0.91; \quad \det \begin{pmatrix} 1 & 0.1 \\ 0.1 & 1 \end{pmatrix} = 0.99;$$

$$\begin{aligned}
\det(\Omega) &= \det \begin{pmatrix} 1 & \rho & 0.3 \\ \rho & 1 & 0.1 \\ 0.3 & 0.1 & 1 \end{pmatrix} \\
&= 1 + 0.1 \cdot 0.3 \cdot \rho + 0.1 \cdot 0.3 \cdot \rho - (0.3)^2 - (0.1)^2 - \rho^2 \\
&= 0.9 + 0.06\rho - \rho^2.
\end{aligned}$$

Thus, the matrix Ω is a symmetric positive semidefinite matrix if and only if

$$\begin{aligned}
1 - \rho^2 &\geq 0; \\
0.9 + 0.06\rho - \rho^2 &\geq 0,
\end{aligned}$$

which is equivalent to $-1 \leq \rho \leq 1$ and

$$\rho^2 - 0.06\rho - 0.9 \leq 0. \tag{7.5}$$

Since the roots of the quadratic equation corresponding to (7.5) are -0.9192 and 0.9792, we conclude that the matrix Ω is symmetric positive semidefinite, and therefore a correlation matrix, if and only if

$$-0.9192 \leq \rho \leq 0.9792. \quad \square$$

Problem 5: Consider three random variables with pairwise correlation ρ, i.e., such that the correlation between any two of the random variables is ρ. What is the smallest possible value for ρ?

Solution: Denote by

$$\Omega = \begin{pmatrix} 1 & \rho & \rho \\ \rho & 1 & \rho \\ \rho & \rho & 1 \end{pmatrix} \tag{7.6}$$

the correlation matrix of the three random variables.

Recall from Sylvester's criterion that a symmetric matrix with diagonal entries equal to 1 is a correlation matrix if and only if the matrix is symmetric positive semidefinite, i.e., if and only if all the principal minors of the matrix are greater than or equal to 0. Thus, we need to find the smallest value of ρ such that all the principal minors of the matrix Ω are greater than or equal to 0.

The principal minors of a matrix are the determinants of all the square matrices obtained by eliminating the same rows and columns from the matrix. In particular, the matrix Ω has the following principal minors:

$$\det(1) = 1; \quad \det \begin{pmatrix} 1 & \rho \\ \rho & 1 \end{pmatrix} = 1 - \rho^2;$$

$$\begin{aligned} \det(\Omega) &= 1 + 2\rho^3 - 3\rho^2 = (1 - \rho)(1 + \rho - 2\rho^2) \\ &= (1 - \rho)^2 (1 + 2\rho). \end{aligned}$$

Thus, the matrix Ω is a symmetric positive semidefinite matrix if and only if

$$\begin{cases} 1 - \rho^2 & \geq 0 \\ (1 - \rho)^2(1 + 2\rho) & \geq 0 \end{cases} \iff \begin{cases} -1 \leq \rho \leq 1 \\ 1 + 2\rho \geq 0 \end{cases} \iff \begin{cases} -1 \leq \rho \leq 1; \\ \rho \geq -\frac{1}{2}. \end{cases}$$

We conclude that $\rho = -\frac{1}{2}$ is the smallest value of ρ such that the matrix Ω is symmetric positive semidefinite, and therefore a correlation matrix.

Note that, in Problem 7 from this chapter, we show that an $n \times n$ matrix with main diagonal entries equal to 1 and with all the other entries equal to ρ is a correlation matrix if and only if $-\frac{1}{n-1} \leq \rho \leq 1$. For $n = 3$, it follows that the smallest value of ρ such that the matrix Ω given by (7.6) is a correlation matrix is $\rho = -\frac{1}{3-1} = -\frac{1}{2}$.

□

Problem 6: Let X_1, X_2, X_3, X_4 be random variables such that

$$\operatorname{corr}(X_1, X_3) = 0.3; \quad \operatorname{corr}(X_1, X_4) = 0.2; \quad \operatorname{corr}(X_2, X_3) = 0.1;$$

$$\operatorname{corr}(X_2, X_4) = -0.1; \quad \operatorname{corr}(X_3, X_4) = -0.2.$$

Find upper and lower bounds for $\rho = \operatorname{corr}(X_1, X_2)$.

Solution: Denote by

$$\Omega = \begin{pmatrix} 1 & \rho & 0.3 & 0.2 \\ \rho & 1 & 0.1 & -0.1 \\ 0.3 & 0.1 & 1 & -0.2 \\ 0.2 & -0.1 & -0.2 & 1 \end{pmatrix} \tag{7.7}$$

the correlation matrix of the random variables X_1, X_2, X_3, X_4.

From Sylvester's criterion, it follows that a symmetric matrix with diagonal entries equal to 1 is a correlation matrix if and only if the matrix is symmetric positive semidefinite, i.e., if and only if all the principal minors of the matrix are greater than or equal to 0. Thus, we need to find all the values of ρ such that all the principal minors of the matrix Ω from (7.7) are greater than or equal to 0.

The principal minors of a matrix are the determinants of all the square matrices obtained by eliminating the same rows and columns from the matrix. Then, the principal minors of the matrix Ω from (7.7) are

$$\det(1) = 1; \quad \det(1) = 1; \quad \det(1) = 1; \quad \det(1) = 1;$$

$$\det \begin{pmatrix} 1 & \rho \\ \rho & 1 \end{pmatrix} = 1 - \rho^2; \quad \det \begin{pmatrix} 1 & 0.3 \\ 0.3 & 1 \end{pmatrix} = 0.91;$$

$$\det \begin{pmatrix} 1 & 0.2 \\ 0.2 & 1 \end{pmatrix} = 0.96; \quad \det \begin{pmatrix} 1 & 0.1 \\ 0.1 & 1 \end{pmatrix} = 0.99;$$

$$\det \begin{pmatrix} 1 & -0.1 \\ -0.1 & 1 \end{pmatrix} = 0.99; \quad \det \begin{pmatrix} 1 & -0.2 \\ -0.2 & 1 \end{pmatrix} = 0.96.$$

$$\det \begin{pmatrix} 1 & \rho & 0.3 \\ \rho & 1 & 0.1 \\ 0.3 & 0.1 & 1 \end{pmatrix} = 1 + 2 \cdot 0.1 \cdot 0.3 \cdot \rho - (0.3)^2 - (0.1)^2 - \rho^2$$

$$= 0.9 + 0.06\rho - \rho^2$$

$$\det \begin{pmatrix} 1 & \rho & 0.2 \\ \rho & 1 & -0.1 \\ 0.2 & -0.1 & 1 \end{pmatrix} = 1 + 2 \cdot (-0.1) \cdot 0.2 \cdot \rho + -(0.2)^2 - (-0.1)^2 - \rho^2$$

$$= 0.95 - 0.04\rho - \rho^2;$$

$$\det \begin{pmatrix} 1 & 0.3 & 0.2 \\ 0.3 & 1 & -0.2 \\ 0.2 & -0.2 & 1 \end{pmatrix} = 0.806;$$

$$\det \begin{pmatrix} 1 & 0.1 & -0.1 \\ 0.1 & 1 & -0.2 \\ -0.1 & -0.2 & 1 \end{pmatrix} = 0.944.$$

$$\det(\Omega)$$

$$= \det \begin{pmatrix} 1 & \rho & 0.3 & 0.2 \\ \rho & 1 & 0.1 & -0.1 \\ 0.3 & 0.1 & 1 & -0.2 \\ 0.2 & -0.1 & -0.2 & 1 \end{pmatrix}$$

$$= 1 \cdot \det \begin{pmatrix} 1 & 0.1 & -0.1 \\ 0.1 & 1 & -0.2 \\ -0.1 & -0.2 & 1 \end{pmatrix} - \rho \cdot \det \begin{pmatrix} \rho & 0.1 & -0.1 \\ 0.3 & 1 & -0.2 \\ 0.2 & -0.2 & 1 \end{pmatrix}$$

$$+ 0.3 \cdot \det \begin{pmatrix} \rho & 1 & -0.1 \\ 0.3 & 0.1 & -0.2 \\ 0.2 & -0.1 & 1 \end{pmatrix} - 0.2 \cdot \det \begin{pmatrix} \rho & 1 & 0.1 \\ 0.3 & 0.1 & 1 \\ 0.2 & -0.1 & -0.2 \end{pmatrix}$$

$$= 0.944 - \rho(0.96\rho - 0.008)$$

$$\quad + 0.3(0.08\rho - 0.335) - 0.2(0.08\rho + 0.255)$$

$$= 0.7925 + 0.016\rho - 0.96\rho^2.$$

Thus, the matrix Ω is a symmetric positive semidefinite matrix if and only if

$$
\begin{cases}
1 - \rho^2 & \geq \ 0 \\
0.9 + 0.06\rho - \rho^2 & \geq \ 0 \\
0.95 - 0.04\rho - \rho^2 & \geq \ 0 \\
0.7925 + 0.016\rho - 0.96\rho^2 & \geq \ 0
\end{cases}
$$

$$
\Longleftrightarrow \quad
\begin{cases}
\rho^2 & \leq \ 1 \\
\rho^2 - 0.06\rho - 0.9 & \leq \ 0 \\
\rho^2 + 0.04\rho - 0.95 & \leq \ 0 \\
0.96\rho^2 - 0.016\rho - 0.7925 & \leq \ 0
\end{cases}
$$

$$
\Longleftrightarrow \quad
\begin{cases}
-1 & \leq \ \rho \ \leq & 1 \\
-0.9192 & \leq \ \rho \ \leq & 0.9792 \\
-0.9949 & \leq \ \rho \ \leq & 0.9549 \\
-0.9003 & \leq \ \rho \ \leq & 0.9170
\end{cases}
$$

$$
\Longleftrightarrow \quad -0.9003 \leq \rho \leq 0.9170,
$$

since, e.g., the roots of the quadratic equation $\rho^2 + 0.04\rho - 0.95$ are -0.9949 and 0.9549.

We conclude that the matrix Ω is symmetric positive semidefinite, and therefore a correlation matrix, if and only if

$$
-0.9003 \leq \rho \leq 0.9170. \quad \square
$$

Problem 7: Assume that all the entries of an $n \times n$ correlation matrix which are not on the main diagonal are equal to ρ. Find upper and lower bounds on the possible values of ρ.

Solution: Recall that a symmetric matrix with diagonal entries equal to 1 is a correlation matrix if and only if the matrix is symmetric positive semidefinite, i.e., if and only if all the eigenvalues of the matrix are greater than or equal to 0.
    Let

$$
\Omega = \begin{pmatrix}
1 & \rho & \cdots & \rho \\
\rho & \ddots & \ddots & \vdots \\
\vdots & \ddots & \ddots & \rho \\
\rho & \cdots & \rho & 1
\end{pmatrix}.
$$

We include two ways to compute the eigenvalues of Ω, which are then used to find the necessary and sufficient conditions for the matrix Ω to be a correlation matrix.

Solution 1: Note that

$$
\Omega = \begin{pmatrix}
1-\rho & 0 & \cdots & 0 \\
0 & \ddots & \ddots & \vdots \\
\vdots & \ddots & \ddots & 0 \\
0 & \cdots & 0 & 1-\rho
\end{pmatrix}
+
\begin{pmatrix}
\rho & \rho & \cdots & \rho \\
\rho & \ddots & \ddots & \vdots \\
\vdots & \ddots & \ddots & \rho \\
\rho & \cdots & \rho & \rho
\end{pmatrix}
$$

$$= (1-\rho) \begin{pmatrix} 1 & 0 & \cdots & 0 \\ 0 & \ddots & \ddots & \vdots \\ \vdots & \ddots & \ddots & 0 \\ 0 & \cdots & 0 & 1 \end{pmatrix} + \rho \begin{pmatrix} 1 & 1 & \cdots & 1 \\ 1 & \ddots & \ddots & \vdots \\ \vdots & \ddots & \ddots & 1 \\ 1 & \cdots & 1 & 1 \end{pmatrix}$$

$$= (1-\rho)I + \rho M,$$

where M is the $n \times n$ matrix with all entries equal to 1, and I is the $n \times n$ identity matrix.

Let λ and $v = (v_i)_{i=1:n}$ be an eigenvalue and a corresponding eigenvector of M, i.e., $Mv = \lambda v$, with $v \neq 0$. Note that $Mv = \lambda v$ can be written as

$$\begin{cases} v_1 + v_2 + \ldots + v_n &= \lambda v_1; \\ v_1 + v_2 + \ldots + v_n &= \lambda v_2; \\ \qquad \vdots & \quad \vdots \\ v_1 + v_2 + \ldots + v_n &= \lambda v_n, \end{cases} \tag{7.8}$$

and therefore

$$\lambda v_1 = \lambda v_2 = \ldots = \lambda v_n$$

Thus, either $\lambda = 0$ or $v_1 = v_2 = \ldots = v_n$. If $v_1 = v_2 = \ldots = v_n$, it follows from (7.8) that $nv_1 = \lambda v_1$, and therefore that $\lambda = n$, since $v = (v_i)_{i=1:n} \neq 0$.

In other words, the eigenvalues of the matrix M are $\lambda = 0$ and $\lambda = n$.

Note that

$$\Omega v = ((1-\rho)I + \rho M)v = (1-\rho)v + \rho M v.$$

Then,

$$Mv = \lambda v \iff \Omega v = (1-\rho)v + \rho \lambda v$$
$$\iff \Omega v = (1-\rho+\rho\lambda)v.$$

Thus, λ and v are an eigenvalue and a corresponding eigenvector of the matrix M if and only if $\mu = 1 - \rho + \rho\lambda$ and v are an eigenvalue and a corresponding eigenvector of Ω.

Since the eigenvalues of M are $\lambda = 0$ and $\lambda = n$, it follows that the eigenvalues of Ω are $\mu = 1 - \rho$, corresponding to $\lambda = 0$, and $\mu = (1-\rho) + n\rho = 1 + (n-1)\rho$, corresponding to $\lambda = n$.[1]

Since Ω is a correlation matrix if and only if all its eigenvalues are greater than or equal to 0, we conclude that the matrix Ω is a correlation matrix if and only if

$$0 \leq 1-\rho \quad \text{and} \quad 0 \leq 1+(n-1)\rho,$$

which is equivalent to

$$-\frac{1}{n-1} \leq \rho \leq 1. \tag{7.9}$$

---

[1] The eigenvalue $1 - \rho$ of Ω has multiplicity $n - 1$, and the eigenvalue $1 + (n-1)\rho$ of Ω has multiplicity 1; see Solution 2 of this problem for details.

Solution 2: Note that

$$\Omega = (1-\rho)I + \rho \begin{pmatrix} 1 & 1 & \cdots & 1 \\ 1 & \ddots & \ddots & \vdots \\ \vdots & \ddots & \ddots & 1 \\ 1 & \cdots & 1 & 1 \end{pmatrix}$$

$$= (1-\rho)I + \rho \begin{pmatrix} 1 \\ \vdots \\ 1 \end{pmatrix} (1 \ \cdots \ 1)$$

$$= (1-\rho)I + \rho \mathbf{1}\mathbf{1}^t$$
$$= (1-\rho)I + \rho A,$$

where I is the $n \times n$ identity matrix, $\mathbf{1} = (1)_{i=1:n}$ is the $n \times 1$ column vector of size n with all entries equal to 1, and $A = \mathbf{1}\mathbf{1}^t$ is an $n \times n$ matrix.

Recall that[2] an $n \times n$ matrix of the form uu^t, where $u = (u_i)_{i=1:n}$ is an $n \times 1$ column vector, has an eigenvalue equal to $\sum_{i=1}^{n} u_i^2$ with multiplicity 1 and another eigenvalue equal to 0 with multiplicity $n - 1$.

Then, the eigenvalues of the matrix $A = \mathbf{1}\mathbf{1}^t$ are:
• $\lambda = \sum_{i=1}^{n} 1 = n$ with multiplicity 1;
• $\lambda = 0$ with multiplicity $n - 1$.

Note that, if λ and v are an eigenvalue and a corresponding eigenvector of A, then $Av = \lambda v$, and therefore

$$\Omega v = (1-\rho)v + \rho Av = (1-\rho)v + \rho \lambda v$$
$$= (1 - \rho + \rho\lambda)v.$$

Thus, $1 - \rho + \rho\lambda$ and v are an eigenvalue and corresponding eigenvector of Ω, and we obtain that the matrix Ω has the following eigenvalues:
• $(1-\rho) + n\rho = 1 + (n-1)\rho$ with multiplicity 1;
• $1 - \rho$ with multiplicity $n - 1$.

As before, since Ω is a correlation matrix if and only if all its eigenvalues are greater than or equal to 0, we conclude that the matrix Ω is a correlation matrix if and only if

$$0 \le 1 + (n-1)\rho \quad \text{and} \quad 0 \le 1 - \rho,$$

which is equivalent to

$$-\frac{1}{n-1} \le \rho \le 1,$$

which is the same as (7.9). □

Problem 8: Show that the matrix $\begin{pmatrix} 1 & 0.1 & 0.2 \\ 0.1 & 1 & -0.3 \\ 0.2 & -0.3 & 1 \end{pmatrix}$ is a correlation matrix, and find its Cholesky factor.

---

[2]See Problem 9 from Chapter 4.

Solution: Let

$$\Omega = \begin{pmatrix} 1 & 0.1 & 0.2 \\ 0.1 & 1 & -0.3 \\ 0.2 & -0.3 & 1 \end{pmatrix}.$$

Recall that a symmetric matrix with diagonal entries equal to 1 is a correlation matrix if and only if the matrix is symmetric positive semidefinite, i.e., if and only if all the principal minors of the matrix are greater than or equal to 0; cf. Sylvester's criterion.

The principal minors of the matrix Ω are

$$\det(1) = 1; \quad \det(1) = 1; \quad \det(1) = 1;$$

$$\det\begin{pmatrix} 1 & 0.1 \\ 0.1 & 1 \end{pmatrix} = 0.99; \quad \det\begin{pmatrix} 1 & 0.2 \\ 0.2 & 1 \end{pmatrix} = 0.96; \quad \det\begin{pmatrix} 1 & -0.3 \\ -0.3 & 1 \end{pmatrix} = 0.91;$$

$$\det(\Omega) = 1 + 2(0.1)(-0.3)(0.2) - (0.2)^2 - (0.1)^2 - (-0.3)^2 = 0.848.$$

Thus, all the principal minors of Ω are greater than or equal to 0, and we conclude that Ω is a correlation matrix.

Moreover, note that the leading principal minors of the matrix Ω, i.e.,

$$\det(1) = 1; \quad \det\begin{pmatrix} 1 & 0.1 \\ 0.1 & 1 \end{pmatrix} = 0.99; \quad \det(\Omega) = 0.848,$$

are positive. Then, from Sylvester's criterion, it follows that Ω is a symmetric positive definite matrix, and therefore Ω has a Cholesky decomposition. By running the Cholesky decomposition algorithm, we find that the Cholesky factor of the matrix Ω is

$$U = \begin{pmatrix} 1 & 0.1 & 0.2 \\ 0 & 0.995 & -0.3216 \\ 0 & 0 & 0.9255 \end{pmatrix}. \quad \square$$

Problem 9: Show that it is not possible to find three random variables on the same probability space with correlations 0.8, 0.7, and -0.5.

Solution: Assume that it would be possible to find three random variables with correlations 0.8, 0.7, and -0.5, and denote by

$$\Omega = \begin{pmatrix} 1 & 0.8 & 0.7 \\ 0.8 & 1 & -0.5 \\ 0.7 & -0.5 & 1 \end{pmatrix} \tag{7.10}$$

the correlation matrix of these random variables.

Solution 1: Recall that any correlation matrix is symmetric positive semidefinite, i.e., all the eigenvalues of the matrix are greater than or equal to 0. Then, the determinant of the matrix, which is equal to the product of all the eigenvalues of the matrix, must be greater than or equal to 0.

However,

$$\det(\Omega) \;=\; \det \begin{pmatrix} 1 & 0.8 & 0.7 \\ 0.8 & 1 & -0.5 \\ 0.7 & -0.5 & 1 \end{pmatrix} \;=\; -0.94. \tag{7.11}$$

Thus, the matrix Ω has a negative determinant and therefore cannot be a correlation matrix.

Solution 2: Recall that any correlation matrix is symmetric positive semidefinite. Then, from Sylvester's criterion, it follows that all the principal minors of a correlation matrix are greater than or equal to 0.

Note that one of the principal minors of a matrix is the determinant of the matrix. Since $\det(\Omega) = -0.94$, see (7.11), we obtain the matrix Ω has at least one negative principal minors, and therefore cannot be a correlation matrix.   □

Problem 10: Consider three random variables given by the following time series data at five data points:

$$\begin{pmatrix} 0.25 & -0.50 & 1.50 \\ 1 & -1 & 1.25 \\ -0.50 & -0.25 & 2 \\ 0 & 0.50 & 0.75 \\ -1 & 0.75 & 1.50 \end{pmatrix}.$$

(i) Show that the time series vectors $\begin{pmatrix} 0.25 \\ 1 \\ -0.50 \\ 0 \\ -1 \end{pmatrix}$, $\begin{pmatrix} -0.50 \\ -1 \\ -0.25 \\ 0.50 \\ 0.75 \end{pmatrix}$, $\begin{pmatrix} 1.50 \\ 1.25 \\ 2 \\ 0.75 \\ 1.50 \end{pmatrix}$ of the three random variables are linearly independent.

(ii) Show that the vectors $\begin{pmatrix} 0.25 \\ 1 \\ -0.50 \\ 0 \\ -1 \end{pmatrix}$, $\begin{pmatrix} -0.50 \\ -1 \\ -0.25 \\ 0.50 \\ 0.75 \end{pmatrix}$, $\begin{pmatrix} 1.50 \\ 1.25 \\ 2 \\ 0.75 \\ 1.50 \end{pmatrix}$, $\begin{pmatrix} 1 \\ 1 \\ 1 \\ 1 \\ 1 \end{pmatrix}$ are not linearly independent.

(iii) Compute the sample covariance matrix of the three random variables and show that it is singular.

Solution: (i) We give a proof by contradiction.

Assume that the vectors $\begin{pmatrix} 0.25 \\ 1 \\ -0.50 \\ 0 \\ -1 \end{pmatrix}$, $\begin{pmatrix} -0.50 \\ -1 \\ -0.25 \\ 0.50 \\ 0.75 \end{pmatrix}$, and $\begin{pmatrix} 1.50 \\ 1.25 \\ 2 \\ 0.75 \\ 1.50 \end{pmatrix}$ are not linearly independent. Then, the matrix

$$T = \begin{pmatrix} 0.25 & -0.50 & 1.50 \\ 1 & -1 & 1.25 \\ -0.50 & -0.25 & 2 \\ 0 & 0.50 & 0.75 \\ -1 & 0.75 & 1.50 \end{pmatrix} \tag{7.12}$$

would have linearly dependent columns. Thus, the matrix

$$T(1:3, 1:3) = \begin{pmatrix} 0.25 & -0.50 & 1.50 \\ 1 & -1 & 1.25 \\ -0.50 & -0.25 & 2 \end{pmatrix}$$

made of the first three rows of the matrix T would also have linearly dependent columns and therefore would be singular and $\det(T(1:3, 1:3)) = 0$. However, by direct computation we find that $\det(T(1:3, 1:3)) = -0.23$, which is a contradiction. We conclude that the vectors $\begin{pmatrix} 0.25 \\ 1 \\ -0.50 \\ 0 \\ -1 \end{pmatrix}$, $\begin{pmatrix} -0.50 \\ -1 \\ -0.25 \\ 0.50 \\ 0.75 \end{pmatrix}$, and $\begin{pmatrix} 1.50 \\ 1.25 \\ 2 \\ 0.75 \\ 1.50 \end{pmatrix}$ are linearly independent.

(ii) Note that

$$\begin{pmatrix} 0.25 \\ 1 \\ -0.50 \\ 0 \\ -1 \end{pmatrix} + \begin{pmatrix} -0.50 \\ -1 \\ -0.25 \\ 0.50 \\ 0.75 \end{pmatrix} + \begin{pmatrix} 1.50 \\ 1.25 \\ 2 \\ 0.75 \\ 1.50 \end{pmatrix} = \begin{pmatrix} 1.25 \\ 1.25 \\ 1.25 \\ 1.25 \\ 1.25 \end{pmatrix}.$$

Thus,

$$\begin{pmatrix} 0.25 \\ 1 \\ -0.50 \\ 0 \\ -1 \end{pmatrix} + \begin{pmatrix} -0.50 \\ -1 \\ -0.25 \\ 0.50 \\ 0.75 \end{pmatrix} + \begin{pmatrix} 1.50 \\ 1.25 \\ 2 \\ 0.75 \\ 1.50 \end{pmatrix} - 1.25 \begin{pmatrix} 1 \\ 1 \\ 1 \\ 1 \\ 1 \end{pmatrix} = 0,$$

and therefore the vectors $\begin{pmatrix} 0.25 \\ 1 \\ -0.50 \\ 0 \\ -1 \end{pmatrix}$, $\begin{pmatrix} -0.50 \\ -1 \\ -0.25 \\ 0.50 \\ 0.75 \end{pmatrix}$, $\begin{pmatrix} 1.50 \\ 1.25 \\ 2 \\ 0.75 \\ 1.50 \end{pmatrix}$, $\begin{pmatrix} 1 \\ 1 \\ 1 \\ 1 \\ 1 \end{pmatrix}$ are not linearly independent.

(iii) The sample means of the three random variables with time series data given by the matrix T from (7.12) are -0.05, -0.10, and 1.40, respectively.[3] By subtracting the sample mean of each column of T, we obtain the following mean–normalized time series matrix of the three random variables:

$$\overline{T} = \begin{pmatrix} 0.25 & -0.50 & 1.50 \\ 1 & -1 & 1.25 \\ -0.50 & -0.25 & 2 \\ 0 & 0.50 & 0.75 \\ -1 & 0.75 & 1.50 \end{pmatrix} - \begin{pmatrix} -0.05 & -0.10 & 1.40 \\ -0.05 & -0.10 & 1.40 \\ -0.05 & -0.10 & 1.40 \\ -0.05 & -0.10 & 1.40 \\ -0.05 & -0.10 & 1.40 \end{pmatrix}$$

[3] For example, for the first random variable, the sample mean is $\frac{0.25+1-0.50+0-1}{5} = -0.05$.

$$
= \begin{pmatrix}
0.30 & -0.40 & 0.10 \\
1.05 & -0.90 & -0.15 \\
-0.45 & -0.15 & 0.60 \\
0.05 & 0.60 & -0.65 \\
-0.95 & 0.85 & 0.10
\end{pmatrix}.
\tag{7.13}
$$

The sample covariance matrix of the three random variables with $N = 5$ time series data points is

$$
\begin{aligned}
\Sigma &= \frac{1}{N-1} \overline{T}^t \overline{T} = \frac{1}{4} \overline{T}^t \overline{T} \\
&= \begin{pmatrix}
0.5750 & -0.4438 & -0.1312 \\
-0.4438 & 0.5188 & -0.0750 \\
-0.1312 & -0.0750 & 0.2062
\end{pmatrix},
\end{aligned}
$$

where the matrix $\overline{T}$ is given by (7.13).

By direct computation, we obtain that $\det(\Sigma) = 0$, and conclude that the sample covariance matrix of the three random variables is singular.

This exercise shows that random variables with linearly independent time series data do not necessarily have a nonsingular sample covariance matrix; the necessary and sufficient condition for the sample covariance matrix to be nonsingular is that the time series data column vectors and the column vector of the same size with all entries equal to 1 are linearly independent.   □

Problem 11: The file *data-DJ30-july2011-june2013.xlsx* from fepress.org/nla-primer contains the end of week and end of month closing prices for eight financial and technology Dow Jones components (AXP; BAC; JPM; CSCO; HPQ; IBM; INTC; MSFT) between July 1, 2011, and June 30, 2013.

(i) Compute the weekly and monthly log returns of these eight stocks. Recall that the log return between t_1 and t_2 of an asset with price $S(t)$ is $\ln\left(\frac{S(t_2)}{S(t_1)}\right)$.

(ii) Compute the sample covariance matrix and the sample correlation matrix of the weekly and monthly log returns of the stocks, and compare them with the sample covariance and correlation matrices of the weekly and monthly percentage returns computed in an example in the corresponding chapter of Stefanica [3].

Solution: (i) Let $n = 8$ be the number of stocks. The number of monthly log returns and the number of weekly log returns that can be computed from the given data are $N_m = 23$ and $N_w = 104$, respectively. The 23×8 matrix $T_{\mathbf{x},m}^{log}$ of the monthly log returns of the eight stocks can be found below; the 104×8 matrix $T_{\mathbf{x},w}^{log}$ of the weekly log returns is too large to be included here.

$$
\begin{pmatrix}
0.0127 & 0.0596 & 0.0334 & -0.0091 & -0.0212 & 0.0848 & 0.0021 & 0.0104 \\
-0.1014 & -0.1041 & -0.1076 & -0.1425 & -0.1702 & -0.0313 & -0.0229 & -0.0599 \\
-0.0169 & -0.0106 & -0.0385 & -0.0087 & 0.1460 & 0.0517 & -0.0922 & -0.1458 \\
-0.0820 & -0.0813 & 0.0303 & -0.0019 & -0.1748 & -0.0602 & -0.0449 & -0.0286 \\
-0.0552 & 0.0071 & -0.0389 & -0.0142 & -0.1989 & 0.0069 & -0.0034 & -0.0209 \\
-0.0263 & 0.0253 & -0.0747 & -0.0454 & -0.1470 & -0.0584 & -0.0201 & -0.0274 \\
-0.0280 & -0.1647 & -0.0679 & -0.0384 & -0.1024 & -0.0078 & -0.0523 & -0.0034 \\
0.0013 & -0.0554 & 0.0146 & -0.1052 & 0.0640 & 0.0188 & 0.0897 & 0.0614 \\
0.0124 & -0.0541 & -0.0367 & 0.1003 & 0.2084 & 0.0643 & 0.0466 & 0.0418 \\
0.0250 & -0.1015 & -0.0859 & -0.0011 & -0.0181 & -0.0626 & 0.0917 & 0.0352 \\
-0.0103 & -0.0852 & -0.0312 & -0.1793 & 0.0775 & 0.0015 & 0.0258 & -0.0515 \\
0.0053 & 0.1088 & -0.0158 & 0.0693 & 0.0978 & -0.0020 & 0.0362 & 0.0374 \\
-0.0417 & -0.1074 & -0.0750 & -0.0501 & 0.1145 & -0.0138 & -0.0309 & -0.0469 \\
0.0756 & 0.0975 & 0.2596 & 0.2106 & 0.0874 & 0.0668 & 0.0874 & 0.0859 \\
-0.0433 & 0.1652 & 0.0611 & 0.0442 & -0.0379 & 0.0075 & -0.0100 & 0.0074 \\
-0.0898 & -0.1827 & -0.1585 & -0.0623 & 0.0550 & -0.0588 & -0.0453 & -0.0162 \\
-0.0535 & -0.1134 & -0.0507 & -0.0116 & 0.1005 & -0.0251 & -0.0249 & -0.0788 \\
-0.0647 & -0.2489 & -0.1221 & -0.0867 & -0.0826 & -0.0463 & -0.0859 & -0.1289 \\
0.0183 & -0.0220 & -0.0710 & 0.0307 & 0.0770 & 0.0221 & 0.0271 & -0.0146 \\
0.0523 & 0.2261 & 0.1153 & -0.0056 & -0.0491 & -0.0221 & -0.0240 & 0.0329 \\
-0.1241 & -0.1091 & -0.1519 & -0.1828 & -0.1701 & -0.0543 & -0.1395 & -0.0678 \\
0.1018 & 0.2885 & 0.2208 & 0.0107 & 0.1426 & -0.0170 & -0.0586 & 0.0666 \\
0.0066 & 0.1720 & 0.0740 & 0.0192 & 0.3011 & 0.0518 & 0.0939 & 0.0231
\end{pmatrix}
$$

(ii) Let $\overline{T}^{log}_{\mathbf{x},w}$ be the mean normalized time series matrix corresponding to the matrix $T^{log}_{\mathbf{x},w}$ of the weekly log returns of the assets. The sample covariance matrix $\widehat{\Sigma}^{log}_{\mathbf{x},w}$ and the sample correlation matrix $\widehat{\Omega}^{log}_{\mathbf{x},w}$ of the weekly log returns of the assets are

$$
\widehat{\Sigma}^{log}_{\mathbf{x},w} = \frac{1}{N_w - 1} \, (\overline{T}^{log}_{\mathbf{x},w})^t \, \overline{T}^{log}_{\mathbf{x},w} ;
$$

$$
\widehat{\Omega}^{log}_{\mathbf{x},w} = \mathrm{diag}\left(\frac{1}{\sqrt{\widehat{\Sigma}^{log}_{\mathbf{x},w}(k)}} \right)_{k=1:n} (\overline{T}^{log}_{\mathbf{x},w})^t \, \overline{T}^{log}_{\mathbf{x},w} \, \mathrm{diag}\left(\frac{1}{\sqrt{\widehat{\Sigma}^{log}_{\mathbf{x},w}(k)}} \right)_{k=1:n} ,
$$

and are given by

$$
\widehat{\Sigma}^{log}_{\mathbf{x},w} =
\begin{pmatrix}
0.0010 & 0.0013 & 0.0010 & 0.0005 & 0.0007 & 0.0004 & 0.0006 & 0.0005 \\
0.0013 & 0.0036 & 0.0021 & 0.0010 & 0.0012 & 0.0009 & 0.0009 & 0.0009 \\
0.0010 & 0.0021 & 0.0019 & 0.0009 & 0.0011 & 0.0007 & 0.0007 & 0.0007 \\
0.0005 & 0.0010 & 0.0009 & 0.0015 & 0.0009 & 0.0006 & 0.0005 & 0.0005 \\
0.0007 & 0.0012 & 0.0011 & 0.0009 & 0.0036 & 0.0009 & 0.0010 & 0.0006 \\
0.0004 & 0.0009 & 0.0007 & 0.0006 & 0.0009 & 0.0009 & 0.0005 & 0.0004 \\
0.0006 & 0.0009 & 0.0007 & 0.0005 & 0.0010 & 0.0005 & 0.0011 & 0.0005 \\
0.0005 & 0.0009 & 0.0007 & 0.0005 & 0.0006 & 0.0004 & 0.0005 & 0.0009
\end{pmatrix}
$$

$$
\widehat{\Omega}^{log}_{\mathbf{x},w} =
\begin{pmatrix}
1 & 0.6895 & 0.6993 & 0.3721 & 0.3482 & 0.4750 & 0.5258 & 0.5570 \\
0.6895 & 1 & 0.8075 & 0.4232 & 0.3206 & 0.4856 & 0.4575 & 0.4864 \\
0.6993 & 0.8075 & 1 & 0.5515 & 0.4215 & 0.5272 & 0.4802 & 0.5149 \\
0.3721 & 0.4232 & 0.5515 & 1 & 0.4020 & 0.5205 & 0.4237 & 0.4509 \\
0.3482 & 0.3206 & 0.4215 & 0.4020 & 1 & 0.4953 & 0.5191 & 0.3260 \\
0.4750 & 0.4856 & 0.5272 & 0.5205 & 0.4953 & 1 & 0.5409 & 0.4813 \\
0.5258 & 0.4575 & 0.4802 & 0.4237 & 0.5191 & 0.5409 & 1 & 0.5129 \\
0.5570 & 0.4864 & 0.5149 & 0.4509 & 0.3260 & 0.4813 & 0.5129 & 1
\end{pmatrix}
$$

Let $\overline{T}^{log}_{\mathbf{x},m}$ be the mean normalized time series matrix corresponding to the matrix $T^{log}_{\mathbf{x},m}$ of the monthly log returns of the assets. The sample covariance matrix $\widehat{\Sigma}^{log}_{\mathbf{x},m}$ and the sample correlation matrix $\widehat{\Omega}^{log}_{\mathbf{x},m}$ of the monthly log returns of the assets are

$$\widehat{\Sigma}^{log}_{\mathbf{x},m} = \frac{1}{N_m - 1} (\overline{T}^{log}_{\mathbf{x},m})^t \, \overline{T}^{log}_{\mathbf{x},m};$$

$$\widehat{\Omega}^{log}_{\mathbf{x},m} = \text{diag}\left(\frac{1}{\sqrt{\widehat{\Sigma}^{log}_{\mathbf{x},m}(k)}}\right)_{k=1:n} (\overline{T}^{log}_{\mathbf{x},m})^t \, \overline{T}^{log}_{\mathbf{x},m} \, \text{diag}\left(\frac{1}{\sqrt{\widehat{\Sigma}^{log}_{\mathbf{x},m}(k)}}\right)_{k=1:n},$$

and are given by

$$\widehat{\Sigma}^{log}_{\mathbf{x},m} = \begin{pmatrix} 0.0031 & 0.0051 & 0.0045 & 0.0029 & 0.0040 & 0.0012 & 0.0018 & 0.0022 \\ 0.0051 & 0.0188 & 0.0116 & 0.0059 & 0.0056 & 0.0024 & 0.0022 & 0.0047 \\ 0.0045 & 0.0116 & 0.0113 & 0.0057 & 0.0047 & 0.0022 & 0.0024 & 0.0040 \\ 0.0029 & 0.0059 & 0.0057 & 0.0077 & 0.0048 & 0.0020 & 0.0025 & 0.0028 \\ 0.0040 & 0.0056 & 0.0047 & 0.0048 & 0.0183 & 0.0033 & 0.0037 & 0.0021 \\ 0.0012 & 0.0024 & 0.0022 & 0.0020 & 0.0033 & 0.0020 & 0.0012 & 0.0008 \\ 0.0018 & 0.0022 & 0.0024 & 0.0025 & 0.0037 & 0.0012 & 0.0039 & 0.0025 \\ 0.0022 & 0.0047 & 0.0040 & 0.0028 & 0.0021 & 0.0008 & 0.0025 & 0.0035 \end{pmatrix}$$

$$\widehat{\Omega}^{log}_{\mathbf{x},m} = \begin{pmatrix} 1 & 0.6738 & 0.7596 & 0.6020 & 0.5365 & 0.4907 & 0.5272 & 0.6693 \\ 0.6738 & 1 & 0.8007 & 0.4891 & 0.3024 & 0.3839 & 0.2623 & 0.5745 \\ 0.7596 & 0.8007 & 1 & 0.6163 & 0.3290 & 0.4642 & 0.3572 & 0.6353 \\ 0.6020 & 0.4891 & 0.6163 & 1 & 0.4018 & 0.5031 & 0.4563 & 0.5417 \\ 0.5365 & 0.3024 & 0.3290 & 0.4018 & 1 & 0.5532 & 0.4420 & 0.2631 \\ 0.4907 & 0.3839 & 0.4642 & 0.5031 & 0.5532 & 1 & 0.4365 & 0.2890 \\ 0.5272 & 0.2623 & 0.3572 & 0.4563 & 0.4420 & 0.4365 & 1 & 0.6729 \\ 0.6693 & 0.5745 & 0.6353 & 0.5417 & 0.2631 & 0.2890 & 0.6729 & 1 \end{pmatrix}$$

As expected, the sample covariance and sample correlation matrices of the weekly and monthly log returns of the stocks are close to the sample covariance and sample correlation matrices of the weekly and monthly percentage returns computed in the example from Stefanica [3].

A more detailed analysis shows that the relative error of the entries of the sample correlation matrices is smaller than the relative error of the entries of the sample covariance matrices. Also, the sample covariance and sample correlation matrices corresponding to weekly log returns are closer to the sample covariance and sample correlation matrices corresponding to weekly percentage returns than is the case for the matrices corresponding to monthly returns.   □

Problem 12: Let X_1, X_2, ..., X_n be nonconstant random variables, and let Y_1, Y_2, ..., Y_n be the random variables given by $Y_i = d_i X_i$, $i = 1 : n$, where $d_i \neq 0$, $i = 1 : n$, are constants. Denote by $\Sigma_{\mathbf{X}}$, $\Omega_{\mathbf{X}}$ and $\Sigma_{\mathbf{Y}}$, $\Omega_{\mathbf{Y}}$ the covariance and correlation matrices of X_1, X_2, ..., X_n and of Y_1, Y_2, ..., Y_n, respectively, and let $D = \text{diag}(d_i)_{i=1:n}$.

(i) Show that $\Sigma_{\mathbf{Y}} = D\Sigma_{\mathbf{X}} D$.

(ii) If $d_i > 0$ for all $i = 1 : n$, show that $\Omega_{\mathbf{Y}} = \Omega_{\mathbf{X}}$.

Solution: (i) Let $\mathbf{Y} = (Y_i)_{i=1:n}$ and $\mathbf{X} = (X_i)_{i=1:n}$. Then,

$$\mathbf{Y} = D\mathbf{X},$$

where $D = \text{diag}(d_i)_{i=1:n}$, and, from the Linear Transformation Property, it follows that

$$\Sigma_{\mathbf{Y}} = D\Sigma_{\mathbf{X}}D^t. \tag{7.14}$$

Note that $D^t = D$, since D is a diagonal matrix. Then, from (7.14), we obtain that

$$\Sigma_{\mathbf{Y}} = D\Sigma_{\mathbf{X}}D. \tag{7.15}$$

(ii) Recall that

$$\Sigma_{\mathbf{X}} = D_{\sigma_{\mathbf{X}}}\Omega_{\mathbf{X}}D_{\sigma_{\mathbf{X}}}; \tag{7.16}$$
$$\Sigma_{\mathbf{Y}} = D_{\sigma_{\mathbf{Y}}}\Omega_{\mathbf{Y}}D_{\sigma_{\mathbf{Y}}}, \tag{7.17}$$

where

$$D_{\sigma_{\mathbf{X}}} = \text{diag}(\sigma(X_i))_{i=1:n}; \quad D_{\sigma_{\mathbf{Y}}} = \text{diag}(\sigma(Y_i))_{i=1:n}.$$

Since $Y_i = d_iX_i$ and $d_i > 0$ for $i = 1:n$, we obtain that

$$\sigma(Y_i) = d_i\sigma(X_i), \quad \forall\, i = 1:n,$$

and therefore

$$
\begin{aligned}
DD_{\sigma_{\mathbf{X}}} &= \text{diag}(d_i)_{i=1:n}\,\text{diag}(\sigma(X_i))_{i=1:n} \\
&= \text{diag}(d_i\sigma(X_i))_{i=1:n} \\
&= \text{diag}(\sigma(Y_i))_{i=1:n} \\
&= D_{\sigma_{\mathbf{Y}}}; \tag{7.18}
\end{aligned}
$$

$$
\begin{aligned}
D_{\sigma_{\mathbf{X}}}D &= \text{diag}(\sigma(X_i))_{i=1:n}\,\text{diag}(d_i)_{i=1:n} \\
&= \text{diag}(d_i\sigma(X_i))_{i=1:n} \\
&= \text{diag}(\sigma(Y_i))_{i=1:n} \\
&= D_{\sigma_{\mathbf{Y}}}; \tag{7.19}
\end{aligned}
$$

From (7.18) and (7.19), it follows that

$$DD_{\sigma_{\mathbf{X}}} = D_{\sigma_{\mathbf{X}}}D = D_{\sigma_{\mathbf{Y}}}. \tag{7.20}$$

From (7.15–7.20), we conclude that

$$
\begin{aligned}
&\Sigma_{\mathbf{Y}} = D\Sigma_{\mathbf{X}}D \\
\Longleftrightarrow\ & D_{\sigma_{\mathbf{Y}}}\Omega_{\mathbf{Y}}D_{\sigma_{\mathbf{Y}}} = DD_{\sigma_{\mathbf{X}}}\Omega_{\mathbf{X}}D_{\sigma_{\mathbf{X}}}D \\
\Longleftrightarrow\ & D_{\sigma_{\mathbf{Y}}}\Omega_{\mathbf{Y}}D_{\sigma_{\mathbf{Y}}} = D_{\sigma_{\mathbf{Y}}}\Omega_{\mathbf{X}}D_{\sigma_{\mathbf{Y}}} \\
\Longleftrightarrow\ & (D_{\sigma_{\mathbf{Y}}})^{-1}D_{\sigma_{\mathbf{Y}}}\Omega_{\mathbf{Y}}D_{\sigma_{\mathbf{Y}}}(D_{\sigma_{\mathbf{Y}}})^{-1} = (D_{\sigma_{\mathbf{Y}}})^{-1}D_{\sigma_{\mathbf{Y}}}\Omega_{\mathbf{X}}D_{\sigma_{\mathbf{Y}}}(D_{\sigma_{\mathbf{Y}}})^{-1} \\
\Longleftrightarrow\ & \Omega_{\mathbf{Y}} = \Omega_{\mathbf{X}};
\end{aligned}
$$

note that $D_{\sigma_{\mathbf{Y}}}$ is nonsingular: $d_i \neq 0$ and $\sigma(X_i) \neq 0$ since X_i is a nonconstant random variable and therefore $\sigma(Y_i) = d_i\sigma(X_i) \neq 0$.

Thus, $\Omega_{\mathbf{Y}} = \Omega_{\mathbf{X}}$, which is what we wanted to show.   □

Problem 13: Let

$$\begin{pmatrix} X_1 \\ X_2 \\ X_3 \end{pmatrix} \sim N \left(\begin{pmatrix} 1 \\ -2 \\ 1 \end{pmatrix}, \begin{pmatrix} 1 & -1 & 0 \\ -1 & 3 & -2 \\ 0 & -2 & 3 \end{pmatrix} \right) \tag{7.21}$$

be a 3–dimensional multivariate normal random variable.
   Find the probability that $X_1 + 2X_2 + 2X_3$ is positive.

Solution: Let $\mathbf{X} = \begin{pmatrix} X_1 \\ X_2 \\ X_3 \end{pmatrix}$. Denote by $\mu_{\mathbf{X}}$ and $\Sigma_{\mathbf{X}}$ the mean vector and the covariance matrix of $\mathbf{X}$. From (7.21), we find that

$$\mu_{\mathbf{X}} = \begin{pmatrix} 1 \\ -2 \\ 1 \end{pmatrix}; \quad \Sigma_{\mathbf{X}} = \begin{pmatrix} 1 & -1 & 0 \\ -1 & 3 & -2 \\ 0 & -2 & 3 \end{pmatrix}. \tag{7.22}$$

Note that

$$X_1 + 2X_2 + 2X_3 = (1 \ 2 \ 2) \begin{pmatrix} X_1 \\ X_2 \\ X_3 \end{pmatrix} = C^t \mathbf{X}, \tag{7.23}$$

where $C = \begin{pmatrix} 1 \\ 2 \\ 2 \end{pmatrix}$.

Since

$$E[C^t \mathbf{X}] = C^t \mu_{\mathbf{X}}; \tag{7.24}$$
$$\text{var}(C^t \mathbf{X}) = C^t \Sigma_{\mathbf{X}} C, \tag{7.25}$$

we obtain from (7.22–7.25) that

$$E[X_1 + 2X_2 + 2X_3] = E[C^t \mathbf{X}] = C^t \mu_{\mathbf{X}} = (1 \ 2 \ 2) \begin{pmatrix} 1 \\ -2 \\ 1 \end{pmatrix}$$

$$= -1; \tag{7.26}$$
$$\text{var}(X_1 + 2X_2 + 2X_3) = \text{var}(C^t \mathbf{X}) = C^t \Sigma_{\mathbf{X}} C$$

$$= (1 \ 2 \ 2) \begin{pmatrix} 1 & -1 & 0 \\ -1 & 3 & -2 \\ 0 & -2 & 3 \end{pmatrix} \begin{pmatrix} 1 \\ 2 \\ 2 \end{pmatrix}$$

$$= 5. \tag{7.27}$$

Recall that a multivariate random variable is multivariate normal if and only if any linear combination of its components is a normal random variable.

Then, since $\mathbf{X} = \begin{pmatrix} X_1 \\ X_2 \\ X_3 \end{pmatrix}$ is a multivariate normal random variable, it follows that $X_1 + 2X_2 + 2X_3$ is a normal random variable. From (7.26) and (7.27), we find that $X_1 + 2X_2 + 2X_3$ is a normal random variable of mean -1 and variance 5, and therefore with standard deviation $\sqrt{5}$. In other words,

$$X_1 + 2X_2 + 2X_3 = -1 + \sqrt{5}Z,$$

where Z is a standard normal random variable. Thus,

$$
\begin{aligned}
P(X_1 + 2X_2 + 2X_3 < 0) &= P(-1 + \sqrt{5}Z < 0) \\
&= P\left(Z < \frac{1}{\sqrt{5}}\right) \\
&= 0.6726. \quad \square
\end{aligned}
$$

Problem 14: Let

$$
\begin{pmatrix} X_1 \\ X_2 \\ X_3 \end{pmatrix} \sim N\left(\begin{pmatrix} 2 \\ -1 \\ -2 \end{pmatrix}, \begin{pmatrix} 2 & 0 & 0.5 \\ 0 & 1 & -0.25 \\ 0.5 & -0.25 & 1 \end{pmatrix} \right) \tag{7.28}
$$

be a 3–dimensional multivariate normal random variable, and let

$$Y_1 = X_1; \quad Y_2 = X_1 + X_2; \quad Y_3 = X_1 + X_2 + X_3.$$

(i) What are the mean vector and the covariance matrix of $\begin{pmatrix} Y_1 \\ Y_2 \\ Y_3 \end{pmatrix}$?

(ii) Find the probability that $Y_1 + Y_2 + Y_3$ is negative.

Solution: Let $\mathbf{X} = \begin{pmatrix} X_1 \\ X_2 \\ X_3 \end{pmatrix}$ and $\mathbf{Y} = \begin{pmatrix} Y_1 \\ Y_2 \\ Y_3 \end{pmatrix}$. Then,

$$
\begin{aligned}
\mathbf{Y} = \begin{pmatrix} Y_1 \\ Y_2 \\ Y_3 \end{pmatrix} &= \begin{pmatrix} X_1 \\ X_1 + X_2 \\ X_1 + X_2 + X_3 \end{pmatrix} = \begin{pmatrix} 1 & 0 & 0 \\ 1 & 1 & 0 \\ 1 & 1 & 1 \end{pmatrix} \begin{pmatrix} X_1 \\ X_2 \\ X_3 \end{pmatrix} \\
&= M\mathbf{X},
\end{aligned}
$$

where

$$
M = \begin{pmatrix} 1 & 0 & 0 \\ 1 & 1 & 0 \\ 1 & 1 & 1 \end{pmatrix}.
$$

Let $\mu_\mathbf{X}$, $\Sigma_\mathbf{X}$ and $\mu_\mathbf{Y}$, $\Sigma_\mathbf{Y}$ be the mean vector and the covariance matrix of the multivariate random variables $\mathbf{X}$ and $\mathbf{Y}$, respectively. From (7.28), we find that

$$
\mu_\mathbf{X} = \begin{pmatrix} 2 \\ -1 \\ -2 \end{pmatrix}; \quad \Sigma_\mathbf{X} = \begin{pmatrix} 2 & 0 & 0.5 \\ 0 & 1 & -0.25 \\ 0.5 & -0.25 & 1 \end{pmatrix}.
$$

Since $\mathbf{Y} = M\mathbf{X}$, we obtain that

$$\mu_\mathbf{Y} = M\mu_\mathbf{X} = \begin{pmatrix} 1 & 0 & 0 \\ 1 & 1 & 0 \\ 1 & 1 & 1 \end{pmatrix} \begin{pmatrix} 2 \\ -1 \\ -2 \end{pmatrix} = \begin{pmatrix} 2 \\ 1 \\ -1 \end{pmatrix}, \qquad (7.29)$$

and, from the Linear Transformation Property, it follows that

$$\Sigma_\mathbf{Y} = M\Sigma_\mathbf{X}M^t = \begin{pmatrix} 1 & 0 & 0 \\ 1 & 1 & 0 \\ 1 & 1 & 1 \end{pmatrix} \begin{pmatrix} 2 & 0 & 0.5 \\ 0 & 1 & -0.25 \\ 0.5 & -0.25 & 1 \end{pmatrix} \begin{pmatrix} 1 & 1 & 1 \\ 0 & 1 & 1 \\ 0 & 0 & 1 \end{pmatrix}$$

$$= \begin{pmatrix} 2 & 2 & 2.5 \\ 2 & 3 & 3.25 \\ 2.5 & 3.25 & 4.5 \end{pmatrix}. \qquad (7.30)$$

(ii) Note that

$$Y_1 + Y_2 + Y_3 = \begin{pmatrix} 1 & 1 & 1 \end{pmatrix} \begin{pmatrix} Y_1 \\ Y_2 \\ Y_3 \end{pmatrix} = \mathbf{1}^t\mathbf{Y}, \qquad (7.31)$$

where $\mathbf{1} = \begin{pmatrix} 1 \\ 1 \\ 1 \end{pmatrix}$. Since, for any 3×1 constant vector C,

$$E[C^t\mathbf{Y}] = C^t\mu_\mathbf{Y}; \qquad (7.32)$$
$$\mathrm{var}(C^t\mathbf{Y}) = C^t\Sigma_\mathbf{Y}C, \qquad (7.33)$$

we obtain from (7.31–7.33) that

$$E[Y_1 + Y_2 + Y_3] = E[\mathbf{1}^t\mathbf{Y}] = \mathbf{1}^t\mu_\mathbf{Y} = \begin{pmatrix} 1 & 1 & 1 \end{pmatrix} \begin{pmatrix} 2 \\ 1 \\ -1 \end{pmatrix}$$

$$= 2; \qquad (7.34)$$

$$\mathrm{var}(Y_1 + Y_2 + Y_3) = \mathrm{var}(\mathbf{1}^t\mathbf{Y}) = \mathbf{1}^t\Sigma_\mathbf{Y}\mathbf{1}$$

$$= \begin{pmatrix} 1 & 1 & 1 \end{pmatrix} \begin{pmatrix} 2 & 2 & 2.5 \\ 2 & 3 & 3.25 \\ 2.5 & 3.25 & 4.5 \end{pmatrix} \begin{pmatrix} 1 \\ 1 \\ 1 \end{pmatrix}$$

$$= 25. \qquad (7.35)$$

Recall that a multivariate random variable is multivariate normal if and only if any linear combination of its components is a normal random variable. Then, since $\mathbf{X} = \begin{pmatrix} X_1 \\ X_2 \\ X_3 \end{pmatrix}$ is a multivariate normal random variable, it follows that

$$Y_1 + Y_2 + Y_3 = X_1 + (X_1 + X_2) + (X_1 + X_2 + X_3) = 3X_1 + 2X_2 + X_3$$

is a normal random variable. From (7.34) and (7.35), we find that $Y_1 + Y_2 + Y_3$ is a normal random variable of mean 2 and variance 25, and therefore with standard deviation 5.

In other words,
$$Y_1 + Y_2 + Y_3 = 2 + 5Z,$$
where Z is a standard normal random variable. Thus,

$$P(Y_1 + Y_2 + Y_3 > 0) = P(2 + 5Z > 0) = P\left(Z > -\frac{2}{5}\right)$$
$$= 1 - P\left(Z < -\frac{2}{5}\right)$$
$$= 0.6554. \quad \square$$

Problem 15: Let

$$\begin{pmatrix} X_1 \\ X_2 \\ X_3 \end{pmatrix} \sim N\left(\begin{pmatrix} 1 \\ 0 \\ -2 \end{pmatrix}, \begin{pmatrix} 2 & 1 & 0.5 \\ 1 & 3 & -1.5 \\ 0.5 & -1.5 & 4 \end{pmatrix} \right) \tag{7.36}$$

be a 3–dimensional multivariate normal random variable.

(i) What are the expected value and the covariance matrix of $\begin{pmatrix} X_1 - X_2 \\ X_2 - X_3 \\ X_3 - X_1 \end{pmatrix}$?

(ii) What is the probability density function of $\begin{pmatrix} X_1 - X_2 \\ X_2 - X_3 \\ X_3 - X_1 \end{pmatrix}$?

Solution: Let $\mathbf{X} = \begin{pmatrix} X_1 \\ X_2 \\ X_3 \end{pmatrix}$ and let

$$\mathbf{Y} = \begin{pmatrix} X_1 - X_2 \\ X_2 - X_3 \\ X_3 - X_1 \end{pmatrix} = \begin{pmatrix} 1 & -1 & 0 \\ 0 & 1 & -1 \\ -1 & 0 & 1 \end{pmatrix} \begin{pmatrix} X_1 \\ X_2 \\ X_3 \end{pmatrix}$$
$$= M\mathbf{X},$$

where

$$M = \begin{pmatrix} 1 & -1 & 0 \\ 0 & 1 & -1 \\ -1 & 0 & 1 \end{pmatrix}.$$

Let $\mu_\mathbf{X}$, $\Sigma_\mathbf{X}$ and $\mu_\mathbf{Y}$, $\Sigma_\mathbf{Y}$ be the mean vector and the covariance matrix of the multivariate random variables $\mathbf{X}$ and $\mathbf{Y}$, respectively. From (7.36), we find that

$$\mu_\mathbf{X} = \begin{pmatrix} 1 \\ 0 \\ -2 \end{pmatrix}; \quad \Sigma_\mathbf{X} = \begin{pmatrix} 2 & 1 & 0.5 \\ 1 & 3 & -1.5 \\ 0.5 & -1.5 & 4 \end{pmatrix}.$$

Since $\mathbf{Y} = M\mathbf{X}$, we obtain that

$$\mu_\mathbf{Y} = M\mu_\mathbf{X} = \begin{pmatrix} 1 & -1 & 0 \\ 0 & 1 & -1 \\ -1 & 0 & 1 \end{pmatrix} \begin{pmatrix} 1 \\ 0 \\ -2 \end{pmatrix} = \begin{pmatrix} 1 \\ 2 \\ -3 \end{pmatrix}, \tag{7.37}$$

and, from the Linear Transformation Property, it follows that

$$\Sigma_\mathbf{Y} = M\Sigma_\mathbf{X}M^t$$

$$= \begin{pmatrix} 1 & -1 & 0 \\ 0 & 1 & -1 \\ -1 & 0 & 1 \end{pmatrix} \begin{pmatrix} 2 & 1 & 0.5 \\ 1 & 3 & -1.5 \\ 0.5 & -1.5 & 4 \end{pmatrix} \begin{pmatrix} 1 & 0 & -1 \\ -1 & 1 & 0 \\ 0 & -1 & 1 \end{pmatrix}$$

$$= \begin{pmatrix} 3 & -4 & 1 \\ -4 & 10 & -6 \\ 1 & -6 & 5 \end{pmatrix}. \tag{7.38}$$

(ii) Recall that the probability density function $f : \mathbb{R}^n \to \mathbb{R}$ of a non-degenerate multivariate normal variable $\mathbf{Y} \sim N(\mu_\mathbf{Y}, \Sigma_\mathbf{Y})$ with non-singular covariance matrix $\Sigma_\mathbf{Y}$ is given by

$$f(x) = \frac{1}{(2\pi)^{n/2}\sqrt{\det(\Sigma_\mathbf{Y})}} \exp\left(-\frac{1}{2}(x-\mu_\mathbf{Y})^t(\Sigma_\mathbf{Y})^{-1}(x-\mu_\mathbf{Y})\right). \tag{7.39}$$

However, $\det(\Sigma_\mathbf{Y}) = 0.^4$ Thus, $\mathbf{Y}$ is a degenerate multivariate normal variable and its probability density function cannot be expressed using (7.39). □

Problem 16: Let Z_1, Z_2, Z_3 be three independent standard normal variables, and let X_1, X_2, X_3 and Y_1, Y_2, Y_3 be normal random variables given by

$$X_1 = Z_3; \quad X_2 = Z_1 + 2Z_2 - 2Z_3; \quad X_3 = -3Z_1 + Z_2 + Z_3;$$

$$Y_1 = Z_1; \quad Y_2 = -2Z_1 + \sqrt{5}Z_2; \quad Y_3 = Z_1 - \frac{1}{\sqrt{5}}Z_2 + \sqrt{9.8}Z_3.$$

Show that the covariance matrix of X_1, X_2, X_3 and the covariance matrix of Y_1, Y_2, Y_3 are equal.

Solution: Let $\mathbf{Z} = \begin{pmatrix} Z_1 \\ Z_2 \\ Z_3 \end{pmatrix}$ and let

$$\mathbf{X} = \begin{pmatrix} X_1 \\ X_2 \\ X_3 \end{pmatrix} = \begin{pmatrix} Z_3 \\ Z_1 + 2Z_2 - 2Z_3 \\ -3Z_1 + Z_2 + Z_3 \end{pmatrix}$$

$$= \begin{pmatrix} 0 & 0 & 1 \\ 1 & 2 & -2 \\ -3 & 1 & 1 \end{pmatrix} \begin{pmatrix} Z_1 \\ Z_2 \\ Z_3 \end{pmatrix}$$

$$= M_1 \mathbf{Z}; \tag{7.40}$$

$$\mathbf{Y} = \begin{pmatrix} Y_1 \\ Y_2 \\ Y_3 \end{pmatrix} = \begin{pmatrix} Z_1 \\ -2Z_1 + \sqrt{5}Z_2 \\ Z_1 - \frac{1}{\sqrt{5}}Z_2 + \sqrt{9.8}Z_3 \end{pmatrix}$$

---

^{4}Since $\Sigma_\mathbf{Y} = M\Sigma_\mathbf{X}M^t$, we find that $\det(\Sigma_\mathbf{Y}) = \det(M)\det(\Sigma_\mathbf{X})\det(M^t)$. Using the fact that $\det(M) = \det\begin{pmatrix} 1 & -1 & 0 \\ 0 & 1 & -1 \\ -1 & 0 & 1 \end{pmatrix} = 0$, we obtain that $\det(\Sigma_\mathbf{Y}) = 0$.

$$= \begin{pmatrix} 1 & 0 & 0 \\ -2 & \sqrt{5} & 0 \\ 1 & -\frac{1}{\sqrt{5}} & \sqrt{9.8} \end{pmatrix} \begin{pmatrix} Z_1 \\ Z_2 \\ Z_3 \end{pmatrix}$$

$$= M_2 \mathbf{Z}, \tag{7.41}$$

where

$$M_1 = \begin{pmatrix} 0 & 0 & 1 \\ 1 & 2 & -2 \\ -3 & 1 & 1 \end{pmatrix};$$

$$M_2 = \begin{pmatrix} 1 & 0 & 0 \\ -2 & \sqrt{5} & 0 \\ 1 & -\frac{1}{\sqrt{5}} & \sqrt{9.8} \end{pmatrix}.$$

Note that the covariance matrix $\Sigma_{\mathbf{Z}}$ of Z_1, Z_2, Z_3 is the identity matrix, i.e.,

$$\Sigma_{\mathbf{Z}} = I, \tag{7.42}$$

since Z_1, Z_2, Z_3 are independent standard normal variables.

Let $\Sigma_{\mathbf{X}}$ and $\Sigma_{\mathbf{Y}}$ be the covariance matrices of $\mathbf{X}$ and $\mathbf{Y}$, respectively. Since $\mathbf{X} = M_1 \mathbf{Z}$ and $\mathbf{Y} = M_2 \mathbf{Z}$, see (7.40) and (7.41), and using the Linear Transformation Property and (7.42), we obtain that

$$\Sigma_{\mathbf{X}} = M_1 \Sigma_{\mathbf{Z}} M_1^t = M_1 M_1^t = \begin{pmatrix} 0 & 0 & 1 \\ 1 & 2 & -2 \\ -3 & 1 & 1 \end{pmatrix} \begin{pmatrix} 0 & 1 & -3 \\ 0 & 2 & 1 \\ 1 & -2 & 1 \end{pmatrix}$$

$$= \begin{pmatrix} 1 & -2 & 1 \\ -2 & 9 & -3 \\ 1 & -3 & 11 \end{pmatrix}; \tag{7.43}$$

$$\Sigma_{\mathbf{Y}} = M_2 \Sigma_{\mathbf{Z}} M_2^t = M_2 M_2^t = \begin{pmatrix} 1 & 0 & 0 \\ -2 & \sqrt{5} & 0 \\ 1 & -\frac{1}{\sqrt{5}} & \sqrt{9.8} \end{pmatrix} \begin{pmatrix} 1 & -2 & 1 \\ 0 & \sqrt{5} & -\frac{1}{\sqrt{5}} \\ 0 & 0 & \sqrt{9.8} \end{pmatrix}$$

$$= \begin{pmatrix} 1 & -2 & 1 \\ -2 & 9 & -3 \\ 1 & -3 & 11 \end{pmatrix}. \tag{7.44}$$

We conclude from (7.43) and (7.44) that $\Sigma_{\mathbf{X}} = \Sigma_{\mathbf{Y}}$, i.e., the covariance matrices of X_1, X_2, X_3 and of Y_1, Y_2, Y_3 are equal. $\square$

Problem 17: Given Z_1 and Z_2 independent standard normal variables, find two normal random variables X_1 and X_2 with covariance matrix $\begin{pmatrix} 4 & -1 \\ -1 & 9 \end{pmatrix}$.

Solution: Let $\Sigma = \begin{pmatrix} 4 & -1 \\ -1 & 9 \end{pmatrix}$. The Cholesky factor of the matrix Σ is[5]

$$U = \begin{pmatrix} 2 & -\frac{1}{2} \\ 0 & \frac{\sqrt{35}}{2} \end{pmatrix},$$

and therefore

$$U^t U = \Sigma. \tag{7.46}$$

Let $\mathbf{Z} = \begin{pmatrix} Z_1 \\ Z_2 \end{pmatrix}$, and let

$$\mathbf{X} = \begin{pmatrix} X_1 \\ X_2 \end{pmatrix} = U^t \mathbf{Z} = \begin{pmatrix} 2 & 0 \\ -\frac{1}{2} & \frac{\sqrt{35}}{2} \end{pmatrix} \begin{pmatrix} Z_1 \\ Z_2 \end{pmatrix}$$

$$= \begin{pmatrix} 2Z_1 \\ -\frac{1}{2}Z_1 + \frac{\sqrt{35}}{2}Z_2 \end{pmatrix}. \tag{7.47}$$

Note that $X_1 = 2Z_1$ and $X_2 = -\frac{1}{2}Z_1 + \frac{\sqrt{35}}{2}Z_2$ are normal random variables since they are linear combinations of independent normal random variables.

Then, from the Linear Transformation Property and using (7.46) and the fact that the covariance matrix $\Sigma_{\mathbf{Z}}$ of the independent standard normal variables Z_1 and Z_2 is $\Sigma_{\mathbf{Z}} = I$, it follows that the covariance matrix $\Sigma_{\mathbf{X}}$ of X_1 and X_2 is

$$\Sigma_{\mathbf{X}} = U^t \Sigma_{\mathbf{Z}} U = U^t U = \Sigma = \begin{pmatrix} 4 & -1 \\ -1 & 9 \end{pmatrix}. \tag{7.48}$$

From (7.47) and (7.48), we conclude that the covariance matrix of the normal random variables

$$X_1 = 2Z_1 \quad \text{and} \quad X_2 = -\frac{1}{2}Z_1 + \frac{\sqrt{35}}{2}Z_2$$

is $\begin{pmatrix} 4 & -1 \\ -1 & 9 \end{pmatrix}$. $\square$

Problem 18: Given Z_1 and Z_2 independent standard normal variables, find two normal random variables X_1 and X_2 with correlation matrix $\begin{pmatrix} 1 & 0.25 \\ 0.25 & 1 \end{pmatrix}$.

Solution: Let $\Omega = \begin{pmatrix} 1 & 0.25 \\ 0.25 & 1 \end{pmatrix}$. The Cholesky factor of the matrix Ω is

$$U = \begin{pmatrix} 1 & \frac{1}{4} \\ 0 & \frac{\sqrt{15}}{4} \end{pmatrix},$$

---

[5]Recall that the Cholesky factor of the 2×2 symmetric positive definite matrix $\begin{pmatrix} a & b \\ b & d \end{pmatrix}$ is

$$\begin{pmatrix} \sqrt{a} & \frac{b}{\sqrt{a}} \\ 0 & \sqrt{\frac{ad-b^2}{a}} \end{pmatrix}. \tag{7.45}$$

see, e.g., (7.45), and therefore

$$U^tU = \Omega. \tag{7.49}$$

Let $\mathbf{Z} = \begin{pmatrix} Z_1 \\ Z_2 \end{pmatrix}$, and let

$$\mathbf{X} = \begin{pmatrix} X_1 \\ X_2 \end{pmatrix} = U^t\mathbf{Z} = \begin{pmatrix} 1 & 0 \\ \frac{1}{4} & \frac{\sqrt{15}}{4} \end{pmatrix} \begin{pmatrix} Z_1 \\ Z_2 \end{pmatrix}$$

$$= \begin{pmatrix} Z_1 \\ \frac{1}{4}Z_1 + \frac{\sqrt{15}}{4}Z_2 \end{pmatrix}. \tag{7.50}$$

Note that $X_1 = Z_1$ and $X_2 = \frac{1}{4}Z_1 + \frac{\sqrt{15}}{4}Z_2$ are normal random variables since they are linear combinations of independent normal random variables.

Then, from the Linear Transformation Property and using (7.49) and the fact that the covariance matrix $\Sigma_{\mathbf{z}}$ of the independent standard normal variables Z_1 and Z_2 is $\Sigma_{\mathbf{z}} = I$, it follows that the covariance matrix $\Sigma_{\mathbf{x}}$ of X_1 and X_2 is

$$\Sigma_{\mathbf{x}} = U^t\Sigma_{\mathbf{z}}U = U^tU = \Omega = \begin{pmatrix} 1 & 0.25 \\ 0.25 & 1 \end{pmatrix}. \tag{7.51}$$

Note that the standard deviations σ_1 and σ_2 of X_1 and X_2, respectively, are equal to 1 since

$$\sigma_1 = \sqrt{\Sigma_{\mathbf{x}}(1,1)} = 1; \quad \sigma_2 = \sqrt{\Sigma_{\mathbf{x}}(2,2)} = 1. \tag{7.52}$$

Then, from the relationship

$$\Sigma_{\mathbf{x}} = \begin{pmatrix} \sigma_1 & 0 \\ 0 & \sigma_2 \end{pmatrix} \Omega_{\mathbf{x}} \begin{pmatrix} \sigma_1 & 0 \\ 0 & \sigma_2 \end{pmatrix}$$

between the covariance matrix $\Sigma_{\mathbf{x}}$ and the correlation matrix $\Omega_{\mathbf{x}}$ of $\mathbf{X}$, and using (7.51) and (7.52), we obtain that

$$\Sigma_{\mathbf{x}} = \Omega_{\mathbf{x}} = \begin{pmatrix} 1 & 0.25 \\ 0.25 & 1 \end{pmatrix}. \tag{7.53}$$

From (7.50) and (7.53), we conclude that the correlation matrix of the normal random variables

$$X_1 = Z_1 \quad \text{and} \quad X_2 = \frac{1}{4}Z_1 + \frac{\sqrt{15}}{4}Z_2$$

is $\begin{pmatrix} 1 & 0.25 \\ 0.25 & 1 \end{pmatrix}$.   □

Problem 19: Given Z_1, Z_2, Z_3 independent standard normal variables, find three normal random variables X_1, X_2, X_3 with covariance matrix $\begin{pmatrix} 1 & 1 & 0.5 \\ 1 & 4 & -2 \\ 0.5 & -2 & 9 \end{pmatrix}$.

Solution: Let $\Sigma = \begin{pmatrix} 1 & 1 & 0.5 \\ 1 & 4 & -2 \\ 0.5 & -2 & 9 \end{pmatrix}$. By running the Cholesky decomposition algorithm, we find that the Cholesky factor of Σ is

$$U = \begin{pmatrix} 1 & 1 & 0.5 \\ 0 & 1.7321 & -1.4434 \\ 0 & 0 & 2.5820 \end{pmatrix};$$

note that

$$U^t U = \Sigma. \tag{7.54}$$

Let $\mathbf{Z} = \begin{pmatrix} Z_1 \\ Z_2 \\ Z_3 \end{pmatrix}$, and let

$$\mathbf{X} = \begin{pmatrix} X_1 \\ X_2 \\ X_3 \end{pmatrix} = U^t \mathbf{Z} = \begin{pmatrix} 1 & 0 & 0 \\ 1 & 1.7321 & \\ 0.5 & -1.4434 & 2.5820 \end{pmatrix} \begin{pmatrix} Z_1 \\ Z_2 \\ Z_3 \end{pmatrix}$$

$$= \begin{pmatrix} Z_1 \\ Z_1 + 1.7321 Z_2 \\ 0.5 Z_1 - 1.4434 Z_2 + 2.5820 Z_3 \end{pmatrix}. \tag{7.55}$$

Note that $X_1 = Z_1$, $X_2 = Z_1 + 1.7321 Z_2$, and $X_3 = 0.5 Z_1 - 1.4434 Z_2 + 2.5820 Z_3$ are normal random variables since they are linear combinations of independent normal random variables.

Then, from the Linear Transformation Property and using (7.54) and the fact that the covariance matrix $\Sigma_{\mathbf{Z}}$ of the independent standard normal variables Z_1, Z_2, Z_3 is $\Sigma_{\mathbf{Z}} = I$, it follows that the covariance matrix $\Sigma_{\mathbf{X}}$ of X_1, X_2, and X_3 is

$$\Sigma_{\mathbf{X}} = U^t \Sigma_{\mathbf{Z}} U = U^t U = \Sigma = \begin{pmatrix} 1 & 1 & 0.5 \\ 1 & 4 & -2 \\ 0.5 & -2 & 9 \end{pmatrix}. \tag{7.56}$$

From (7.55) and (7.56), we conclude that the covariance matrix of the normal random variables

$$X_1 = Z_1; \quad X_2 = Z_1 + 1.7321 Z_2; \quad X_3 = 0.5 Z_1 - 1.4434 Z_2 + 2.5820 Z_3$$

is $\begin{pmatrix} 1 & 1 & 0.5 \\ 1 & 4 & -2 \\ 0.5 & -2 & 9 \end{pmatrix}$. $\square$

Problem 20: Given Z_1, Z_2, Z_3 independent standard normal variables, find three normal random variables X_1, X_2, X_3 with correlation matrix $\begin{pmatrix} 1 & 0.3 & 0.4 \\ 0.3 & 1 & 0.5 \\ 0.4 & 0.5 & 1 \end{pmatrix}$.

Solution: Let $\Omega = \begin{pmatrix} 1 & 0.3 & 0.4 \\ 0.3 & 1 & 0.5 \\ 0.4 & 0.5 & 1 \end{pmatrix}$. By running the Cholesky decomposition algorithm, we find that the Cholesky factor of the matrix Ω is

$$U = \begin{pmatrix} 1 & 0.3 & 0.4 \\ 0 & 0.9539 & 0.3983 \\ 0 & 0 & 0.8254 \end{pmatrix};$$

note that

$$U^t U = \Omega. \tag{7.57}$$

Let $\mathbf{Z} = \begin{pmatrix} Z_1 \\ Z_2 \\ Z_3 \end{pmatrix}$, and let

$$
\mathbf{X} = \begin{pmatrix} X_1 \\ X_2 \\ X_3 \end{pmatrix} = U^t \mathbf{Z} = \begin{pmatrix} 1 & 0 & 0 \\ 0.3 & 0.9539 & 0 \\ 0.4 & 0.3983 & 0.8254 \end{pmatrix} \begin{pmatrix} Z_1 \\ Z_2 \\ Z_3 \end{pmatrix}
$$

$$
= \begin{pmatrix} Z_1 \\ 0.3Z_1 + 0.9539Z_2 \\ 0.4Z_1 + 0.3983Z_2 + 0.8254Z_3 \end{pmatrix}. \tag{7.58}
$$

Note that $X_1 = Z_1$, $X_2 = 0.3Z_1 + 0.9539Z_2$, and $X_3 = 0.4Z_1 + 0.3983Z_2 + 0.8254Z_3$ are normal random variables since they are linear combinations of independent normal random variables.

Then, from the Linear Transformation Property and using (7.57) and the fact that the covariance matrix $\Sigma_{\mathbf{Z}}$ of the independent standard normal variables Z_1, Z_2, Z_3 is $\Sigma_{\mathbf{Z}} = I$, it follows that the covariance matrix $\Sigma_{\mathbf{X}}$ of X_1, X_2, X_3 is

$$\Sigma_{\mathbf{X}} = U^t \Sigma_{\mathbf{Z}} U = U^t U = \Omega = \begin{pmatrix} 1 & 0.3 & 0.4 \\ 0.3 & 1 & 0.5 \\ 0.4 & 0.5 & 1 \end{pmatrix}. \tag{7.59}$$

Note that the standard deviations σ_1, σ_2, and σ_3 of X_1, X_2, and X_3, respectively, are equal to 1, since

$$\sigma_1 = \sqrt{\Sigma_{\mathbf{X}}(1,1)} = 1; \quad \sigma_2 = \sqrt{\Sigma_{\mathbf{X}}(2,2)} = 1; \quad \sigma_3 = \sqrt{\Sigma_{\mathbf{X}}(3,3)} = 1. \tag{7.60}$$

Then, from the relationship

$$\Sigma_{\mathbf{X}} = \begin{pmatrix} \sigma_1 & 0 & 0 \\ 0 & \sigma_2 & 0 \\ 0 & 0 & \sigma_3 \end{pmatrix} \Omega_{\mathbf{X}} \begin{pmatrix} \sigma_1 & 0 & 0 \\ 0 & \sigma_2 & 0 \\ 0 & 0 & \sigma_3 \end{pmatrix}$$

between the covariance matrix $\Sigma_{\mathbf{X}}$ and the correlation matrix $\Omega_{\mathbf{X}}$ of $\mathbf{X}$, and using (7.59) and (7.60), we obtain that

$$\Sigma_{\mathbf{X}} = \Omega_{\mathbf{X}} = \begin{pmatrix} 1 & 0.3 & 0.4 \\ 0.3 & 1 & 0.5 \\ 0.4 & 0.5 & 1 \end{pmatrix}. \tag{7.61}$$

From (7.58) and (7.61), we conclude that the correlation matrix of the normal random variables

$$X_1 \; = \; Z_1; \quad X_2 = 0.3Z_1 + 0.9539Z_2; \quad X_3 = 0.4Z_1 + 0.3983Z_2 + 0.8254Z_3$$

is $\begin{pmatrix} 1 & 0.3 & 0.4 \\ 0.3 & 1 & 0.5 \\ 0.4 & 0.5 & 1 \end{pmatrix}$.   □

Chapter 8

Ordinary least squares (OLS). Linear regression.

8.1 Exercises

1. Denote by x^* the solution to the ordinary least squares problem $y \approx Ax$, where A is an $m \times n$ matrix with linearly independent columns and with $m > n$, y is an $m \times 1$ column vector, and x is an $n \times 1$ column vector.

   Show that Ax^* is the projection of the vector y onto the space generated by the columns of A, or, equivalently, show that the vector $y - Ax^*$ is orthogonal to the space generated by the columns of A. In other words, show that

   $$(y - Ax^*, Az) = 0, \quad \forall z \in \mathbb{R}^n.$$

   Note: Recall that the space generated by the columns of A is $\{Az \mid z \in \mathbb{R}^n\}$, and is also called the range of the matrix A.

2. The mid prices on March 9, 2012, of the S&P 500 options maturing on December 22, 2012, can be found in Table 8.1.

   The spot price of the index corresponding to these option prices was $1,370$. For the options above, the market estimates for the annualized continuous dividend yield of the S&P 500 index and for the risk–free rate were $q = 0.0193 = 1.93\%$ and $r = 0.0015 = 0.15\%$, respectively.

   (i) Use Newton's method to compute the implied volatilities for each of the options.

   (ii) How do these values compare to the implied volatilities obtained by using ordinary least squares to compute the discount factor and the present value of the forward price?

   (iii) How do the implied volatilities of calls and puts with the same strike compare to each other?

Table 8.1: SPX options prices

Call Price	Strike	Put Price
225.40	1175	46.60
205.55	1200	51.55
186.20	1225	57.15
167.50	1250	63.30
149.15	1275	70.15
131.70	1300	77.70
115.25	1325	86.20
99.55	1350	95.30
84.90	1375	105.30
71.10	1400	116.55
58.70	1425	129.00
47.25	1450	143.20
29.25	1500	173.95
15.80	1550	210.80
11.10	1575	230.90
7.90	1600	252.40

Table 8.2: SPX options prices

Bid Price Call	Ask Price Call	Strike	Bid Price Put	Ask Price Put
431.20	434.40	1450	8.90	10.20
384.10	387.40	1500	11.60	13.10
337.90	341.20	1550	15.20	16.90
292.80	296.20	1600	19.90	21.80
228.30	231.10	1675	29.70	31.40
207.50	210.10	1700	33.60	35.50
167.60	170.00	1750	43.50	45.40
148.90	151.30	1775	49.40	51.50
130.60	132.80	1800	56.50	58.10
113.20	115.30	1825	63.60	65.70
96.80	98.60	1850	71.90	74.20
81.50	83.20	1875	81.40	83.50
67.10	69.00	1900	92.10	94.50
54.20	55.90	1925	103.80	106.30
32.80	34.50	1975	132.40	134.70
24.50	25.90	2000	149.20	151.90
12.40	13.60	2050	186.90	189.70
5.60	6.50	2100	230.00	232.40

3. On May 22, 2014, The bid and ask prices on May 22, 2014, of the S&P 500 options maturing on January 17, 2015, can be found in Table 8.2. The spot price of the index corresponding to these option prices was 1,894.

(i) Compute the mid prices of the options, i.e., the average of the bid price and ask price of the options. Use ordinary least squares to compute the present value of the forward price PVF and the discount factor $disc$ corresponding to the mid prices of the options.

(ii) Compute the implied volatilities of these options. How do the implied volatilities of calls and puts with the same strike compare to each other?

4. Recall that the vega of an option measures the sensitivity of the price of the option to changes in volatility. In other words, for call and put options, $\text{vega}(C) = \frac{\partial C}{\partial \sigma}$ and $\text{vega}(P) = \frac{\partial P}{\partial \sigma}$.

   Use the Put–Call parity
   $$C - P = Se^{-qT} - Ke^{-rT}$$
   to show that
   $$\text{vega}(C) = \text{vega}(P).$$

5. The goal of this exercise is to derive the Black–Scholes formulas for the vegas of plain vanilla European calls and puts.

   In the Black–Scholes framework, an underlying asset with spot price S follows a lognormal distribution with volatility σ and pays continuous dividends at rate q. Denote by r be the risk–free interest rate, assumed to be constant. The Black–Scholes values of a European call option and of a European put option with strike K and maturity T are

   $$C_{BS}(S, K, T, \sigma, r, q) = Se^{-qT} N(d_1) - Ke^{-rT} N(d_2);$$
   $$P_{BS}(S, K, T, \sigma, r, q) = Ke^{-rT} N(-d_2) - Se^{-qT} N(-d_1),$$

   respectively, where
   $$N(z) = \frac{1}{\sqrt{2\pi}} \int_{-\infty}^{z} e^{-\frac{x^2}{2}} \, dx$$

   denotes the cumulative distribution of the standard normal variable, and

   $$d_1 = \frac{\ln\left(\frac{S}{K}\right) + \left(r - q + \frac{\sigma^2}{2}\right) T}{\sigma\sqrt{T}}; \quad d_2 = d_1 - \sigma\sqrt{T}.$$

   (i) Use Chain Rule to show that

   $$\text{vega}(C) = \frac{\partial C}{\partial \sigma} = Se^{-qT} N'(d_1) \frac{\partial d_1}{\partial \sigma} - Ke^{-rT} N'(d_2) \frac{\partial d_2}{\partial \sigma}.$$

   (ii) Show that
   $$N'(z) = \frac{1}{\sqrt{2\pi}} e^{-\frac{z^2}{2}},$$

   and use this fact to prove that

   $$Se^{-qT} N'(d_1) = Ke^{-rT} N'(d_2).$$

(iii) Show that

$$\text{vega}(C) \;=\; \frac{1}{\sqrt{2\pi}} \, Se^{-qT} \, e^{-\frac{d_1^2}{2}} \, \sqrt{T};$$

$$\text{vega}(P) \;=\; \frac{1}{\sqrt{2\pi}} \, Se^{-qT} \, e^{-\frac{d_1^2}{2}} \, \sqrt{T}.$$

6. The goal of this exercise is to show that the theoretical values of the implied volatilities of plain vanilla European call and put options with the same strike and maturity are equal.

Denote by $C_{BS}(S, K, T, \sigma, r, q)$ and $P_{BS}(S, K, T, \sigma, r, q)$ the Black–Scholes values of a European call option and of a European put option, respectively, with strike K and maturity T on an underlying asset following a lognormal distribution with volatility σ, with spot price S, and paying dividends continuously at the rate q, if interest rates are constant and equal to r. Let C_m and P_m be the market prices of a call and of a put option, respectively, with parameters S, K, T, r, and q.

The implied volatility $\sigma_{imp,C}$ corresponding to the price C is the solution to

$$C_{BS}(\sigma_{imp,C}) \;=\; C_m.$$

The implied volatility $\sigma_{imp,P}$ corresponding to the price P is the solution to

$$P_{BS}(\sigma_{imp,P}) \;=\; P_m.$$

Note that, here and below, $C_{BS}(\sigma)$ and $P_{BS}(\sigma)$ are shorthand notations for $C_{BS}(S, K, T, \sigma, r, q)$ and $P_{BS}(S, K, T, \sigma, r, q)$, respectively.

(i) Use the facts that the Black–Scholes values of put and call options satisfy the Put–Call parity for any value $\sigma > 0$ of the volatility, i.e.,

$$C_{BS}(\sigma) - P_{BS}(\sigma) \;=\; Se^{-qT} - Ke^{-rT},$$

and that the market prices of put and call options satisfy the Put–Call parity as well, i.e.,

$$C_m - P_m \;=\; Se^{-qT} - Ke^{-rT},$$

to conclude that

$$P_{BS}(\sigma_{imp,P}) \;=\; P_{BS}(\sigma_{imp,C}).$$

(ii) Show that the Black–Scholes value of a put option is a strictly increasing function of volatility, and conclude that the implied volatilities corresponding to put and call options with the same strike and maturity on the same asset must be equal, i.e.,

$$\sigma_{imp,P} = \sigma_{imp,C}.$$

7. Recall the example from the book where for 15 consecutive trading days, the yields of the 2-year, 3-year, 5-year, and 10-year treasury bonds were, respectively:

2-year	3-year	5-year	10-year
4.69	4.58	4.57	4.63
4.81	4.71	4.69	4.73
4.81	4.72	4.70	4.74
4.79	4.78	4.77	4.81
4.79	4.77	4.77	4.80
4.83	4.75	4.73	4.79
4.81	4.71	4.72	4.76
4.81	4.72	4.74	4.77
4.83	4.76	4.77	4.80
4.81	4.73	4.75	4.77
4.82	4.75	4.77	4.80
4.82	4.75	4.76	4.80
4.80	4.73	4.75	4.78
4.78	4.71	4.72	4.73
4.79	4.71	4.71	4.73

Denote by T_2, T_3, T_5, and T_{10} the time series data vectors corresponding to the yield of the 2-year, 3-year, 5-year, and 10-year treasury bonds, respectively.

(i) Find the coefficients a, b_1, b_2, b_3 of the linear regression for the yield of the 3-year bond in terms of the yields of the 2-year, 5-year, and 10-year bonds, i.e., find a, b_1, b_2, b_3 corresponding to the solution to the ordinary least squares problem

$$T_3 \approx a\mathbf{1} + b_1 T_2 + b_2 T_5 + b_3 T_{10},$$

where $\mathbf{1}$ is the 15×1 column vector with all entries equal to 1. Let

$$T_{3,LR} = a\mathbf{1} + b_1 T_2 + b_2 T_5 + b_3 T_{10}.$$

Find the approximation error

$$\text{error}_{LR} = ||T_3 - T_{3,LR}||$$

of the linear regression.

(ii) Compute the linear interpolation values of the 3-year yield by doing linear interpolation between the 2-year yield and the 5-year yield at each data point. Denote by $T_{3,linear\_interp}$ the time series vector of these values. In other words,

$$T_{3,linear\_interp} = \frac{2}{3}T_2 + \frac{1}{3}T_5.$$

Find the approximation error

$$\text{error}_{linear\_interp} = ||T_3 - T_{3,linear\_interp}||$$

of the linear interpolation.

(iii) Compute the cubic interpolation values of the 3-year yield by doing cubic spline interpolation between the 2-year, 5-year, and 10-year yield at each data point. Denote by $T_{3,cubic\_interp}$ the time series vector of these values.

Find the approximation error

$$\text{error}_{cubic\_interp} = ||T_3 - T_{3,cubic\_interp}||$$

of the cubic interpolation.

(iv) Compare the approximation errors from (i), (ii), and (iii), and comment on the results.

8. The file *financials2012.xlsx* from www.fepress.org/nla-primer contains the end of week adjusted closing prices for the stocks of the following financial companies: JPM; GS; MS; BAC (Bank of America); RBS; CS; UBS; RY (RBC); BCS (Barclays) between January 11, 2012, and October 15, 2012.

(i) Compute the weekly percentage returns of these stocks.

(ii) Find the linear regression of the JPM returns with respect to the returns of the other stocks. What is the approximation error of this linear regression?

(iii) Find the linear regression of the JPM returns with respect to the returns of the other American financial companies, i.e., with respect to GS, MS, and BAC. What is the approximation error of this linear regression?

(iv) Find the linear regression of the JPM stock prices with respect to the prices of the other stocks. What is the approximation error of this linear regression? How does it compare with the approximation error of the linear regression of the JPM returns computed at (ii)?

8.2 Solutions to Chapter 8 Exercises

Problem 1: Denote by x^* the solution to the ordinary least squares problem $y \approx Ax$, where A is an $m \times n$ matrix with linearly independent columns and with $m > n$, y is an $m \times 1$ column vector, and x is an $n \times 1$ column vector.

Show that Ax^* is the projection of the vector y onto the space generated by the columns of A, or, equivalently, show that the vector $y - Ax^*$ is orthogonal to the space generated by the columns of A. In other words, show that

$$(y - Ax^*, Az) = 0, \quad \forall z \in \mathbb{R}^n. \tag{8.1}$$

Solution: Recall that the solution to the ordinary least squares problem $y \approx Ax$ is

$$x^* = (A^t A)^{-1} A^t y. \tag{8.2}$$

From (8.2), we obtain that

$$
\begin{aligned}
(y - Ax^*, Az) &= (Az)^t (y - Ax^*) \\
&= z^t A^t (y - Ax^*) \\
&= z^t A^t y - z^t A^t Ax^* \\
&= z^t A^t y - z^t A^t A \cdot (A^t A)^{-1} A^t y \\
&= z^t A^t y - z^t A^t y \\
&= 0, \quad \forall z \in \mathbb{R}^n,
\end{aligned}
$$

since $A^t A \cdot (A^t A)^{-1} = I$.

Note that $\{Az \mid z \in \mathbb{R}^n\}$ is the space generated by the columns of A. Thus, (8.1) says that Ax^* is the projection of the vector y on the space generated by the columns of the matrix A. □

Problem 2: The mid prices on March 9, 2012, of the S&P 500 options maturing on December 22, 2012, can be found in Table 8.3.

The spot price of the index corresponding to these option prices was $1,370$. For these options, the market estimates for the annualized continuous dividend yield of the S&P 500 index and for the risk–free rate were $q = 0.0193 = 1.93\%$ and $r = 0.0015 = 0.15\%$, respectively.

(i) Use Newton's method to compute the implied volatilities for each of the options.

(ii) How do these values compare to the implied volatilities obtained by using ordinary least squares to compute the discount factor and the present value of the forward price?

(iii) How do the implied volatilities of calls and puts with the same strike compare to each other?

Solution: (i) The maturity of every option from Table 8.3 is $T = 199/252$. Using Newton's method corresponding to a spot index price of $S = 1370$, a continuous dividend yield $q = 0.0193$, and a risk–free rate $r = 0.0015$, we obtain the implied volatilities from Table 8.4.

Table 8.3: SPX options prices

Call Price	Strike	Put Price
225.40	1175	46.60
205.55	1200	51.55
186.20	1225	57.15
167.50	1250	63.30
149.15	1275	70.15
131.70	1300	77.70
115.25	1325	86.20
99.55	1350	95.30
84.90	1375	105.30
71.10	1400	116.55
58.70	1425	129.00
47.25	1450	143.20
29.25	1500	173.95
15.80	1550	210.80
11.10	1575	230.90
7.90	1600	252.40

Table 8.4: Implied volatiles using given dividend yield and risk–free rate

Strike	Implied Vol Call	Implied Vol Put	Strike	Implied Vol Call	Implied Vol Put
1175	26.31%	25.46%	1375	20.02%	19.24%
1200	25.51%	24.66%	1400	19.24%	18.48%
1225	24.70%	23.88%	1425	18.53%	17.73%
1250	23.91%	23.11%	1450	17.78%	17.09%
1275	23.06%	22.34%	1500	16.56%	15.45%
1300	22.27%	21.57%	1550	15.24%	13.90%
1325	21.52%	20.84%	1575	14.65%	12.93%
1350	20.76%	20.05%	1600	14.29%	11.88%

(ii) Recall from the example in the book that, if least squares are used to compute the present value of the forward price of the underlying asset and the discount factor corresponding to time T, the implied volatilities corresponding to the options from Table 8.3 are those from Table 8.5.

The implied volatilities from Table 8.5 were computed using the dividend yield (1.91%) and the risk–free rate (0.45%) embedded in the option prices, while the implied volatilities from Table 8.4 are computed using a given dividend yield (1.93%) and risk–free rate (0.15%) which are determined from market considerations not related to the option prices.

We note that, for each strike, the implied volatilities from Table 8.5 are close to each other and in between the implied volatilities from Table 8.4.

While the historical yield of the S&P500 index is a good approximation for the dividend yield embedded in the option prices (1.93% and 1.91%, respectively), that is not the case for the risk–free rates (0.15% and 0.45%, respectively), which is what

Table 8.5: Implied volatiles for SPX options

Strike	Implied Vol Call	Implied Vol Put	Strike	Implied Vol Call	Implied Vol Put
1175	25.73%	25.72%	1375	19.69%	19.66%
1200	24.96%	24.92%	1400	18.94%	18.94%
1225	24.19%	24.16%	1425	18.26%	18.25%
1250	23.44%	23.40%	1450	17.53%	17.68%
1275	22.63%	22.65%	1500	16.34%	16.24%
1300	21.86%	21.91%	1550	15.05%	15.08%
1325	21.15%	21.20%	1575	14.48%	14.47%
1350	20.41%	20.43%	1600	14.13%	14.02%

generates the discrepancy in the implied volatilities from Table 8.4 and Table 8.5.

(iii) According to Put–Call parity, the implied volatilities of calls and puts with the same strike should be the same. However, the implied volatilities from Table 8.4 computed using a given dividend yield and risk–free rate are different by at least half of a percentage point and up to over two percentage points from each other, which makes them wrong estimates for implied volatilities.

The implied volatilities from Table 8.5 obtained by using the dividend yield and the risk–free rate embedded in the option prices are very accurate estimates for implied volatilities.

We note that how close the implied volatilities for calls and puts with the same strike are to each other is very sensitive to changes in the interest rate and the discount factor that are used in the Newton's method routine for computing implied volatilities. Even a change of 10 basis points in the interest rate could result in differences of one percentage point or more for the implied volatilities corresponding to a call and to a put option with the same strike.

Using the interest rate and the dividend yield embedded in the option prices is the optimal way to compute implied volatilities in practice.   □

Problem 3: The bid and ask prices on May 22, 2014, of the S&P 500 options maturing on January 17, 2015, can be found in Table 8.6.

The spot price of the index corresponding to these option prices was $1,894$.

(i) Compute the mid prices of the options, i.e., the average of the bid price and ask price of the options. Use ordinary least squares to compute the present value of the forward price PVF and the discount factor $disc$ corresponding to the mid prices of the options.

(ii) Compute the implied volatilities of these options. How do the implied volatilities of calls and puts with the same strike compare to each other?

Table 8.6: SPX options prices

Bid Price Call	Ask Price Call	Strike	Bid Price Put	Ask Price Put
431.20	434.40	1450	8.90	10.20
384.10	387.40	1500	11.60	13.10
337.90	341.20	1550	15.20	16.90
292.80	296.20	1600	19.90	21.80
228.30	231.10	1675	29.70	31.40
207.50	210.10	1700	33.60	35.50
167.60	170.00	1750	43.50	45.40
148.90	151.30	1775	49.40	51.50
130.60	132.80	1800	56.50	58.10
113.20	115.30	1825	63.60	65.70
96.80	98.60	1850	71.90	74.20
81.50	83.20	1875	81.40	83.50
67.10	69.00	1900	92.10	94.50
54.20	55.90	1925	103.80	106.30
32.80	34.50	1975	132.40	134.70
24.50	25.90	2000	149.20	151.90
12.40	13.60	2050	186.90	189.70
5.60	6.50	2100	230.00	232.40

Solution: (i) The mid prices of the call and put options from Table 8.6 are

$$
C_{mid} = \begin{pmatrix} 432.80 \\ 385.75 \\ 339.55 \\ 294.50 \\ 229.70 \\ 208.80 \\ 168.80 \\ 150.10 \\ 131.70 \\ 114.25 \\ 97.70 \\ 82.35 \\ 68.05 \\ 55.05 \\ 33.65 \\ 25.20 \\ 13.00 \\ 6.05 \end{pmatrix} \; ; \quad P_{mid} = \begin{pmatrix} 9.55 \\ 12.35 \\ 16.05 \\ 20.85 \\ 30.55 \\ 34.55 \\ 44.45 \\ 50.45 \\ 57.30 \\ 64.65 \\ 73.05 \\ 82.45 \\ 93.30 \\ 105.05 \\ 133.55 \\ 150.55 \\ 188.30 \\ 231.20 \end{pmatrix} . \tag{8.3}
$$

Recall that, if C and P are the values of a European call and of a European put with strike K and maturity T on an underlying asset with spot price S paying dividends continuously at rate q, and if the risk–free interest rate is constant and equal to r, the Put–Call parity is the following model–independent relationship:

$$
C - P = Se^{-qT} - Ke^{-rT}. \tag{8.4}
$$

Also, recall that (8.4) can be written in terms of the discount factor $disc = e^{-rT}$ and the present value of the forward price of the asset $PVF = Fe^{-rT}$, where $F = Se^{(r-q)T}$ is the forward price of the asset at time T, as follows:

$$C - P = PVF - K \cdot disc. \tag{8.5}$$

The data from Table 8.6 provides call and put options values for 18 different strikes. From (8.5), it follows that the values of PVF and $disc$ can be obtained by solving a least squares problem $y \approx Ax$ with $x = \begin{pmatrix} PVF \\ disc \end{pmatrix}$ and with the following 18×2 matrix A and the following 18×1 column vector y corresponding to $C_{mid} - P_{mid}$ for each strike, see (8.3):

$$A = \begin{pmatrix} 1 & -1450 \\ 1 & -1500 \\ 1 & -1550 \\ 1 & -1600 \\ 1 & -1675 \\ 1 & -1700 \\ 1 & -1750 \\ 1 & -1775 \\ 1 & -1800 \\ 1 & -1825 \\ 1 & -1850 \\ 1 & -1875 \\ 1 & -1900 \\ 1 & -1925 \\ 1 & -1975 \\ 1 & -2000 \\ 1 & -2050 \\ 1 & -2100 \end{pmatrix} ; \quad y = \begin{pmatrix} 423.25 \\ 373.40 \\ 323.50 \\ 273.65 \\ 199.15 \\ 174.25 \\ 124.35 \\ 99.65 \\ 74.40 \\ 49.60 \\ 24.65 \\ -0.10 \\ -25.25 \\ -50 \\ -99.90 \\ -125.35 \\ -175.30 \\ -225.15 \end{pmatrix}.$$

The solution $x = (A^t A)^{-1} A^t y$ to this least squares problem is computed as $x =$ linear_solve_cholesky$(A^t A, A^t y)$, and we find that

$$x = \begin{pmatrix} PVF \\ disc \end{pmatrix} = \begin{pmatrix} 1869.40 \\ 0.9972 \end{pmatrix}.$$

Thus, $PVF = 1869.40$ and $disc = 0.9972$.

(ii) There are 166 trading days between May 22, 2014, and January 17, 2015, and 252 trading days in a year, and therefore the maturity of the options from Table 8.6 is $T = \frac{166}{252}$. The implied volatilities corresponding to the options from Table 8.6 are included in Table 8.7 and are obtained by using Newton's method for the form of the Black–Scholes formulas involving the discount factor $disc$ (obtained above to be $disc = 0.9972$) and the present value of the forward price of the asset PVF (obtained above to be $PVF = 1869.40$); see also Section 8.1.1 from Stefanica [3].

The implied volatilities from Table 8.7 corresponding to calls and puts with the same strike are nearly identical, i.e., within at most 0.15% of each other, and even closer for options near at–the–money. This conforms to the theoretical result derived from the Put–Call parity that the implied volatilities of calls and puts with the same strike are equal. □

Table 8.7: Implied volatiles for SPX options

Strike	Implied Vol Call	Implied Vol Put	Strike	Implied Vol Call	Implied Vol Put
1450	2187%	21.97%	1825	14.71%	14.69%
1500	2089%	20.96%	1850	14.21%	14.19%
1550	19.93%	20.00%	1875	13.73%	13.68%
1600	19.00%	19.06%	1900	13.23%	13.21%
1675	17.64%	17.61%	1925	12.73%	12.69%
1700	17.14%	17.11%	1975	11.83%	11.79%
1750	16.16%	16.14%	2000	11.40%	11.47%
1775	15.73%	15.67%	2050	10.67%	10.78%
1800	15.21%	15.21%	2100	10.12%	10.28%

Problem 4: Recall that the vega of an option measures the sensitivity of the price of the option to changes in volatility. In other words, for call and put options, $\text{vega}(C) = \frac{\partial C}{\partial \sigma}$ and $\text{vega}(P) = \frac{\partial P}{\partial \sigma}$.
Use the Put–Call parity

$$C - P = Se^{-qT} - Ke^{-rT} \tag{8.6}$$

to show that

$$\text{vega}(C) = \text{vega}(P).$$

Solution: By differentiating the Put–Call parity (8.6) with respect to σ, we find that

$$\frac{\partial C}{\partial \sigma} - \frac{\partial P}{\partial \sigma} = \frac{\partial Se^{-qT}}{\partial \sigma} - \frac{\partial Ke^{-rT}}{\partial \sigma} = 0,$$

since $\frac{\partial}{\partial \sigma}\left(Se^{-qT}\right) = 0$ and $\frac{\partial}{\partial \sigma}\left(Ke^{-rT}\right) = 0$.
Thus, $\frac{\partial C}{\partial \sigma} = \frac{\partial P}{\partial \sigma}$, and therefore $\text{vega}(C) = \text{vega}(P)$.   □

Problem 5: The goal of this exercise is to derive the Black–Scholes formulas for the vegas of plain vanilla European calls and puts.

In the Black–Scholes framework, an underlying asset with spot price S follows a lognormal distribution with volatility σ and pays continuous dividends at rate q. Denote by r be the risk–free interest rate, assumed to be constant. The Black–Scholes values of a European call option and of a European put option with strike K and maturity T are

$$\begin{aligned}C_{BS}(S, K, T, \sigma, r, q) &= Se^{-qT}N(d_1) - Ke^{-rT}N(d_2); \\ P_{BS}(S, K, T, \sigma, r, q) &= Ke^{-rT}N(-d_2) - Se^{-qT}N(-d_1),\end{aligned}$$

respectively, where

$$N(z) = \frac{1}{\sqrt{2\pi}} \int_{-\infty}^{z} e^{-\frac{x^2}{2}} \, dx$$

denotes the cumulative distribution of the standard normal variable, and

$$d_1 = \frac{\ln\left(\frac{S}{K}\right) + \left(r - q + \frac{\sigma^2}{2}\right)T}{\sigma\sqrt{T}}; \quad d_2 = d_1 - \sigma\sqrt{T}. \tag{8.7}$$

(i) Use Chain Rule to show that

$$\text{vega}(C) = \frac{\partial C}{\partial \sigma} = Se^{-qT} N'(d_1) \frac{\partial d_1}{\partial \sigma} - Ke^{-rT} N'(d_2) \frac{\partial d_2}{\partial \sigma}. \tag{8.8}$$

(ii) Show that

$$N'(z) = \frac{1}{\sqrt{2\pi}} e^{-\frac{z^2}{2}},$$

and use this fact to prove that

$$Se^{-qT} N'(d_1) = Ke^{-rT} N'(d_2). \tag{8.9}$$

(iii) Show that

$$\text{vega}(C) = \frac{1}{\sqrt{2\pi}} Se^{-qT} e^{-\frac{d_1^2}{2}} \sqrt{T}; \tag{8.10}$$

$$\text{vega}(P) = \frac{1}{\sqrt{2\pi}} Se^{-qT} e^{-\frac{d_1^2}{2}} \sqrt{T}. \tag{8.11}$$

Solution: (i) Note that d_1 and d_2 given by (8.7) are functions of σ. Thus, $N(d_1)$ and $N(d_2)$ are functions of σ and therefore, by applying the Chain Rule, we obtain that

$$\frac{\partial}{\partial \sigma}(N(d_1)) = N'(d_1) \frac{\partial d_1}{\partial \sigma}; \quad \frac{\partial}{\partial \sigma}(N(d_2)) = N'(d_2) \frac{\partial d_2}{\partial \sigma}.$$

Then,

$$\begin{aligned}
\text{vega}(C) &= \frac{\partial C}{\partial \sigma} = \frac{\partial}{\partial \sigma}\left(Se^{-qT} N(d_1) - Ke^{-rT} N(d_2)\right) \\
&= Se^{-qT} \frac{\partial}{\partial \sigma}(N(d_1)) - Ke^{-rT} \frac{\partial}{\partial \sigma}(N(d_2)) \\
&= Se^{-qT} N'(d_1) \frac{\partial d_1}{\partial \sigma} - Ke^{-rT} N'(d_2) \frac{\partial d_2}{\partial \sigma}.
\end{aligned}$$

(ii) Recall that

$$N(0) = \frac{1}{\sqrt{2\pi}} \int_{-\infty}^{0} e^{-\frac{x^2}{2}} dx = \frac{1}{2},$$

since the probability density function $\frac{1}{\sqrt{2\pi}} e^{-\frac{x^2}{2}}$ of the standard normal distribution is an even function. Then,

$$\begin{aligned}
N(z) &= \frac{1}{\sqrt{2\pi}} \int_{-\infty}^{z} e^{-\frac{x^2}{2}} dx \\
&= \frac{1}{\sqrt{2\pi}} \int_{-\infty}^{0} e^{-\frac{x^2}{2}} dx + \frac{1}{\sqrt{2\pi}} \int_{0}^{z} e^{-\frac{x^2}{2}} dx \\
&= \frac{1}{2} + \frac{1}{\sqrt{2\pi}} \int_{0}^{z} e^{-\frac{x^2}{2}} dx.
\end{aligned}$$

Thus,

$$
\begin{aligned}
N'(z) &= \frac{1}{\sqrt{2\pi}} \frac{d}{dz} \left(\int_0^z e^{-\frac{x^2}{2}} \, dx \right) \\
&= \frac{1}{\sqrt{2\pi}} e^{-\frac{z^2}{2}},
\end{aligned}
\tag{8.12}
$$

since

$$
\frac{d}{dz} \left(\int_0^z f(x) \, dx \right) = f(z)
\tag{8.13}
$$

for any continuous function $f(x)$.[1]

From (8.12), we obtain that

$$
N'(d_1) = \frac{1}{\sqrt{2\pi}} e^{-\frac{d_1^2}{2}}; \quad N'(d_2) = \frac{1}{\sqrt{2\pi}} e^{-\frac{d_2^2}{2}},
\tag{8.14}
$$

and therefore

$$
Se^{-qT} N'(d_1) = Ke^{-rT} N'(d_2)
\tag{8.15}
$$

$$
\begin{aligned}
&\Longleftrightarrow \quad Se^{-qT} e^{-\frac{d_1^2}{2}} = Ke^{-rT} e^{-\frac{d_2^2}{2}} \\
&\Longleftrightarrow \quad \frac{Se^{(r-q)T}}{K} = \exp\left(\frac{d_1^2 - d_2^2}{2} \right),
\end{aligned}
\tag{8.16}
$$

where $\exp(x) = e^x$. From (8.7), we obtain that

$$
\begin{aligned}
\frac{d_1^2 - d_2^2}{2} &= \frac{d_1^2 - \left(d_1 - \sigma\sqrt{T} \right)^2}{2} \\
&= d_1 \sigma \sqrt{T-t} - \frac{\sigma^2 T}{2} \\
&= \ln\left(\frac{S}{K} \right) + (r-q)T \\
&= \ln\left(\frac{Se^{(r-q)T}}{K} \right),
\end{aligned}
$$

and therefore

$$
\exp\left(\frac{d_1^2 - d_2^2}{2} \right) = \frac{Se^{(r-q)T}}{K},
$$

which is the same as (8.16). From (8.15) and (8.16), we conclude that

$$
Se^{-qT} N'(d_1) = Ke^{-rT} N'(d_2).
$$

---

[1] Note that (8.13) is a particular form of the following result for differentiating definite integrals:

$$
\frac{d}{dt} \left(\int_{a(t)}^{b(t)} f(x) \, dx \right) = f(b(t))b'(t) - f(a(t))a'(t),
$$

where $f(t)$ is a continuous function and $a(t)$ and $b(t)$ are continuously differentiable functions.

(iii) From (8.8) and (8.9), we obtain that

$$
\begin{aligned}
\text{vega}(C) &= Se^{-qT} N'(d_1) \frac{\partial d_1}{\partial \sigma} - Ke^{-rT} N'(d_2) \frac{\partial d_2}{\partial \sigma} \\
&= Se^{-qT} N'(d_1) \frac{\partial d_1}{\partial \sigma} - Se^{-qT} N'(d_1) \frac{\partial d_2}{\partial \sigma} \\
&= Se^{-qT} N'(d_1) \left(\frac{\partial d_1}{\partial \sigma} - \frac{\partial d_2}{\partial \sigma} \right).
\end{aligned}
\tag{8.17}
$$

Since $d_2 = d_1 - \sigma\sqrt{T}$, see (8.7), we find that $d_1 - d_2 = \sigma\sqrt{T}$ and thus

$$
\frac{\partial d_1}{\partial \sigma} - \frac{\partial d_2}{\partial \sigma} = \sqrt{T}.
\tag{8.18}
$$

From (8.17) and (8.18), and using the fact that $N'(d_1) = \frac{1}{\sqrt{2\pi}} e^{\frac{-d_1^2}{2}}$, see (8.14), we conclude that

$$
\begin{aligned}
\text{vega}(C) &= Se^{-qT} N'(d_1)\sqrt{T} \\
&= \frac{1}{\sqrt{2\pi}} Se^{-qT} e^{-\frac{d_1^2}{2}} \sqrt{T}.
\end{aligned}
\tag{8.19}
$$

To compute the vega of the put option, we differentiate the Put–Call parity

$$
C - P = Se^{-qT} - Ke^{-rT}
$$

with respect to σ and obtain that

$$
\frac{\partial C}{\partial \sigma} - \frac{\partial P}{\partial \sigma} = 0.
$$

In other words,

$$
\text{vega}(P) = \frac{\partial P}{\partial \sigma} = \frac{\partial C}{\partial \sigma} = \text{vega}(C).
\tag{8.20}
$$

From (8.19) and (8.20), we conclude that

$$
\text{vega}(P) = \frac{1}{\sqrt{2\pi}} Se^{-qT} e^{-\frac{d_1^2}{2}} \sqrt{T}. \quad \square
$$

Problem 6: The goal of this exercise is to show that the theoretical values of the implied volatilities of plain vanilla European call and put options with the same strike and maturity are equal.

Denote by $C_{BS}(S, K, T, \sigma, r, q)$ and $P_{BS}(S, K, T, \sigma, r, q)$ the Black–Scholes values of a European call option and of a European put option, respectively, with strike K and maturity T on an underlying asset following a lognormal distribution with volatility σ, with spot price S, and paying dividends continuously at the rate q, if interest rates are constant and equal to r. Let C_m and P_m be the market prices of a call and of a put option, respectively, with parameters S, K, T, r, and q.

The implied volatility $\sigma_{imp,C}$ corresponding to the price C is the solution to

$$C_{BS}(\sigma_{imp,C}) = C_m. \tag{8.21}$$

The implied volatility $\sigma_{imp,P}$ corresponding to the price P is the solution to

$$P_{BS}(\sigma_{imp,P}) = P_m. \tag{8.22}$$

Here and below, $C_{BS}(\sigma)$ and $P_{BS}(\sigma)$ are shorthand notations for $C_{BS}(S, K, T, \sigma, r, q)$ and $P_{BS}(S, K, T, \sigma, r, q)$, respectively.

(i) Use the facts that the Black–Scholes values of put and call options satisfy the Put–Call parity for any value $\sigma > 0$ of the volatility, i.e.,

$$C_{BS}(\sigma) - P_{BS}(\sigma) = Se^{-qT} - Ke^{-rT}, \tag{8.23}$$

and that the market prices of put and call options satisfy the Put–Call parity as well, i.e.,

$$C_m - P_m = Se^{-qT} - Ke^{-rT}, \tag{8.24}$$

to conclude that

$$P_{BS}(\sigma_{imp,P}) = P_{BS}(\sigma_{imp,C}).$$

(ii) Show that the Black–Scholes value of a put option is a strictly increasing function of volatility, and conclude that the implied volatilities corresponding to put and call options with the same strike and maturity on the same asset must be equal, i.e.,

$$\sigma_{imp,P} = \sigma_{imp,C}.$$

Solution: (i) Let $\sigma = \sigma_{imp,C}$ in (8.23). Then,

$$C_{BS}(\sigma_{imp,C}) - P_{BS}(\sigma_{imp,C}) = Se^{-qT} - Ke^{-rT}, \tag{8.25}$$

which is the same as

$$C_m - P_{BS}(\sigma_{imp,C}) = Se^{-qT} - Ke^{-rT}, \tag{8.26}$$

since $C_{BS}(\sigma_{imp,C}) = C_m$; see (8.21).
From (8.24) and (8.26), it follows that

$$P_m = P_{BS}(\sigma_{imp,C}). \tag{8.27}$$

Then, from (8.27) and the definition (8.22) of the implied volatility $\sigma_{imp,P}$, we conclude that

$$P_{BS}(\sigma_{imp,P}) = P_{BS}(\sigma_{imp,C}). \tag{8.28}$$

(ii) Recall from (8.11) that

$$\text{vega}(P) = \frac{1}{\sqrt{2\pi}} Se^{-qT} e^{-\frac{d_1^2}{2}} \sqrt{T} > 0,$$

where $\text{vega}(P) = \frac{\partial P}{\partial \sigma}$. Thus, $\frac{\partial P}{\partial \sigma} > 0$ for all $\sigma > 0$ and therefore the Black–Scholes value of a put option is a strictly increasing function of volatility. Since

$P_{BS}(\sigma_{imp,P}) = P_{BS}(\sigma_{imp,C})$, see (8.28), we conclude that $\sigma_{imp,P} = \sigma_{imp,C}$, i.e., the implied volatilities corresponding to put and call options with the same strike and maturity on the same asset are equal. □

Problem 7: Recall the example from the book where for 15 consecutive trading days, the yields of the 2-year, 3-year, 5-year, and 10-year treasury bonds were, respectively:

2-year	3-year	5-year	10-year
4.69	4.58	4.57	4.63
4.81	4.71	4.69	4.73
4.81	4.72	4.70	4.74
4.79	4.78	4.77	4.81
4.79	4.77	4.77	4.80
4.83	4.75	4.73	4.79
4.81	4.71	4.72	4.76
4.81	4.72	4.74	4.77
4.83	4.76	4.77	4.80
4.81	4.73	4.75	4.77
4.82	4.75	4.77	4.80
4.82	4.75	4.76	4.80
4.80	4.73	4.75	4.78
4.78	4.71	4.72	4.73
4.79	4.71	4.71	4.73

Denote by T_2, T_3, T_5, and T_{10} the time series data vectors corresponding to the yield of the 2-year, 3-year, 5-year, and 10-year treasury bonds, respectively.

(i) Find the coefficients a, b_1, b_2, b_3 of the linear regression for the yield of the 3-year bond in terms of the yields of the 2-year, 5-year, and 10-year bonds, i.e., find a, b_1, b_2, b_3 corresponding to the solution to the ordinary least squares problem

$$T_3 \approx a\mathbf{1} + b_1 T_2 + b_2 T_5 + b_3 T_{10}, \qquad (8.29)$$

where $\mathbf{1}$ is the 15×1 column vector with all entries equal to 1. Let

$$T_{3,LR} = a\mathbf{1} + b_1 T_2 + b_2 T_5 + b_3 T_{10}.$$

Find the approximation error

$$\text{error}_{LR} = ||T_3 - T_{3,LR}||$$

of the linear regression.

(ii) Compute the linear interpolation values of the 3-year yield by doing linear interpolation between the 2-year yield and the 5-year yield at each data point. Denote by $T_{3,linear\_interp}$ the time series vector of these values. In other words,

$$T_{3,linear\_interp} = \frac{2}{3}T_2 + \frac{1}{3}T_5. \qquad (8.30)$$

Find the approximation error

$$\text{error}_{linear\_interp} = ||T_3 - T_{3,linear\_interp}||$$

of the linear interpolation.

(iii) Compute the cubic interpolation values of the 3-year yield by doing cubic spline interpolation between the 2-year, 5-year, and 10-year yield at each data point. Denote by $T_{3,cubic\_interp}$ the time series vector of these values. Find the approximation error

$$error_{cubic\_interp} \; = \; ||T_3 \; - \; T_{3,cubic\_interp}||$$

of the cubic interpolation.

(iv) Compare the approximation errors from (i), (ii), and (iii), and comment on the results.

Solution: (i) The ordinary least squares problem (8.29), i.e.,

$$T_3 \; \approx \; a\mathbf{1} \; + \; b_1 T_2 \; + \; b_2 T_5 \; + \; b_3 T_{10},$$

can be written in least squares form as $y \approx Ax$, with

$$x = \begin{pmatrix} a \\ b_1 \\ b_2 \\ b_3 \end{pmatrix} ; \; y = T_3 = \begin{pmatrix} 4.58 \\ 4.71 \\ 4.72 \\ 4.78 \\ 4.77 \\ 4.75 \\ 4.71 \\ 4.72 \\ 4.76 \\ 4.73 \\ 4.75 \\ 4.75 \\ 4.73 \\ 4.71 \\ 4.71 \end{pmatrix} ; \; A = (\mathbf{1} \; T_2 \; T_5 \; T_{10}) = \begin{pmatrix} 1 & 4.69 & 4.57 & 4.63 \\ 1 & 4.81 & 4.70 & 4.74 \\ 1 & 4.79 & 4.77 & 4.81 \\ 1 & 4.79 & 4.77 & 4.80 \\ 1 & 4.83 & 4.73 & 4.79 \\ 1 & 4.81 & 4.72 & 4.76 \\ 1 & 4.81 & 4.74 & 4.77 \\ 1 & 4.83 & 4.77 & 4.80 \\ 1 & 4.81 & 4.75 & 4.77 \\ 1 & 4.82 & 4.77 & 4.80 \\ 1 & 4.82 & 4.76 & 4.80 \\ 1 & 4.80 & 4.75 & 4.78 \\ 1 & 4.78 & 4.72 & 4.73 \\ 1 & 4.79 & 4.71 & 4.73 \end{pmatrix}$$

The solution to the least squares problem $y \approx Ax$ is $x = (A^t A)^{-1} A^t y$. By using the linear Cholesky solver linear_solve_cholesky, we obtain that

$$x = \text{linear\_solve\_cholesky}(A^t A, A^t y) = \begin{pmatrix} a \\ b_1 \\ b_2 \\ b_3 \end{pmatrix} = \begin{pmatrix} 0.0123 \\ 0.1272 \\ 0.3340 \\ 0.5298 \end{pmatrix}.$$

Thus, the ordinary least square linear regression for the yield of the 3-year bond in terms of the yields of the 2-year, 5-year, and 10-year bonds is

$$T_3 \; \approx \; 0.0123 \cdot \mathbf{1} \; + \; 0.1272 \; T_2 \; + \; 0.3340 \; T_5 \; + \; 0.5298 \; T_{10} \; = \; T_{3,LR}.$$

The approximation error of the linear regression is

$$error_{LR} \; = \; ||T_3 \; - \; T_{3,LR}|| \; = \; 0.0430.$$

(ii) From (8.30), we find that the time series vector $T_{3,linear\_interp}$ of the 3-year yield obtained by doing a linear interpolation between the 2-year yield and the 5-year yield at each data point is

$$T_{3,linear\_interp} = \frac{2}{3}T_2 + \frac{1}{3}T_5 = \begin{pmatrix} 4.6500 \\ 4.7700 \\ 4.7733 \\ 4.7833 \\ 4.7833 \\ 4.7967 \\ 4.7800 \\ 4.7867 \\ 4.8100 \\ 4.7900 \\ 4.8033 \\ 4.8000 \\ 4.7833 \\ 4.7600 \\ 4.7633 \end{pmatrix}.$$

The approximation error of the linear interpolation is

$$\text{error}_{linear\_interp} = ||T_3 - T_{3,linear\_interp}|| = 0.2066.$$

(iii) The time series vector $T_{3,cubic\_interp}$ of the 3-year yield obtained by doing cubic spline interpolation between the 2-year, 5-year, and 10-year yields at each data point is

$$T_{3,cubic\_interp} = \begin{pmatrix} 4.6413 \\ 4.7620 \\ 4.7659 \\ 4.7809 \\ 4.7812 \\ 4.7891 \\ 4.7737 \\ 4.7818 \\ 4.8057 \\ 4.7860 \\ 4.7996 \\ 4.7953 \\ 4.7796 \\ 4.7563 \\ 4.7582 \end{pmatrix}.$$

The corresponding approximation error is

$$\text{error}_{cubic\_interp} = ||T_3 - T_{3,cubic\_interp}|| = 0.1864.$$

(iv) The approximation errors corresponding to the linear regression of the entire time series data, the linear interpolation at each data point, and the cubic spline interpolation at each data point are

$$\text{error}_{LR} = 0.0430;$$
$$\text{error}_{linear\_interp} = 0.2066;$$
$$\text{error}_{cubic\_interp} = 0.1864.$$

As expected, the linear regression gives the most accurate results and does so by a rather wide margin, while the cubic spline interpolation has a smaller error than the linear interpolation.   □

Problem 8: The file *financials2012.xlsx* from www.fepress.org/nla-primer contains the end of week adjusted closing prices for the stocks of the following financial companies: JPM; GS; MS; BAC (Bank of America); RBS; CS; UBS; RY (RBC); BCS (Barclays) between January 11, 2012, and October 15, 2012.

(i) Compute the weekly percentage returns of these stocks.

(ii) Find the linear regression of the JPM returns with respect to the returns of the other stocks. What is the approximation error of this linear regression?

(iii) Find the linear regression of the JPM returns with respect to the returns of the other American financial companies, i.e., with respect to GS, MS, and BAC. What is the approximation error of this linear regression?

(iv) Find the linear regression of the JPM stock prices with respect to the prices of the other stocks. What is the approximation error of this linear regression? How does it compare with the approximation error of the linear regression of the JPM returns computed at (ii)?

Solution: (i) It is important to note that the data from the file *financials2012.xlsx* is reverse chronological and must be sorted from oldest to newest prior before the returns are computed.

Denote by M_weekly_prices the 41×9 matrix of the weekly prices of the stocks from *financials2012.xlsx* after they are ordered from oldest to newest, and denote by M_weekly_returns the 40×9 matrix of the weekly percentage returns of JPM; GS; MS; BAC; RBS; CS; UBS; RY; BCS, in this column order, e.g., column 3 corresponds to the weekly returns of MS. For example, the percentage return of MS during week 5 corresponds to M_weekly_returns$(5, 3)$ and is given by

$$
\begin{aligned}
\text{M\_weekly\_returns}(5,3) \;&=\; \frac{\text{M\_weekly\_prices}(6,3) - \text{M\_weekly\_prices}(5,3)}{\text{M\_weekly\_prices}(5,3)} \\
&=\; \frac{19.03 - 19.53}{19.53} \\
&=\; -0.0256;
\end{aligned}
$$

the percentage return of GS during week 12 corresponds to M_weekly_returns$(12, 2)$ and is given by

$$
\begin{aligned}
\text{M\_weekly\_returns}(12,2) \;&=\; \frac{\text{M\_weekly\_prices}(13,2) - \text{M\_weekly\_prices}(12,2)}{\text{M\_weekly\_prices}(12,2)} \\
&=\; \frac{116.92 - 123.24}{123.24} \\
&=\; -0.0513.
\end{aligned}
$$

The weekly percentage return matrix M_weekly_prices can be found below:

0.0401	0.0988	0.1063	0.0699	0.1671	0.1507	0.1479	0.0442	0.1273
−0.0041	0.0279	0.0115	0.0312	0.0140	0.0283	0.0258	−0.0049	0.0103
0.0289	0.0515	0.0944	0.0758	0.0436	0.0441	0.0438	0.0248	0.0718
−0.0176	−0.0290	−0.0317	0.0294	−0.0352	−0.0768	−0.0523	−0.0034	−0.0190
0.0231	0.0157	−0.0256	−0.0062	0.0091	0.0422	0.0305	−0.0013	0.0704
−0.0050	−0.0003	−0.0347	−0.0175	0.0215	0.0312	−0.0021	0.0235	−0.0006
0.0614	0.0384	0.0201	0.0331	−0.0232	−0.0138	−0.0303	0.0443	0.0323
0.0098	−0.0223	−0.0261	−0.0099	−0.0668	−0.0303	−0.0284	0.0050	−0.0625
0.0862	0.0481	0.0630	0.2167	0.0910	0.1165	0.0719	0.0214	0.0667
0.0133	0.0264	0.0407	0.0051	−0.0022	−0.0140	−0.0140	−0.0087	−0.0212
0.0181	−0.0143	−0.0337	−0.0275	−0.0145	−0.0238	−0.0149	−0.0004	−0.0421
−0.0294	−0.0513	−0.0636	−0.0356	−0.0939	−0.0655	−0.0655	−0.0137	−0.0827
−0.0254	−0.0246	−0.0608	−0.0597	−0.0187	−0.0342	−0.0477	−0.0212	−0.0218
−0.0113	−0.0230	0.0117	−0.0370	−0.0267	0.0133	0.0016	0.0283	0.0045
0.0145	0.0175	−0.0276	−0.0132	0.0392	−0.0549	0.0161	0.0186	0.0636
−0.0368	−0.0474	−0.0557	−0.0620	−0.0088	−0.0883	−0.0214	−0.0506	−0.0709
−0.1146	−0.0630	−0.0659	−0.0246	−0.0698	−0.0277	−0.0114	−0.0229	−0.0404
−0.0940	−0.0650	−0.1068	−0.0704	−0.1446	−0.0707	−0.0755	−0.0606	−0.1342
0.0003	0.0127	−0.0075	0.0186	0.0399	0.0072	0.0266	−0.0356	0.0243
−0.0470	−0.0375	−0.0394	−0.0168	−0.0521	−0.0401	−0.0277	−0.0162	−0.0633
0.0550	0.0206	0.0773	0.0770	0.1294	0.0692	0.0418	0.0200	0.1136
0.0401	0.0118	0.0425	0.0450	0.1218	−0.0697	0.0239	0.0243	0.0649
0.0273	−0.0212	−0.0105	0.0051	−0.0268	−0.0042	−0.0033	0.0101	−0.0071
−0.0073	0.0238	0.0312	0.0303	−0.1076	−0.0245	−0.0201	0.0102	−0.1834
−0.0432	−0.0041	−0.0303	−0.0636	−0.0779	−0.0262	−0.0589	0.0124	−0.0029
0.0642	0.0205	−0.0064	0.0209	0.0319	−0.0213	−0.0299	−0.0023	−0.0029
−0.0603	−0.0335	−0.0907	−0.0960	−0.0124	−0.0287	−0.0505	0	−0.0324
0.0883	0.0795	0.0597	0.0340	0.0814	0.0543	0.0798	0.0066	0.0660
−0.0216	−0.0065	0.0215	0.0164	−0.0145	−0.0364	−0.0173	−0.0002	0.0105
0.0243	0.0202	0.0602	0.0418	0.0338	0.0221	0.0121	0.0037	0.0858
0.0003	0.0057	−0.0014	0.0336	0.0355	0.0381	0.0128	0.0557	0.0477
0.0052	0.0131	−0.0021	0.0200	−0.0274	0.0526	0.0145	−0.0062	−0.0174
−0.0008	0.0117	0.0302	−0.0209	0.0141	0.0026	−0.0045	0.0330	−0.0194
0.0580	0.1004	0.1387	0.1028	0.0932	0.0987	0.1076	0.0261	0.1324
0.0579	0.0432	0.0679	0.0852	0.1463	0.0922	0.0915	0.0056	0.1245
−0.0167	−0.0382	−0.0636	−0.0461	−0.0111	−0.0100	−0.0415	−0.0066	−0.0263
−0.0099	−0.0260	−0.0199	−0.0307	−0.0662	−0.0756	−0.0573	0	−0.0381
0.0381	0.0495	0.0454	0.0555	0.0168	0.0671	0.0509	0.0223	0.0454
−0.0022	0.0075	−0.0109	−0.0215	0.0213	−0.0062	−0.0148	−0.0123	0.0193
0.0168	0.0285	0.0127	0.0351	0.0347	0.0473	0.0349	0.0162	0.0088

(ii) The linear regression of the JPM returns with respect to the returns of the other stocks, and including a 40×1 vector denoted by $\mathbf{1}_{40}$ with all entries equal to 1 corresponding to a constant weekly return, can be found by solving the least squares problem $y \approx Ax$, where x is the 9×1 vector of the regression coefficients, y is the 40×1 vector of the JPM returns corresponding to the first column of M_weekly_returns, and A is the 40×9 matrix given by a constant vector of with entry 1 and the returns of the other eight stocks corresponding to the columns 2, 3, ..., 9 of M_weekly_returns, i.e.,

$$y = \text{M\_weekly\_returns}(:,1);$$
$$A = [\ \mathbf{1}_{40}\ \text{M\_weekly\_returns}(:,2:9)\ ].$$

By using the Cholesky linear solver linear_solve_cholesky, we obtain that

$$x = (A^t A)^{-1}(A^t y) = \text{linear\_solve\_cholesky}(A^t A, A^t y)$$

$$= \begin{pmatrix} -0.0034 \\ 0.7663 \\ -0.0778 \\ 0.2980 \\ 0.2812 \\ -0.0576 \\ -0.4140 \\ 0.1476 \\ -0.0088 \end{pmatrix}.$$

The approximation error of this linear regression is

$$\text{error\_JPM\_weekly\_returns} = 0.1295.$$

(iii) The linear regression of the JPM returns with respect to the returns of GS, MS, and BAC, and including the 40×1 vector $\mathbf{1}_{40}$ with all entries equal to 1 corresponding to a constant weekly return, can be found by solving the least squares problem $y \approx A_s x_s$, where x_s is the 4×1 vector of the regression coefficients, y is the 40×1 vector of the JPM returns corresponding to the first column of M_weekly_returns, and A_s is the 40×4 matrix given by a constant vector of with entry 1 and the returns of GS, MS, and BAC corresponding to the columns 2, 3, and 4 of M_weekly_returns, i.e.,

$$y = \text{M\_weekly\_returns}(:, 1);$$
$$A_s = [\ \mathbf{1}_{40} \ \ \text{M\_weekly\_returns}(:, 2:4)\ ].$$

By using the Cholesky linear solver linear_solve_cholesky, we obtain that

$$x_s = (A_s^t A_s)^{-1}(A_s^t y) = \text{linear\_solve\_cholesky}(A_s^t A_s, A_s^t y)$$

$$= \begin{pmatrix} -0.0014 \\ 0.6537 \\ -0.0375 \\ 0.2613 \end{pmatrix}.$$

The approximation error of this linear regression is

$$\text{error\_JPM\_weekly\_returns\_Amer} = 0.1499.$$

The approximation error error_JPM_weekly_returns_Amer of the linear regression of the JPM weekly returns with respect to the weekly returns of the American financial stocks is larger than the approximation error error_JPM_weekly_returns $= 0.1312$ of the linear regression of the JPM weekly returns with respect to the weekly returns of all the financial stocks.

(iv) The linear regression of the JPM prices with respect to the prices of the other stocks, and including a 41×1 vector denoted by $\mathbf{1}_{41}$ with all entries equal to 1 corresponding to a constant weekly price, can be found by solving the least squares problem $y \approx Ax$, where x is the 9×1 vector of the regression coefficients, y is the 41×1 vector of the JPM weekly prices, and A is the 41×9 matrix given by a

constant vector with all entries equal to 1 and the weekly prices of the other eight
stocks corresponding to the columns 2, 3, ..., 9 of M_weekly_prices, i.e.,

$$y = \text{M\_weekly\_prices}(:, 1);$$
$$A = [\ \mathbf{1}_{41}\ \ \text{M\_weekly\_prices}(:, 2:9)\ ].$$

By using the Cholesky linear solver linear_solve_cholesky, we obtain that

$$x = (A^t A)^{-1}(A^t y) = \text{linear\_solve\_cholesky}(A^t A, A^t y)$$

$$= \begin{pmatrix} 8.8166 \\ 0.0166 \\ 0.0668 \\ 1.4713 \\ 1.2074 \\ 0.8999 \\ -2.5373 \\ 0.3502 \\ -0.1705 \end{pmatrix}.$$

The approximation error of this linear regression is

$$\text{error\_JPM\_weekly\_prices} = 6.5079.$$

As expected, this approximation is on a different scale than the approximation
error of the linear regression of the JPM returns computed at (ii), since stock prices
are on a different scale than stock returns.

Note that we could do a linear regression of the JPM prices with respect to the
prices of the other stocks without including the 41×1 vector $\mathbf{1}_{41}$ with all entries
equal to 1. In this case, the least squares problem $y \approx Ax$ that needs to be solved
corresponds to $y = \text{M\_weekly\_prices}(:, 1)$, i.e., the 41×1 vector of the JPM weekly
prices, and $A = \text{M\_weekly\_prices}(:, 2:9)$, i.e., the 41×8 matrix given by the weekly
prices of the other eight stocks. The 8×1 vector x of the regression coefficients is

$$x = \text{linear\_solve\_cholesky}(A^t A, A^t y) = \begin{pmatrix} -0.0012 \\ -0.0625 \\ 1.3119 \\ 1.0633 \\ 0.8533 \\ -1.7220 \\ 0.5252 \\ -0.4148 \end{pmatrix}.$$

The approximation error of the linear regression of the weekly prices of JPM over
the weekly prices of the other financial stocks without including a constant vector is

$$\text{error\_JPM\_weekly\_prices\_no\_const} = 6.6395,$$

which is very similar to the approximation error error_JPM_weekly_prices = 6.5079
for the linear regression including a constant vector.   □

Chapter 9

Efficient portfolios. Value at Risk. Portfolio VaR.

9.1 Exercises

1. Two stocks trade at $100 and $60, respectively. Their three-months returns have expected values of 8% and 4%, respectively, and standard deviation of 20% and 15%, respectively. The correlation of the returns is 25%.

   (i) Consider a portfolio made of 150 shares of the first stock and 500 shares of the second stock. What are the weights of each stock in this portfolio?

   (ii) Assume that you have $500,000 to invest. Find a portfolio made of the two stocks that has a 9% expected return.

   (iii) Identify the two portfolios fully invested in the two assets that have a 14.5% standard deviation of return. What are the expected returns of the two portfolios?

2. Assume that the asset allocation for a $100 million maximum return portfolio invested in three assets is $30 million in the first asset, $10 million in the second asset, $40 million in the third asset, and $20 million in cash. What is the asset allocation of the tangency portfolio made of these three assets?

3. Consider three assets with the following expected values, standard deviations, and correlations of their returns:

$$\mu_1 = 0.08; \quad \sigma_1 = 0.25; \quad \rho_{1,2} = -0.25;$$
$$\mu_2 = 0.12; \quad \sigma_2 = 0.25; \quad \rho_{2,3} = -0.25;$$
$$\mu_3 = 0.16; \quad \sigma_3 = 0.30; \quad \rho_{1,3} = 0.25.$$

The risk–free interest rate is 4%.

   (i) Find the asset allocation corresponding to the tangency portfolio. What are the expected value and the standard deviation of the return of the tangency portfolio?

(ii) Find the asset allocation corresponding to the minimum variance portfolio. What are the expected value and the standard deviation of the return of the minimum variance portfolio?

4. Assume that you invest $10 million in two different assets and cash. The three-months returns of the two assets have expected values of 8% and 12%, respectively, and standard deviations of 15% and 20%, respectively. The correlation of the returns of the two assets is 25%. The risk–free interest rate is 5%.

(i) Find the asset allocation for the tangency portfolio.

(ii) Find the asset allocation for a minimum variance portfolio with 7% expected return, and the standard deviation of the return of this portfolio.

(iii) Find the asset allocation for a minimum variance portfolio with 11% expected return, and the standard deviation of the return of this portfolio.

(iv) Find the asset allocation for a maximum return portfolio with 12% standard deviation of return, and the expected return of this portfolio.

(v) Find the asset allocation for a maximum return portfolio with 18% standard deviation of return, and the expected return of this portfolio.

(vi) Assume that the risk–free interest rate changes to 5.25%. How do you adjust the asset allocation of the minimum variance portfolio with 7% expected return in order to maintain a minimum variance portfolio with 7% expected return?

5. Consider two assets with three-months returns with expected values of 6% and 10%, respectively, and standard deviations of 25% and 35%, respectively. The correlation of the three-months returns of the two assets is -25%. The risk–free interest rate is 3%. Assume that you invest $100 million in the two assets.

(i) Find the minimum variance of the return of a portfolio made of the two assets. What is the expected return of this portfolio?

(ii) What is the 10–day 98% VaR of a $100 million portfolio, if it is invested in the first asset, invested in the second asset, or invested in the minimum variance portfolio, respectively?

6. Consider three assets with the following expected values, standard deviations, and correlations of their returns:

$$\mu_1 = 0.06; \quad \sigma_1 = 0.18; \quad \rho_{1,2} = -0.50;$$
$$\mu_2 = 0.09; \quad \sigma_2 = 0.20; \quad \rho_{2,3} = -0.25;$$
$$\mu_3 = 0.12; \quad \sigma_3 = 0.24; \quad \rho_{1,3} = 0.15.$$

The risk–free interest rate is 3%.

(i) Find the asset allocation for the tangency portfolio. What are the expected return and the standard deviation of the return of the tangency portfolio?

(ii) Find the asset allocation for a minimum variance portfolio with 10% expected return, and the standard deviation of the return of this portfolio.

(iii) Find the asset allocation for a maximum return portfolio with 20% standard deviation of return, and the expected return of this portfolio.

7. Consider five assets with the following expected values, standard deviations, and correlations of their returns:

$$\mu_1 = 0.08; \quad \sigma_1 = 0.25;$$
$$\mu_2 = 0.12; \quad \sigma_2 = 0.25;$$
$$\mu_3 = 0.16; \quad \sigma_3 = 0.30;$$
$$\mu_4 = 0.18; \quad \sigma_4 = 0.32;$$
$$\mu_5 = 0.21; \quad \sigma_5 = 0.35.$$

$$\rho_{1,2} = -0.25; \quad \rho_{1,3} = -0.25; \quad \rho_{1,4} = 0.35; \quad \rho_{1,5} = -0.10;$$
$$\rho_{2,3} = 0.30; \quad \rho_{2,4} = -0.50; \quad \rho_{2,5} = 0.10;$$
$$\rho_{3,4} = -0.30; \quad \rho_{3,5} = -0.35; \quad \rho_{4,5} = 0.65;$$

The risk–free interest rate is 5%.

(i) Find the asset allocation for a minimum variance portfolio with 17% expected return, and the standard deviation of the return of this portfolio;

(ii) Find the asset allocation for a maximum return portfolio with 30% standard deviation of return, and the expected return of this portfolio.

8. Consider four assets with the following expected returns over a fixed time period:

$$\mu_1 = 4\%; \quad \mu_2 = 3.5\%; \quad \mu_3 = 5\%; \quad \mu_4 = 3.4\%,$$

and with the following covariance matrix of their returns over the same time period:

$$\begin{pmatrix} 0.09 & 0.01 & 0.03 & -0.015 \\ 0.01 & 0.0625 & -0.02 & -0.01 \\ 0.03 & -0.02 & 0.1225 & 0.02 \\ -0.015 & -0.01 & 0.02 & 0.0576 \end{pmatrix}.$$

Assume that the risk–free interest rate is 1%.

(i) Find the asset allocation for the tangency portfolio. Find the expected value and the standard deviation of the return of the tangency portfolio. What is the Sharpe ratio of the tangency portfolio?

(ii) Find the asset allocation for a minimum variance portfolio with 3% expected return, and the standard deviation of the return of this portfolio. What is the Sharpe ratio of this portfolio?

(iii) Find the asset allocation for a maximum return portfolio with 27% standard deviation of return, and the expected return of this portfolio. What is the Sharpe ratio of this portfolio?

(iv) Find the asset allocation for the minimum variance portfolio fully invested in the assets (i.e., with no cash position). What is the Sharpe ratio of this portfolio?

9. Consider three assets with the following expected values, standard deviations, and correlations of their returns:

$$
\begin{array}{lll}
\mu_1 = 0.05; & \sigma_1 = 0.15; & \rho_{1,2} = -0.25; \\
\mu_2 = 0.09; & \sigma_2 = 0.20; & \rho_{2,3} = 0.25; \\
\mu_3 = 0.10; & \sigma_3 = 0.25; & \rho_{1,3} = 0.50.
\end{array}
$$

The risk–free interest rate is 2%.

(i) Find the asset allocation for a minimum variance portfolio with 8% expected return, and the standard deviation of the return of this portfolio.

(ii) Assume that the returns of the three assets have a joint multivariate normal distribution. Find the probability density function of the return of the minimum variance portfolio with 8% expected return.

(iii) Find the probability that the return of the minimum variance portfolio with 8% expected return is between 7% and 9%. Also, find the probability that the return of this portfolio is below 5%, and the probability that the return of this portfolio is above 10%.

(iv) Consider a portfolio equally invested in each of the three assets. Note that the expected return of this portfolio is 8%. Find the probabilities that the return of this portfolio is between 7% and 9%, is below 5%, and is above 10%, respectively.

10. Let μ_1, σ_1 and μ_2, σ_2 be the expected values and the standard deviations of the returns of two assets over a fixed time period, respectively, and let $\rho_{1,2}$ be the correlation of the returns of the two assets over the same time period. Recall that the minimum variance portfolio fully invested in two assets is obtained by allocating

$$
w_1 = \frac{\sigma_2(\sigma_2 - \rho_{1,2}\sigma_1)}{\sigma_1^2 - 2\sigma_1\sigma_2\rho_{1,2} + \sigma_2^2}
$$

of the portfolio value to the first asset and $w_2 = 1 - w_1$ of the portfolio value to the second asset.

Show that the minimum variance portfolio has only long asset positions, i.e., $0 \le w_1, w_2 \le 1$, if and only if

$$
\rho_{1,2} \le \min\left(\frac{\sigma_1}{\sigma_2}, \frac{\sigma_2}{\sigma_1}\right).
$$

11. (i) Show that the asset allocation for a minimum variance portfolio with expected return μ_P can also be written as follows in terms of the asset weights vector w_T of the tangency portfolio:

- asset weights vector w_{min} given by

$$w_{min} = \frac{\mu_P - r_f}{\bar{\mu}^t w_T} w_T. \tag{9.1}$$

- weight $w_{min,cash}$ of the cash position given by

$$w_{min,cash} = 1 - 1^t w_{min}; \tag{9.2}$$

(ii) Write a pseudocode for computing the asset allocation for a minimum variance portfolio with expected return μ_P using the formulas (9.1) and (9.2).

12. (i) Show that the asset allocation for a maximum return portfolio with variance of return equal to σ_P^2 can also be written as follows in terms of the asset weights vector w_T of the tangency portfolio:

- asset weights vector w_{max} given by

$$w_{max} = \frac{\sigma_P}{\sqrt{w_T^t \Sigma_R w_T}} \cdot \text{sign}(1^t \Sigma_R^{-1} \bar{\mu}) w_T; \tag{9.3}$$

- weight $w_{max,cash}$ of the cash position equal to $1 - 1^t w_{max}$, i.e.,

$$w_{max,cash} = 1 - 1^t w_{max}. \tag{9.4}$$

(ii) Write a pseudocode for computing the asset allocation for maximum return portfolio with variance of return equal to σ_P^2 using the formulas (9.3) and (9.4).

13. Assume that the 99% two day VaR of a portfolio is $10 million. Estimate the two day 95% VaR of the portfolio and the five day 95% VaR of the portfolio.

14. Show that if two portfolios and their combined portfolio have normally distributed returns, then the VaR of the combined portfolio is smaller than the sum of the VaRs of each individual portfolio.

In other words, if V_1, V_2, and $V = V_1 + V_2$ are portfolios with normal returns, show that

$$\text{VaR}_V(N,C) \leq \text{VaR}_{V_1}(N,C) + \text{VaR}_{V_2}(N,C),$$

where $\text{VaR}_V(N,C)$, $\text{VaR}_{V_1}(N,C)$, and $\text{VaR}_{V_2}(N,C)$ are the N day $C\%$ VaRs of V, V_1, and V_2, respectively.

Hint: Recall that, for a portfolio W with normal returns,

$$\text{VaR}_W(N,C) = \sqrt{\frac{N}{252}} z_C \sigma_{R_W} W(0) - \frac{N}{252} \mu_{R_W} W(0),$$

and use the fact that the return R_V of the combined portfolio is the weighted average of the returns R_{V_1} and R_{V_2} of the two portfolios, i.e.,

$$R_V = \frac{V_1(0)}{V(0)} R_{V_1} + \frac{V_2(0)}{V(0)} R_{V_2}.$$

15. Consider three assets with multivariate normal distribution of their returns. The expected values, standard deviations, and correlations of their returns are

$$\mu_1 = 0.08; \quad \sigma_1 = 0.25; \quad \rho_{1,2} = -0.25;$$
$$\mu_2 = 0.12; \quad \sigma_2 = 0.25; \quad \rho_{2,3} = -0.25;$$
$$\mu_3 = 0.16; \quad \sigma_3 = 0.30; \quad \rho_{1,3} = 0.25.$$

(i) What is the 5–day 95% VaR of $100 million portfolios fully invested in the first asset, fully invested in the second asset, and fully invested in the third asset, respectively?

(ii) What is the minimal 5–day 95% VaR of a $100 million portfolio fully invested in the first and second asset, fully invested in the second and third asset, and fully invested in the first and third asset, respectively?

(iii) What is the minimal 5–day 95% VaR of a $100 million portfolio fully invested in all three assets?

9.2   Solutions to Chapter 9 Exercises

Problem 1: Two stocks trade at $100 and $60, respectively. Their three-months returns have expected values of 8% and 4%, respectively, and standard deviation of 20% and 15%, respectively. The correlation of the returns is 25%.

(i) Consider a portfolio made of 150 shares of the first stock and 500 shares of the second stock. What are the weights of each stock in this portfolio?

(ii) Assume that you have $500,000 to invest. Find a portfolio made of the two stocks that has a 9% expected return.

(iii) Identify the two portfolios fully invested in the two assets that have a 14.5% standard deviation of return. What are the expected returns of the two portfolios?

Solution: The expected values, the standard deviations, and the correlation of the returns of the two stocks are:

$$\mu_1 = 0.08; \quad \sigma_1 = 0.20; \quad \rho_{1,2} = 0.25;$$
$$\mu_2 = 0.04; \quad \sigma_2 = 0.15;$$

(i) The value of the portfolio position in the first stock is $150 \cdot \$100 = \$15,000$, the value of the portfolio position in the second stock is $500 \cdot \$60 = \$30,000$, and the total value of the portfolio is $45,000. Thus, the weights w_1 and w_2 of the first and of the second stock in the portfolio are, respectively,[1]

$$w_1 = \frac{15,000}{45,000} = \frac{1}{3}; \quad w_2 = \frac{30,000}{45,000} = \frac{2}{3}.$$

(ii) Let w and $1 - w$ be the weights of the first stock and of the second stock, respectively. The return R of the portfolio is

$$R = wR_1 + (1 - w)R_2, \tag{9.5}$$

where R_1 and R_2 denote the returns of the first and second stock, respectively. The expected value μ_R of the return of the portfolio is given by

$$\mu_R = w\mu_1 + (1 - w)\mu_2 = 0.08w + (1 - w)0.04 = 0.04 + 0.04w. \tag{9.6}$$

By solving (9.6) for $\mu_R = 0.09$, we obtain that the weights of the first and second stock are $w = \frac{0.05}{0.04} = 1.25$ and $1 - w = -0.25$, respectively, which, in a $500,000 portfolio, corresponds to a $625,000 position in the first stock and a $-\$125,000$ position in the second stock.

In other words, a $500,000 portfolio with 9% expected return made of the two stocks can be obtained by taking a $625,000 long position in the first stock and a $125,000 short position in the second stock.

(iii) From (9.5), it follows that the variance of the return R of the portfolio is

$$\sigma_R^2 = w^2\sigma_1^2 + 2w(1 - w)\rho_{1,2}\sigma_1\sigma_2 + (1 - w)^2\sigma_2^2$$
$$= 0.04w^2 + 0.015w(1 - w) + 0.0225(1 - w)^2$$
$$= 0.0475w^2 - 0.03w + 0.0225. \tag{9.7}$$

---

[1]Note that, as expected, $w_1 + w_2 = 1$.

By solving (9.7) for $\sigma_R = 0.145$, we obtain two solutions, $w_1 = 0.5778$ and $w_2 = 0.0537$.

If $w = w_1 = 0.5778$, then the weights of the first and of the second stock in the portfolio are $w_1 = 0.5778$ and $1 - w_1 = 0.4222$, which, in a \$500,000 portfolio, corresponds to a \$288,920 position in the first stock and a \$211,080 position in the second stock. In other words, a \$500,000 portfolio with 14.5% standard deviation of return can be obtained by taking a \$288,920 long position in the first stock and a \$211,080 long position in the second stock. This portfolio has expected return $w_1\mu_1 + (1 - w_1)\mu_2 = 0.0631 = 6.31\%$.

If $w = w_2 = 0.0537$, then the weights of the first and of the second stock in the portfolio are $w_2 = 0.0537$ and $1 - w_2 = 0.9463$, which, in a \$500,000 portfolio, corresponds to a \$26,830 position in the first stock and a \$473,130 position in the second stock. In other words, a \$500,000 portfolio with 14.5% standard deviation of return can be obtained by taking a \$26,830 long position in the first stock and a \$473,130 long position in the second stock. This portfolio has expected return $w_2\mu_1 + (1 - w_2)\mu_2 = 0.0421 = 4.21\%$.   □

Problem 2: Assume that the asset allocation for a \$100 million maximum return portfolio invested in three assets is \$30 million in the first asset, \$10 million in the second asset, \$40 million in the third asset, and \$20 million in cash. What is the asset allocation of the tangency portfolio made of these three assets?

Solution: Recall that any maximum return portfolio consists of a cash position with the rest of the portfolio invested in the tangency portfolio.

In other words, the \$80 million assets position in the given maximum return portfolio corresponds to the tangency portfolio, i.e., the \$80 million position in the tangency portfolio is made of \$30 million in the first asset, \$10 million in the second asset, \$40 million in the third asset. This corresponds to the following asset allocation:

$$\frac{30}{80} = 0.375 \quad = \quad 37.5\% \;\; \text{first asset};$$

$$\frac{10}{80} = 0.125 \quad = \quad 12.5\% \;\; \text{second asset};$$

$$\frac{40}{80} = 0.5 \quad = \quad 50\% \;\; \text{third asset};$$

We conclude that the tangency portfolio is 37.5% invested in the first asset, 12.5% invested in the second asset, and 50% invested in the third asset.   □

Problem 3: Consider three assets with the following expected values, standard deviations, and correlations of their returns:

$$\mu_1 = 0.08; \quad \sigma_1 = 0.25; \quad \rho_{1,2} = -0.25;$$
$$\mu_2 = 0.12; \quad \sigma_2 = 0.25; \quad \rho_{2,3} = -0.25;$$
$$\mu_3 = 0.16; \quad \sigma_3 = 0.30; \quad \rho_{1,3} = 0.25.$$

The risk–free interest rate is 4%.

(i) Find the asset allocation corresponding to the tangency portfolio. What are the expected value and the standard deviation of the return of the tangency portfolio?

(ii) Find the asset allocation corresponding to the minimum variance portfolio. What are the expected value and the standard deviation of the return of the minimum variance portfolio?

Solution: Note that $r_f = 0.04$ and

$$\bar{\mu} = \mu - r_f \mathbf{1} = \begin{pmatrix} 0.04 \\ 0.08 \\ 0.12 \end{pmatrix};$$

$$\Sigma_R = \begin{pmatrix} \sigma_1^2 & \sigma_1\sigma_2\rho_{1,2} & \sigma_1\sigma_3\rho_{1,3} \\ \sigma_1\sigma_2\rho_{1,2} & \sigma_2^2 & \sigma_2\sigma_3\rho_{2,3} \\ \sigma_1\sigma_3\rho_{1,3} & \sigma_2\sigma_3\rho_{2,3} & \sigma_3^2 \end{pmatrix} = \begin{pmatrix} 0.0625 & -0.0156 & 0.0187 \\ -0.0156 & 0.0625 & -0.0187 \\ 0.0187 & -0.0187 & 0.0900 \end{pmatrix}.$$

(i) The asset weights vector of the tangency portfolio is

$$w_T = \frac{1}{\mathbf{1}^t \Sigma_R^{-1} \bar{\mu}} \Sigma_R^{-1} \bar{\mu} = \begin{pmatrix} 0.1538 \\ 0.4615 \\ 0.3846 \end{pmatrix},$$

where the vector $\Sigma_R^{-1} \bar{\mu} = $ linear_solve_cholesky$(\Sigma_R, \bar{\mu})$ was computed by using the Cholesky decomposition to solve the linear system corresponding to the matrix Σ_R and to the right hand side $\bar{\mu}$.

In other words, the tangency portfolio is obtained by investing 15.38% of the portfolio in asset 1, 46.15% of the portfolio in asset 2, and 38.46% of the portfolio in asset 3.

The expected value and the standard deviation of the return of the tangency portfolio are given by

$$\mu_T = r_f + \bar{\mu}^t w_T = 0.1292;$$

$$\sigma_T = \sqrt{w_T^t \Sigma_R w_T} = 0.1465.$$

Thus, the tangency portfolio has a 12.92% expected return with a 14.65% standard deviation of return.

(ii) The asset weights vector of the minimum variance portfolio with no cash position is given by

$$w_m = \frac{1}{\mathbf{1}^t \Sigma_R^{-1} \mathbf{1}} \Sigma_R^{-1} \mathbf{1},$$

where $\mathbf{1} = \begin{pmatrix} 1 \\ 1 \\ 1 \end{pmatrix}$. By computing the vector $\Sigma_R^{-1} \mathbf{1} = $ linear_solve_cholesky$(\Sigma_R, \mathbf{1})$ using the Cholesky decomposition to solve the linear system corresponding to the matrix Σ_R and to the right hand side $\mathbf{1}$, we obtain that

$$w_m = \begin{pmatrix} 0.3339 \\ 0.4417 \\ 0.2244 \end{pmatrix},$$

i.e., the minimum variance portfolio with no cash position is obtained by investing 33.39% of the portfolio in asset 1, 44.17% of the portfolio in asset 2, and 22.44% of the portfolio in asset 3.

The expected value and the standard deviation of the return of the minimum variance portfolio with no cash position are

$$\mu_m \;=\; r_f + \bar{\mu}^t w_m \;=\; 0.1156 \;=\; 11.56\%;$$

$$\sigma_m \;=\; \sqrt{w_m^t \Sigma_R w_m} \;=\; 0.1348 \;=\; 13.48\%. \quad \square$$

Problem 4: Assume that you invest \$10 million in two different assets and cash. The three-months returns of the two assets have expected values of 8% and 12%, respectively, and standard deviations of 15% and 20%, respectively. The correlation of the returns of the two assets is 25%. The risk–free interest rate is 5%.

(i) Find the asset allocation for the tangency portfolio.

(ii) Find the asset allocation for a minimum variance portfolio with 7% expected return, and the standard deviation of the return of this portfolio.

(iii) Find the asset allocation for a minimum variance portfolio with 11% expected return, and the standard deviation of the return of this portfolio.

(iv) Find the asset allocation for a maximum return portfolio with 12% standard deviation of return, and the expected return of this portfolio.

(v) Find the asset allocation for a maximum return portfolio with 18% standard deviation of return, and the expected return of this portfolio.

(vi) Assume that the risk–free interest rate changes to 5.25%. How do you adjust the asset allocation of the minimum variance portfolio with 7% expected return in order to maintain a minimum variance portfolio with 7% expected return?

Solution: Let
$$\mu_1 \;=\; 0.08; \quad \sigma_1 \;=\; 0.15; \quad \rho_{1,2} \;=\; 0.25;$$
$$\mu_2 \;=\; 0.12; \quad \sigma_2 \;=\; 0.20.$$

Note that $r_f = 0.05$ and

$$\bar{\mu} \;=\; \mu - r_f \mathbf{1} \;=\; \begin{pmatrix} 0.03 \\ 0.07 \end{pmatrix};$$

$$\Sigma_R \;=\; \begin{pmatrix} \sigma_1^2 & \sigma_1\sigma_2\rho_{1,2} \\ \sigma_1\sigma_2\rho_{1,2} & \sigma_2^2 \end{pmatrix} \;=\; \begin{pmatrix} 0.0625 & -0.0219 \\ -0.0219 & 0.1225 \end{pmatrix}.$$

(i) The asset weights vector of the tangency portfolio is

$$w_T \;=\; \frac{1}{\mathbf{1}^t \Sigma_R^{-1} \bar{\mu}} \Sigma_R^{-1} \bar{\mu} \;=\; \begin{pmatrix} 0.3333 \\ 0.6667 \end{pmatrix},$$

where the vector $\Sigma_R^{-1} \bar{\mu} = $ linear_solve_cholesky($\Sigma_R, \bar{\mu}$) was computed by using the Cholesky decomposition to solve the linear system corresponding to the matrix Σ_R and to the right hand side $\bar{\mu}$.

In other words, the tangency portfolio is obtained by investing one third of the portfolio in asset 1 and two thirds of the portfolio in asset 2.

(ii) The asset weights vector $w_{min,1}$ and the weight $w_{min,cash,1}$ of the cash position of the minimum variance portfolio with 7% expected return, i.e., corresponding to $\mu_{P,1} = 0.07$ in the formulas below, are given by

$$w_{min,cash,1} = 1 - \frac{\mu_{P,1} - r_f}{\mu^t w_T} = 0.6471;$$

$$w_{min,1} = (1 - w_{min,cash,1})w_T = \begin{pmatrix} 0.1176 \\ 0.2353 \end{pmatrix}.$$

In other words, the minimum variance portfolio with 7% expected return is obtained by holding a cash position equal to 64.71% of the total value of the portfolio, and by investing 11.76% of the portfolio value in asset 1 and 23.53% of the portfolio value in asset 2.

The standard deviation of the return of the minimum variance portfolio with 7% expected return is

$$\sigma_{min,1} = \sqrt{w_{min,1}^t \Sigma_R w_{min,1}} = 0.0542 = 5.42\%.$$

(iii) The asset weights vector $w_{min,2}$ and the weight $w_{min,cash,2}$ of the cash position of the minimum variance portfolio with 11% expected return, i.e., corresponding to $\mu_{P,2} = 0.11$ in the formulas below, are given by

$$w_{min,cash,2} = 1 - \frac{\mu_{P,2} - r_f}{\mu^t w_T} = -0.0588;$$

$$w_{min,2} = (1 - w_{min,cash,2})w_T = \begin{pmatrix} 0.3529 \\ 0.7059 \end{pmatrix}.$$

Thus, the minimum variance portfolio with 11% expected return is obtained by borrowing an amount of cash equal to 5.88% of the total value of the portfolio and by investing 35.29% of the portfolio value in asset 1 and 70.59% of the portfolio value in asset 2.

The standard deviation of the return of the minimum variance portfolio with 11% expected return is

$$\sigma_{min,2} = \sqrt{w_{min,2}^t \Sigma_R w_{min,2}} = 0.1627 = 16.27\%.$$

(iv) The asset weights vector $w_{max,1}$ and the weight $w_{max,cash,1}$ of the cash position of the maximum return portfolio with 12% standard deviation of return, i.e., corresponding to $\sigma_{P,1} = 0.12$ in the formulas below, are given by

$$w_{max,cash,1} = 1 - \frac{\sigma_{P,1}}{\sqrt{w_T^t \Sigma_R w_T}} \cdot \text{sign}(1^t \Sigma_R^{-1} \mu) = 0.2191;$$

$$w_{max,1} = (1 - w_{max,cash,1})w_T = \begin{pmatrix} 0.2603 \\ 0.5206 \end{pmatrix}.$$

In other words, the maximum return portfolio with 12% standard deviation of return is obtained by keeping 21.91% of the total value of the portfolio in cash and investing 26.03% of the portfolio value in asset 1 and 52.06% of the portfolio value in asset 2.

The expected return of the maximum return portfolio with 12% standard deviation of return is

$$\mu_{max} = r_f + \bar{\mu}^t w_{max} = 0.0943 = 9.43\%.$$

(v) The asset weights vector $w_{max,2}$ and the weight $w_{max,cash,2}$ of the cash position of the maximum return portfolio with 18% standard deviation of return, i.e., corresponding to $\sigma_{P,2} = 0.18$ in the formulas below, are given by

$$w_{max,cash,2} = 1 - \frac{\sigma_{P,2}}{\sqrt{w_T^t \Sigma_R w_T}} \cdot \text{sign}(\mathbf{1}^t \Sigma_R^{-1} \bar{\mu}) = -0.1714;$$

$$w_{max,2} = (1 - w_{max,cash,1}) w_T = \begin{pmatrix} 0.3905 \\ 0.7809 \end{pmatrix}.$$

Thus, the maximum return portfolio with 18% standard deviation of return is obtained by borrowing an amount of cash equal to 17.14% of the total value of the portfolio, and by investing 39.05% of the portfolio value in asset 1 and 78.09% of the portfolio value in asset 2.

The expected return of the maximum return portfolio with 18% standard deviation of return is

$$\mu_{max} = r_f + \bar{\mu}^t w_{max} = 0.1164 = 11.64\%.$$

(vi) If the risk–free interest rate changes to $r_{f,3} = 0.0525$, then $\bar{\mu}$ changes to

$$\bar{\mu}_3 = \mu - r_f \mathbf{1} = \begin{pmatrix} 0.0275 \\ 0.0675 \end{pmatrix},$$

and the asset weights vector of the new tangency portfolio is

$$w_{T,3} = \frac{1}{\mathbf{1}^t \Sigma_R^{-1} \bar{\mu}_3} \Sigma_R^{-1} \bar{\mu}_3 = \begin{pmatrix} 0.3115 \\ 0.6885 \end{pmatrix}.$$

Then, the asset weights vector $w_{min,3}$ and the weight $w_{min,cash,3}$ of the cash position of the minimum variance portfolio with 7% expected return are given by

$$w_{min,cash,3} = 1 - \frac{\mu_{P,1} - r_{f,3}}{\bar{\mu}^t w_{T,3}} = 0.6821;$$

$$w_{min,3} = (1 - w_{min,cash,3}) w_{T,3} = \begin{pmatrix} 0.0990 \\ 0.2189 \end{pmatrix},$$

where $\mu_{P,1} = 0.07$.

In other words, the minimum variance portfolio with 7% expected return if the risk–free rate is 5.25% is obtained by holding a cash position equal to 68.21% of the total value of the portfolio, and by investing 9.90% of the portfolio value in asset 1 and 21.89% of the portfolio value in asset 2.

When comparing this portfolio with the minimum variance portfolio with 7% expected return corresponding to a risk–free interest rate of 5%, we note that, if the risk–free rate becomes 5.25%, then the cash position must be increased from 64.71% to 68.21% of the total value of the portfolio, while the position in asset 1 must be reduced from 11.76% to 9.90% of the portfolio value and the position in asset 2 must be reduced from 23.53% to 21.89% of the portfolio value. □

Problem 5: Consider two assets with three-months returns with expected values of 6% and 10%, respectively, and standard deviations of 25% and 35%, respectively. The correlation of the three-months returns of the two assets is -25%. The risk–free interest rate is 3%. Assume that you invest $100 million in the two assets.

(i) Find the minimum variance of the return of a portfolio made of the two assets. What is the expected return of this portfolio?

(ii) What is the 10–day 98% VaR of a $100 million portfolio, if it is invested in the first asset, invested in the second asset, or invested in the minimum variance portfolio, respectively?

Solution: (i) Let

$$\mu_1 = 0.06; \quad \sigma_1 = 0.25; \quad \rho_{1,2} = -0.25;$$
$$\mu_2 = 0.10; \quad \sigma_2 = 0.35.$$

Note that $r_f = 0.03$ and

$$\bar{\mu} = \mu - r_f \mathbf{1} = \begin{pmatrix} 0.03 \\ 0.07 \end{pmatrix};$$

$$\Sigma_R = \begin{pmatrix} \sigma_1^2 & \sigma_1\sigma_2\rho_{1,2} \\ \sigma_1\sigma_2\rho_{1,2} & \sigma_2^2 \end{pmatrix} = \begin{pmatrix} 0.0625 & -0.0219 \\ -0.0219 & 0.1225 \end{pmatrix}.$$

The asset weights vector of the minimum variance portfolio with no cash position is given by

$$w_m = \frac{1}{\mathbf{1}^t \Sigma_R^{-1} \mathbf{1}} \Sigma_R^{-1} \mathbf{1} = \begin{pmatrix} 0.6311 \\ 0.3689 \end{pmatrix},$$

where the vector $\Sigma_R^{-1} \mathbf{1} = \text{linear\_solve\_cholesky}\left(\Sigma_R, \begin{pmatrix} 1 \\ 1 \end{pmatrix}\right)$ was computed using a Cholesky decomposition linear solver.

In other words, the minimum variance portfolio with no cash position is obtained by investing 63.11% of the portfolio in asset 1 and 36.89% of the portfolio in asset 2.

The expected value and the standard deviation of the return of the minimum variance portfolio with no cash position are given by

$$\mu_m = r_f + \bar{\mu}^t w_m = 0.0748 = 7.48\%;$$
$$\sigma_m = \sqrt{w_m^t \Sigma_R w_m} = 0.1771 = 17.71\%.$$

Thus, the minimum variance portfolio with no cash position has a 7.48% expected return with a 17.71% standard deviation of return, which corresponds to a 3.14% variance of the return.

(ii) The N day $C\%$ VaR of a portfolio can be approximated as follows:

$$\text{VaR}(N, C) \approx \sqrt{\frac{N}{252}} \sigma_{R_V} z_C V(0), \tag{9.8}$$

where σ_{R_V} is the annualized standard deviation of the return of the portfolio, z_C is the z–score of the standard normal distribution corresponding to C, and $V(0)$ is the current value of the portfolio.

Let $\mathrm{VaR}_1(10 \text{ days}, 98\%)$, $\mathrm{VaR}_2(10 \text{ days}, 98\%)$, $\mathrm{VaR}_m(10 \text{ days}, 98\%)$ be the 10–day 98% VaR of the $100 million portfolio fully invested in the first asset, fully invested in the second asset, and invested in the minimum variance portfolio fully invested in the two assets, respectively.

From the approximate VaR formula (9.8) with $V(0) = \$100\mathrm{mil}$, $N = 10$, and $z_{98} = 2.0537$, we obtain that

$$\mathrm{VaR}_1(10 \text{ days}, 98\%) \approx \sqrt{\frac{10}{252}} \sigma_1 z_{98} \cdot \$100\mathrm{mil} = \$10.228\mathrm{mil};$$

$$\mathrm{VaR}_2(10 \text{ days}, 98\%) \approx \sqrt{\frac{10}{252}} \sigma_2 z_{98} \cdot \$100\mathrm{mil} = \$14.319\mathrm{mil};$$

$$\mathrm{VaR}_m(10 \text{ days}, 98\%) \approx \sqrt{\frac{10}{252}} \sigma_m z_{98} \cdot \$100\mathrm{mil} = \$7.245\mathrm{mil}.$$

The VaR of the minimum variance portfolio is significantly smaller than the VaR of both the portfolio invested in the first asset, and of the portfolio invested in the second asset, which shows that portfolio diversification can be used to reduce the risk of the portfolio. □

Problem 6: Consider three assets with the following expected values, standard deviations, and correlations of their returns:

$$\begin{aligned}
\mu_1 &= 0.06; & \sigma_1 &= 0.18; & \rho_{1,2} &= -0.50; \\
\mu_2 &= 0.09; & \sigma_2 &= 0.20; & \rho_{2,3} &= -0.25; \\
\mu_3 &= 0.12; & \sigma_3 &= 0.24; & \rho_{1,3} &= 0.15.
\end{aligned}$$

The risk–free interest rate is 3%.

(i) Find the asset allocation for the tangency portfolio. What are the expected return and the standard deviation of the return of the tangency portfolio?

(ii) Find the asset allocation for a minimum variance portfolio with 10% expected return, and the standard deviation of the return of this portfolio.

(iii) Find the asset allocation for a maximum return portfolio with 20% standard deviation of return, and the expected return of this portfolio.

Solution: Note that $r_f = 0.03$ and

$$\bar{\mu} = \mu - r_f\mathbf{1} = \begin{pmatrix} 0.03 \\ 0.06 \\ 0.09 \end{pmatrix};$$

$$\Sigma_R = \begin{pmatrix} \sigma_1^2 & \sigma_1\sigma_2\rho_{1,2} & \sigma_1\sigma_3\rho_{1,3} \\ \sigma_1\sigma_2\rho_{1,2} & \sigma_2^2 & \sigma_2\sigma_3\rho_{2,3} \\ \sigma_1\sigma_3\rho_{1,3} & \sigma_2\sigma_3\rho_{2,3} & \sigma_3^2 \end{pmatrix} = \begin{pmatrix} 0.0324 & -0.0180 & 0.0065 \\ -0.0180 & 0.0400 & -0.0120 \\ 0.0065 & -0.0120 & 0.0576 \end{pmatrix}.$$

(i) The asset weights vector of the tangency portfolio is given by

$$w_T \;=\; \frac{1}{1^t \Sigma_R^{-1} \overline{\mu}} \Sigma_R^{-1} \overline{\mu} \;=\; \begin{pmatrix} 0.3087 \\ 0.4241 \\ 0.2672 \end{pmatrix},$$

where the vector $\Sigma_R^{-1}\overline{\mu} =$ linear_solve_cholesky$(\Sigma_R, \overline{\mu})$ was computed by using the Cholesky decomposition to solve the linear system corresponding to the matrix Σ_R and to the right hand side $\overline{\mu}$.

Thus, the tangency portfolio is obtained by investing 30.87% of the portfolio in asset 1, 42.41% of the portfolio in asset 2, and 26.72% of the portfolio in asset 3.

The expected value and the standard deviation of the return of the tangency portfolio are given by

$$
\begin{aligned}
\mu_T &= r_f + \overline{\mu}^t w_T = 0.0888 = 8.88\%; \\
\sigma_T &= \sqrt{w_T^t \Sigma_R w_T} = 0.0896 = 8.96\%.
\end{aligned}
$$

(ii) The asset weights vector w_{min} and the weight $w_{min,cash}$ of the cash position of the minimum variance portfolio with 10% expected return, i.e., corresponding to $\mu_P = 0.10$ in the formulas below, are given by

$$
\begin{aligned}
w_{min,cash} &= 1 - \frac{\mu_P - r_f}{\overline{\mu}^t w_T} = -0.1914; \\
w_{min} &= (1 - w_{min,cash}) w_T = \begin{pmatrix} 0.3678 \\ 0.5053 \\ 0.3183 \end{pmatrix}.
\end{aligned}
$$

In other words, the minimum variance portfolio with 10% expected return is obtained by borrowing an amount of cash equal to 19.14% of the total value of the portfolio, and by investing 36.78% of the portfolio value in asset 1, 50.53% of the portfolio value in asset 2, and 31.83% of the portfolio value in asset 3.

The standard deviation of the return of the minimum variance portfolio with 10% expected return is

$$\sigma_{min} \;=\; \sqrt{w_{min}^t \Sigma_R w_{min}} \;=\; 0.1068 \;=\; 10.68\%.$$

(iii) The asset weights vector w_{max} and the weight $w_{max,cash}$ of the cash position of the maximum return portfolio with 20% standard deviation of return, i.e., corresponding to $\sigma_P = 0.20$ in the formulas below, are given by

$$
\begin{aligned}
w_{max,cash} &= 1 - \frac{\sigma_P}{\sqrt{w_T^t \Sigma_R w_T}} \cdot \mathrm{sign}(1^t \Sigma_R^{-1} \overline{\mu}) = -1.2319; \\
w_{max} &= (1 - w_{max,cash}) w_T = \begin{pmatrix} 0.6890 \\ 0.9465 \\ 0.5963 \end{pmatrix}.
\end{aligned}
$$

In other words, the maximum return portfolio with 20% standard deviation of return is obtained by borrowing an amount of cash equal to 123.19% of the total value of the portfolio, and by investing 68.90% of the portfolio value in asset 1, 94.65% of the portfolio value in asset 2, and 59.63% of the portfolio value in asset 3.

The expected return of the maximum return portfolio with 20% standard deviation of return is

$$\mu_{max} = r_f + \overline{\mu}^t w_{max} = 0.1611 = 16.11\%. \quad \square$$

Problem 7: Consider five assets with the following expected values, standard deviations, and correlations of their returns:

$$
\begin{aligned}
\mu_1 &= 0.08; & \sigma_1 &= 0.25; \\
\mu_2 &= 0.12; & \sigma_2 &= 0.25; \\
\mu_3 &= 0.16; & \sigma_3 &= 0.30; \\
\mu_4 &= 0.18; & \sigma_4 &= 0.32; \\
\mu_5 &= 0.21; & \sigma_5 &= 0.35.
\end{aligned}
$$

$$
\begin{aligned}
\rho_{1,2} &= -0.25; & \rho_{1,3} &= -0.25; & \rho_{1,4} &= 0.35; & \rho_{1,5} &= -0.10; \\
\rho_{2,3} &= 0.30; & \rho_{2,4} &= -0.50; & \rho_{2,5} &= 0.10; \\
\rho_{3,4} &= -0.30; & \rho_{3,5} &= -0.35; & \rho_{4,5} &= 0.65;
\end{aligned}
$$

The risk–free interest rate is 5%.

(i) Find the asset allocation for a minimum variance portfolio with 17% expected return, and the standard deviation of the return of this portfolio;

(ii) Find the asset allocation for a maximum return portfolio with 30% standard deviation of return, and the expected return of this portfolio.

Solution: Note that $r_f = 0.05$ and

$$
\overline{\mu} = \mu - r_f \mathbf{1} = \begin{pmatrix} 0.03 \\ 0.07 \\ 0.11 \\ 0.13 \\ 0.16 \end{pmatrix};
$$

$$
\Sigma_R = \begin{pmatrix}
0.0625 & -0.0156 & -0.0187 & 0.0280 & -0.0087 \\
-0.0156 & 0.0625 & 0.0225 & -0.0400 & 0.0087 \\
-0.0187 & 0.0225 & 0.0900 & -0.0288 & -0.0367 \\
0.0280 & -0.0400 & -0.0288 & 0.1024 & 0.0728 \\
-0.0087 & 0.0087 & -0.0367 & 0.0728 & 0.1225
\end{pmatrix}.
$$

The asset weights vector of the tangency portfolio is given by

$$
w_T = \frac{1}{\mathbf{1}^t \Sigma_R^{-1} \overline{\mu}} \Sigma_R^{-1} \overline{\mu} = \begin{pmatrix} 0.1464 \\ 0.2005 \\ 0.2937 \\ 0.1944 \\ 0.1650 \end{pmatrix},
$$

where the vector $\Sigma_R^{-1} \overline{\mu} = $ linear_solve_cholesky($\Sigma_R, \overline{\mu}$) was computed by using the Cholesky decomposition to solve the linear system corresponding to the matrix Σ_R and to the right hand side $\overline{\mu}$.

(i) The asset weights vector w_{min} and the weight $w_{min,cash}$ of the cash position of the minimum variance portfolio with 17% expected return, i.e., corresponding to $\mu_P = 0.17$ in the formulas below, are given by

$$w_{min,cash} \quad = \quad 1 - \frac{\mu_P - r_f}{\bar{\mu}^t w_T} \quad = \quad -0.1718;$$

$$w_{min} \quad = \quad (1 - w_{min,cash})w_T \quad = \quad \begin{pmatrix} 0.1716 \\ 0.2350 \\ 0.3441 \\ 0.2278 \\ 0.1934 \end{pmatrix}.$$

In other words, the minimum variance portfolio with 17% expected return is obtained by borrowing an amount of cash equal to 17.18% of the total value of the portfolio, and by investing 17.16% of the portfolio value in asset 1, 23.50% of the portfolio value in asset 2, 34.41% of the portfolio value in asset 3, 22.78% of the portfolio value in asset 4, and 19.34% of the portfolio value in asset 5.

The standard deviation of the return of the minimum variance portfolio with 17% expected return is

$$\sigma_{min} \quad = \quad \sqrt{w_{min}^t \Sigma_R w_{min}} \quad = \quad 0.1454 \quad = \quad 14.54\%.$$

(ii) The asset weights vector w_{max} and the weight $w_{max,cash}$ of the cash position of the maximum return portfolio with 30% standard deviation of return, i.e., corresponding to $\sigma_P = 0.30$ in the formulas below, are given by

$$w_{max,cash} \quad = \quad 1 - \frac{\sigma_P}{\sqrt{w_T^t \Sigma_R w_T}} \cdot \text{sign}(1^t \Sigma_R^{-1} \bar{\mu}) \quad = \quad -1.4181;$$

$$w_{max} \quad = \quad (1 - w_{max,cash})w_T \quad = \quad \begin{pmatrix} 0.3540 \\ 0.4849 \\ 0.7101 \\ 0.4700 \\ 0.3991 \end{pmatrix}.$$

In other words, the maximum return portfolio with 30% standard deviation of return is obtained by borrowing an amount of cash equal to 141.81% of the total value of the portfolio, and by investing 35.40% of the portfolio value in asset 1, 48.49% of the portfolio value in asset 2, 71.01% of the portfolio value in asset 3, 47% of the portfolio value in asset 4, and 39.91% of the portfolio value in asset 5.

The expected return of the maximum return portfolio with 30% standard deviation of return is

$$\mu_{max} \quad = \quad r_f + \bar{\mu}^t w_{max} \quad = \quad 0.2976 \quad = \quad 29.76\%. \quad \square$$

Problem 8: Consider four assets with the following expected returns over a fixed time period:

$$\mu_1 = 4\%; \ \mu_2 = 3.5\%; \ \mu_3 = 5\%; \ \mu_4 = 3.4\%,$$

and with the following covariance matrix of their returns over the same time period:

$$\Sigma_R = \begin{pmatrix} 0.09 & 0.01 & 0.03 & -0.015 \\ 0.01 & 0.0625 & -0.02 & -0.01 \\ 0.03 & -0.02 & 0.1225 & 0.02 \\ -0.015 & -0.01 & 0.02 & 0.0576 \end{pmatrix}. \tag{9.9}$$

Assume that the risk–free interest rate is 1%.

(i) Find the asset allocation for the tangency portfolio. Find the expected value and the standard deviation of the return of the tangency portfolio. What is the Sharpe ratio of the tangency portfolio?

(ii) Find the asset allocation for a minimum variance portfolio with 3% expected return, and the standard deviation of the return of this portfolio. What is the Sharpe ratio of this portfolio?

(iii) Find the asset allocation for a maximum return portfolio with 27% standard deviation of return, and the expected return of this portfolio. What is the Sharpe ratio of this portfolio?

(iv) Find the asset allocation for the minimum variance portfolio fully invested in the assets (i.e., with no cash position). What is the Sharpe ratio of this portfolio?

Solution: (i) Note that $r_f = 0.01$ and

$$\bar{\mu} = \mu - r_f \mathbf{1} = \begin{pmatrix} 0.03 \\ 0.025 \\ 0.04 \\ 0.024 \end{pmatrix}.$$

The asset weights vector of the tangency portfolio is given by

$$w_T = \frac{1}{\mathbf{1}^t \Sigma_R^{-1} \bar{\mu}} \Sigma_R^{-1} \bar{\mu} = \begin{pmatrix} 0.1737 \\ 0.3381 \\ 0.1736 \\ 0.3146 \end{pmatrix},$$

the vector $\Sigma_R^{-1}\bar{\mu} = $ linear_solve_cholesky$(\Sigma_R, \bar{\mu})$ was computed by using the Cholesky decomposition to solve the linear system corresponding to the matrix Σ_R given by (9.9) and to the right hand side $\bar{\mu}$.

Thus, the tangency portfolio is obtained by investing 17.37% of the portfolio in asset 1, 33.81% of the portfolio in asset 2, 17.36% of the portfolio in asset 3, and 31.46% of the portfolio in asset 4. The expected value and the standard deviation of the return of the tangency portfolio and its Sharpe ratio are given by

$$\mu_T = r_f + \bar{\mu}^t w_T = 0.0382 = 3.82\%;$$

$$\sigma_T = \sqrt{w_T^t \Sigma_R w_T} = 0.1353 = 13.53\%;$$

$$\text{Sharpe}_T = \frac{\mu_T - r_f}{\sigma_T} = 0.2081.$$

(ii) The asset weights vector w_{min} and the weight $w_{min,cash}$ of the cash position of the minimum variance portfolio with 3% expected return, i.e., corresponding to

$\mu_P = 0.03$ in the formulas below, are given by

$$w_{min,cash} = 1 - \frac{\mu_P - r_f}{\overline{\mu}^t w_T} = 0.2897;$$

$$w_{min} = (1 - w_{min,cash})w_T = \begin{pmatrix} 0.1234 \\ 0.2402 \\ 0.1233 \\ 0.2234 \end{pmatrix}.$$

In other words, the minimum variance portfolio with 3% expected return is obtained by holding a cash position equal to 28.97% of the total value of the portfolio, and by investing 12.34% of the portfolio value in asset 1, 24.02% of the portfolio value in asset 2, 12.33% of the portfolio value in asset 3, and 22.34% of the portfolio value in asset 4.

The standard deviation of the return of the minimum variance portfolio with 3% expected return is

$$\sigma_{min} = \sqrt{w_{min}^t \Sigma_R w_{min}} = 0.0961 = 9.61\%$$

and the Sharpe ratio is 0.2081.

(iii) The asset weights vector w_{max} and the weight $w_{max,cash}$ of the cash position of the maximum return portfolio with 27% standard deviation of return, i.e., corresponding to $\sigma_P = 0.27$ in the formulas below, are given by

$$w_{max,cash} = 1 - \frac{\sigma_P}{\sqrt{w_T^t \Sigma_R w_T}} \cdot \operatorname{sign}(1^t \Sigma_R^{-1} \overline{\mu}) = -0.9955;$$

$$w_{max} = (1 - w_{max,cash})w_T = \begin{pmatrix} 0.3466 \\ 0.6748 \\ 0.3464 \\ 0.6277 \end{pmatrix}.$$

In other words, the maximum return portfolio with 27% standard deviation of return is obtained by borrowing an amount of cash equal to 99.55% of the total value of the portfolio, and by investing 34.66% of the portfolio value in asset 1, 67.48% of the portfolio value in asset 2, 34.64% of the portfolio value in asset 3, and 62.77% of the portfolio value in asset 4.

The expected return of the maximum return portfolio with 27% standard deviation of return is

$$\mu_{max} = r_f + \overline{\mu}^t w_{max} = 0.0662 = 6.62\%$$

and the Sharpe ratio is 0.2081.

(iv) The asset weights vector of the minimum variance portfolio with no cash position is given by

$$w_m = \frac{1}{1^t \Sigma_R^{-1} 1} \Sigma_R^{-1} 1,$$

where $1 = \begin{pmatrix} 1 \\ 1 \\ 1 \\ 1 \end{pmatrix}$. By computing the vector $\Sigma_R^{-1} 1 = \text{linear\_solve\_cholesky}(\Sigma_R, 1)$

using the Cholesky decomposition to solve the linear system corresponding to the matrix Σ_R and to the right hand side $\mathbf{1}$, we obtain that

$$w_m \;=\; \begin{pmatrix} 0.1905 \\ 0.3390 \\ 0.0893 \\ 0.3812 \end{pmatrix},$$

i.e., the minimum variance portfolio with no cash position is obtained by investing 19.05% of the portfolio in asset 1, 33.90% of the portfolio in asset 2, 8.93% of the portfolio in asset 3, and 38.12% of the portfolio in asset 4.

The expected value, the standard deviation, and the Sharpe ratio of the return of the minimum variance portfolio with no cash position are

$$\mu_m \;=\; r_f + \overline{\mu}^t w_m \;=\; 0.0369 \;=\; 3.69\%;$$

$$\sigma_m \;=\; \sqrt{w_m^t \Sigma_R w_m} \;=\; 0.1323 \;=\; 13.23\%;$$

$$\text{Sharpe}_m \;=\; \frac{\mu_m - r_f}{\sigma_m} \;=\; 0.2034. \quad \square$$

Problem 9: Consider three assets with the following expected values, standard deviations, and correlations of their returns:

$$\begin{aligned}
\mu_1 &= 0.05; & \sigma_1 &= 0.15; & \rho_{1,2} &= -0.25; \\
\mu_2 &= 0.09; & \sigma_2 &= 0.20; & \rho_{2,3} &= 0.25; \\
\mu_3 &= 0.10; & \sigma_3 &= 0.25; & \rho_{1,3} &= 0.50.
\end{aligned}$$

The risk–free interest rate is 2%.

(i) Find the asset allocation for a minimum variance portfolio with 8% expected return, and the standard deviation of the return of this portfolio.

(ii) Assume that the returns of the three assets have a joint multivariate normal distribution. Find the probability density function of the return of the minimum variance portfolio with 8% expected return.

(iii) Find the probability that the return of the minimum variance portfolio with 8% expected return is between 7% and 9%. Also, find the probability that the return of this portfolio is below 5%, and the probability that the return of this portfolio is above 10%.

(iv) Consider a portfolio equally invested in each of the three assets. Note that the expected return of this portfolio is 8%. Find the probabilities that the return of this portfolio is between 7% and 9%, is below 5%, and is above 10%, respectively.

Solution: (i) Note that $r_f = 0.02$ and

$$\overline{\mu} \;=\; \mu - r_f \mathbf{1} \;=\; \begin{pmatrix} 0.03 \\ 0.07 \\ 0.08 \end{pmatrix};$$

$$\Sigma_R = \begin{pmatrix} \sigma_1^2 & \sigma_1\sigma_2\rho_{1,2} & \sigma_1\sigma_3\rho_{1,3} \\ \sigma_1\sigma_2\rho_{1,2} & \sigma_2^2 & \sigma_2\sigma_3\rho_{2,3} \\ \sigma_1\sigma_3\rho_{1,3} & \sigma_2\sigma_3\rho_{2,3} & \sigma_3^2 \end{pmatrix} = \begin{pmatrix} 0.0225 & -0.0075 & 0.0187 \\ -0.0075 & 0.0400 & 0.0125 \\ 0.0187 & 0.0125 & 0.0625 \end{pmatrix}.$$

The asset weights vector of the tangency portfolio is given by

$$w_T = \frac{1}{1^t \Sigma_R^{-1} \bar{\mu}} \Sigma_R^{-1} \bar{\mu} = \begin{pmatrix} 0.4134 \\ 0.4860 \\ 0.1006 \end{pmatrix},$$

where the vector $\Sigma_R^{-1}\bar{\mu}$ = linear_solve_cholesky($\Sigma_R, \bar{\mu}$) is computed by using the Cholesky decomposition to solve the linear system corresponding to the matrix Σ_R and to the right hand side $\bar{\mu}$.

The asset weights vector w_{min} and the weight $w_{min,cash}$ of the cash position of the minimum variance portfolio with 8% expected return, i.e., corresponding to $\mu_P = 0.08$ in the formulas below, are given by

$$w_{min,cash} = 1 - \frac{\mu_P - r_f}{\bar{\mu}^t w_T} = -0.1015;$$

$$w_{min} = (1 - w_{min,cash}) w_T = \begin{pmatrix} 0.4554 \\ 0.5354 \\ 0.1108 \end{pmatrix}.$$

In other words, the minimum variance portfolio with 8% expected return is obtained by borrowing an amount of cash equal to 10.15% of the total value of the portfolio, and by investing 45.54% of the portfolio value in asset 1, 53.54% of the portfolio value in asset 2, and 11.08% of the portfolio value in asset 3.

The standard deviation of the return of the minimum variance portfolio with 8% expected return is

$$\sigma_{min} = \sqrt{w_{min}^t \Sigma_R w_{min}} = 0.1289 = 12.89\%.$$

(ii) If the returns of the three assets have a joint multivariate normal distribution, then the return of any portfolio made of these assets and cash is normal, since it is a linear combination of the returns of the assets and the risk–free rate.

In particular, the return R_P of the minimum variance portfolio with 8% expected return is normally distributed with mean $\mu_P = 0.08$ and standard deviation $\sigma_{min} = 0.1289$, i.e.,

$$R_P = \mu_P + \sigma_{min} Z = 0.08 + 0.1289 Z, \tag{9.10}$$

where Z denotes the standard normal distribution, and the probability density function of R_P is given by

$$f(x) = \frac{1}{\sqrt{2\pi\sigma_{min}^2}} \exp\left(-\frac{(x - \mu_P)^2}{2\sigma_{min}^2}\right).$$

(iii) Using (9.10), we find that the probability that the return R_P of the minimum variance portfolio with 8% expected return is between 7% and 9% is

$$
\begin{aligned}
P(0.07 \le R_P \le 0.09) &= P(0.07 \le \mu_P + \sigma_{min} Z \le 0.09) \\
&= P\left(\frac{0.07 - \mu_P}{\sigma_{min}} \le Z \le \frac{0.09 - \mu_P}{\sigma_{min}}\right) \\
&= P\left(Z \le \frac{0.09 - \mu_P}{\sigma_{min}}\right) - P\left(Z \le \frac{0.07 - \mu_P}{\sigma_{min}}\right) \\
&= P(Z \le 0.0776) - P(Z \le -0.0776) \\
&= 0.0618.
\end{aligned}
$$

The probability that the return R_P of the minimum variance portfolio with 8% expected return is below 5% is

$$
\begin{aligned}
P(R_P \le 0.05) &= P(\mu_P + \sigma_{min} Z \le 0.05) \\
&= P\left(Z \le \frac{0.05 - \mu_P}{\sigma_{min}}\right) = P(Z \le -0.2327) \\
&= 0.4080.
\end{aligned}
$$

The probability that the return R_P of the minimum variance portfolio with 8% expected return is above 10% is

$$
\begin{aligned}
P(0.10 \le R_P) &= P(0.10 \le \mu_P + \sigma_{min} Z) \\
&= P\left(\frac{0.10 - \mu_P}{\sigma_{min}} \le Z\right) \\
&= P(0.1552 \le Z) = P(Z \le -0.1552) \\
&= 0.4383.
\end{aligned}
$$

(iv) The asset weights vector of the portfolio equally invested in each of the three assets is $w_e = \begin{pmatrix} \frac{1}{3} \\ \frac{1}{3} \\ \frac{1}{3} \end{pmatrix}$. Then, the return of a portfolio equally invested in each of the three assets has expected value

$$
\mu_e = w_e^t \mu = 0.08
$$

and standard deviation

$$
\sigma_e = \sqrt{w_e^t \Sigma_R w_e} = 0.1384.
$$

Thus, the return R_e of the portfolio equally invested in each of the three assets is normally distributed with mean μ_e and standard deviation σ_e, i.e.,

$$
R_e = \mu_e + \sigma_e Z = 0.08 + 0.1384Z. \tag{9.11}
$$

Using (9.11) and following the same steps as in (iii), we find that the probabilities that the return R_e of the portfolio equally invested in each of the three assets is between 7% and 9%, is below 5%, and is above 10% are, respectively,

$$
\begin{aligned}
P(0.07 \le R_e \le 0.09) &= P(0.07 \le \mu_e + \sigma_e Z \le 0.09) \\
&= P\left(\frac{0.07 - \mu_e}{\sigma_e} \le Z \le \frac{0.09 - \mu_e}{\sigma_e}\right) \\
&= P(-0.0722 \le Z \le 0.0722) \\
&= 0.0576.
\end{aligned}
$$

$$
\begin{aligned}
P(R_e \le 0.05) &= P(\mu_e + \sigma_e Z \le 0.05) \\
&= P\left(Z \le \frac{0.05 - \mu_e}{\sigma_e}\right) = P(Z \le -0.2167) \\
&= 0.4142.
\end{aligned}
$$

$$
\begin{aligned}
P(0.10 \leq R_e) &= P(0.10 \leq \mu_e + \sigma_e Z) \\
&= P\left(\frac{0.10 - \mu_e}{\sigma_e} \leq Z\right) \\
&= P(0.1445 \leq Z) = P(Z \leq -0.1445) \\
&= 0.4426. \quad \square
\end{aligned}
$$

Problem 10: Let μ_1, σ_1 and μ_2, σ_2 be the expected values and the standard deviations of the returns of two assets over a fixed time period, respectively, and let $\rho_{1,2}$ be the correlation of the returns of the two assets over the same time period. Recall that the minimum variance portfolio fully invested in two assets is obtained by allocating

$$
w_1 = \frac{\sigma_2(\sigma_2 - \rho_{1,2}\sigma_1)}{\sigma_1^2 - 2\sigma_1\sigma_2\rho_{1,2} + \sigma_2^2} \tag{9.12}
$$

of the portfolio value to the first asset and $w_2 = 1 - w_1$ of the portfolio value to the second asset.

Show that the minimum variance portfolio has only long asset positions, i.e., $0 \leq w_1, w_2 \leq 1$, if and only if

$$
\rho_{1,2} \leq \min\left(\frac{\sigma_1}{\sigma_2}, \frac{\sigma_2}{\sigma_1}\right).
$$

Solution: Note that, if $0 \leq w_1 \leq 1$, then $0 \leq w_2 = 1 - w_1 \leq 1$. In other words,

$$
0 \leq w_1, w_2 \leq 1 \iff 0 \leq w_1 \leq 1. \tag{9.13}
$$

Also, note that

$$
\sigma_1^2 - 2\sigma_1\sigma_2\rho_{1,2} + \sigma_2^2 \geq 0, \tag{9.14}
$$

for all $\sigma_1, \sigma_2 > 0$ and $-1 \leq \rho_{1,2} \leq 1$. One way to see this is as follows

$$
\begin{aligned}
\sigma_1^2 - 2\sigma_1\sigma_2\rho_{1,2} + \sigma_2^2 &= \sigma_1^2 - 2\sigma_1\sigma_2 + \sigma_2^2 + 2\sigma_1\sigma_2 - 2\sigma_1\sigma_2\rho_{1,2} \\
&= (\sigma_1 - \sigma_2)^2 + 2\sigma_1\sigma_2(1 - \rho_{1,2}) \\
&\geq 0,
\end{aligned}
$$

since $-1 \leq \rho_{1,2} \leq 1$ and therefore $1 - \rho_{1,2} \geq 0$.

From (9.12) and (9.14) it follows that

$$
\begin{aligned}
0 \leq w_1 &\iff 0 \leq \frac{\sigma_2(\sigma_2 - \rho_{1,2}\sigma_1)}{\sigma_1^2 - 2\sigma_1\sigma_2\rho_{1,2} + \sigma_2^2} \\
&\iff 0 \leq \sigma_2(\sigma_2 - \rho_{1,2}\sigma_1) \\
&\iff \sigma_1\sigma_2\rho_{1,2} \leq \sigma_2^2 \\
&\iff \rho_{1,2} \leq \frac{\sigma_2}{\sigma_1}. \tag{9.15}
\end{aligned}
$$

Moreover, since the denominator from (9.12) is nonnegative, see (9.14), we find from (9.12) that

$$
w_1 \leq 1 \iff \frac{\sigma_2(\sigma_2 - \rho_{1,2}\sigma_1)}{\sigma_1^2 - 2\sigma_1\sigma_2\rho_{1,2} + \sigma_2^2} \leq 1
$$

$$\Longleftrightarrow \quad \sigma_2(\sigma_2 - \rho_{1,2}\sigma_1) \leq \sigma_1^2 - 2\sigma_1\sigma_2\rho_{1,2} + \sigma_2^2$$
$$\Longleftrightarrow \quad \sigma_2^2 - \sigma_1\sigma_2\rho_{1,2} \leq \sigma_1^2 - 2\sigma_1\sigma_2\rho_{1,2} + \sigma_2^2$$
$$\Longleftrightarrow \quad \sigma_1\sigma_2\rho_{1,2} \leq \sigma_1^2$$
$$\Longleftrightarrow \quad \rho_{1,2} \leq \frac{\sigma_1}{\sigma_2}. \tag{9.16}$$

From (9.15) and (9.16), we obtain that

$$0 \leq w_1 \leq 1 \quad \Longleftrightarrow \quad \rho_{1,2} \leq \frac{\sigma_2}{\sigma_1} \quad \text{and} \quad \rho_{1,2} \leq \frac{\sigma_1}{\sigma_2}$$

$$\Longleftrightarrow \quad \rho_{1,2} \leq \min\left(\frac{\sigma_1}{\sigma_2}, \frac{\sigma_2}{\sigma_1}\right). \tag{9.17}$$

From (9.13) and (9.17), we conclude that

$$0 \leq w_1, w_2 \leq 1 \quad \Longleftrightarrow \quad \rho_{1,2} \leq \min\left(\frac{\sigma_1}{\sigma_2}, \frac{\sigma_2}{\sigma_1}\right). \quad \Box$$

Problem 11: (i) Show that the asset allocation for a minimum variance portfolio with expected return μ_P can also be written as follows in terms of the asset weights vector w_T of the tangency portfolio:

• asset weights vector w_{min} given by

$$w_{min} = \frac{\mu_P - r_f}{\bar{\mu}^t w_T} w_T. \tag{9.18}$$

• weight $w_{min,cash}$ of the cash position given by

$$w_{min,cash} = 1 - \mathbf{1}^t w_{min}; \tag{9.19}$$

(ii) Write a pseudocode for computing the asset allocation for a minimum variance portfolio with expected return μ_P using the formulas (9.18) and (9.19).

Solution: (i) Recall that the asset weights vector w_{min} and the weight $w_{min,cash}$ of the cash position for a minimum variance portfolio with expected return μ_P are

$$w_{min,cash} = 1 - \frac{\mu_P - r_f}{\bar{\mu}^t w_T}; \tag{9.20}$$

$$w_{min} = (1 - w_{min,cash})w_T. \tag{9.21}$$

Then, from (9.21), and using the fact that

$$1 - w_{min,cash} = \frac{\mu_P - r_f}{\bar{\mu}^t w_T},$$

see (9.20), we obtain that

$$w_{min} = \frac{\mu_P - r_f}{\bar{\mu}^t w_T} w_T.$$

Since the cash position of any portfolio with asset weights vector w is $w_{cash} = 1 - \mathbf{1}^t w$, it follows that the cash position $w_{min,cash}$ of the minimum variance portfolio is

$$w_{min,cash} = 1 - \mathbf{1}^t w_{min}.$$

(ii) The pseudocode for computing the asset allocation for a minimum variance portfolio with expected return μ_P using the formulas (9.18) and (9.19) can be found in Table 9.1.

Table 9.1: Asset allocation of minimum variance portfolio from tangency portfolio

---
Input:
$\Sigma_R = n \times n$ covariance matrix of the returns of n assets
$\mu = n \times 1$ vector of expected values of the returns of n assets
$r_f =$ risk–free rate
$\mu_P =$ required expected return of the portfolio

Output:
$w_{min} =$ asset weights vector for the minimum variance portfolio
$w_{min,cash} =$ weight of cash position for the minimum variance portfolio
$\sigma_{min} =$ standard deviation of the return of the minimum variance portfolio

$\overline{\mu} = \mu - r_f \mathbf{1}$
$x = \text{linear\_solve\_cholesky}(\Sigma_R, \overline{\mu})$ // compute $x = \Sigma_R^{-1} \overline{\mu}$
$w_T = \frac{1}{\mathbf{1}^t_x x} x$
$w_{min} = \frac{\mu_P - r_f}{\overline{\mu}^t w_T} w_T$
$w_{min,cash} = 1 - \mathbf{1}^t w_{min}$
$\sigma_{min} = \sqrt{w_{min}^t \Sigma_R w_{min}}$

Problem 12: (i) Show that the asset allocation for a maximum return portfolio with variance of return equal to σ_P^2 can also be written as follows in terms of the asset weights vector w_T of the tangency portfolio:

- asset weights vector w_{max} given by

$$w_{max} = \frac{\sigma_P}{\sqrt{w_T^t \Sigma_R w_T}} \cdot \text{sign}(\mathbf{1}^t \Sigma_R^{-1} \overline{\mu}) w_T; \qquad (9.22)$$

- weight $w_{max,cash}$ of the cash position equal to $1 - \mathbf{1}^t w_{max}$, i.e.,

$$w_{max,cash} = 1 - \mathbf{1}^t w_{max}. \qquad (9.23)$$

(ii) Write a pseudocode for computing the asset allocation for maximum return portfolio with variance of return equal to σ_P^2 using the formulas (9.22) and (9.23).

Solution: (i) Recall that the asset weights vector w_{max} and the weight $w_{max,cash}$ of the cash position for a maximum return portfolio with variance of return equal to

σ_P^2 are

$$w_{max,cash} = 1 - \frac{\sigma_P}{\sqrt{w_T^t \Sigma_R w_T}} \cdot \text{sign}(\mathbf{1}^t \Sigma_R^{-1} \bar{\mu}); \qquad (9.24)$$

$$w_{max} = (1 - w_{max,cash}) w_T. \qquad (9.25)$$

Then, from (9.24), and using the fact that $1 - w_{max,cash} = \frac{\sigma_P}{\sqrt{w_T^t \Sigma_R w_T}} \cdot \text{sign}(\mathbf{1}^t \Sigma_R^{-1} \bar{\mu})$, we obtain that

$$w_{max} = \frac{\sigma_P}{\sqrt{w_T^t \Sigma_R w_T}} \cdot \text{sign}(\mathbf{1}^t \Sigma_R^{-1} \bar{\mu}) w_T.$$

Since the cash position of any portfolio with asset weights vector w is

$$w_{cash} = 1 - \mathbf{1}^t w,$$

it follows that the cash position $w_{max,cash}$ of the maximum return portfolio is

$$w_{max,cash} = 1 - \mathbf{1}^t w_{max}.$$

(ii) The pseudocode for computing the asset allocation for maximum return portfolio with variance of return equal to σ_P^2 using the formulas (9.22) and (9.23) can be found in Table 9.2.

Table 9.2: Asset allocation of maximum return portfolio from tangency portfolio

Input:
$\Sigma_R = n \times n$ covariance matrix of the returns of n assets
$\mu = n \times 1$ vector of expected values of the returns of n assets
$r_f = $ risk–free rate
$\sigma_P = $ required standard deviation of the return of the portfolio

Output:
$w_{max} = $ asset weights vector for the maximum return portfolio
$w_{max,cash} = $ weight of cash position for the maximum return portfolio
$\mu_{max} = $ expected return of the maximum return portfolio

$\bar{\mu} = \mu - r_f \mathbf{1}$
$x = $ linear_solve_cholesky$(\Sigma_R, \bar{\mu})$ // compute $x = \Sigma_R^{-1} \bar{\mu}$
$w_T = \frac{1}{\mathbf{1}^t x} x$
if $\mathbf{1}^t \Sigma_R^{-1} \bar{\mu} > 0$
$\quad w_{max} = \frac{\sigma_P}{\sqrt{w_T^t \Sigma_R w_T}} w_T$
else
$\quad w_{max} = -\frac{\sigma_P}{\sqrt{w_T^t \Sigma_R w_T}} w_T$
end
$w_{max,cash} = 1 - \mathbf{1}^t w_{max}$
$\mu_{max} = r_f + \bar{\mu}^t w_{max}$

Problem 13: Assume that the 99% two day VaR of a portfolio is \$10 million. Estimate the two day 95% VaR of the portfolio and the five day 95% VaR of the portfolio.

Solution: The Value at Risk of a portfolio scales to different time horizons and confidence levels as follows:

$$\text{VaR}(N_2, C_2) \approx \frac{z_{C_2}\sqrt{N_2}}{z_{C_1}\sqrt{N_1}} \text{VaR}(N_1, C_1), \tag{9.26}$$

where $\text{VaR}(N_1, C_1)$ is the N_1 day C_1% VaR of the portfolio, $\text{VaR}(N_2, C_2)$ is the N_2 day C_2% VaR of the portfolio, and z_C denotes the z-score of the standard normal distribution corresponding to C, i.e., $P(Z \leq z_C) = C$.

Since

$$\text{VaR}(2 \text{ days}, 99\%) = \$10\text{mil},$$

and using the facts that

$$z_{99} = 2.3263; \quad z_{95} = 1.6449,$$

we obtain from (9.26) that the two day 95% VaR of the portfolio and the five day 95% VaR of the portfolio can be estimated as follows:

$$\text{VaR}(2 \text{ days}, 95\%) \approx \frac{z_{95}\sqrt{2}}{z_{99}\sqrt{2}}\text{VaR}(2 \text{ days}, 99\%) = \frac{1.6449}{2.3263} \cdot \$10\text{mil}$$

$$\approx \$7.07\text{mil};$$

$$\text{VaR}(5 \text{ days}, 95\%) \approx \frac{z_{95}\sqrt{5}}{z_{99}\sqrt{2}}\text{VaR}(2 \text{ days}, 99\%) = \frac{1.6449\sqrt{5}}{2.3263\sqrt{2}} \cdot \$10\text{mil}$$

$$\approx \$11.18\text{mil}. \quad \square$$

Problem 14: Show that if two portfolios and their combined portfolio have normally distributed returns, then the VaR of the combined portfolio is smaller than the sum of the VaRs of each individual portfolio.

In other words, if V_1, V_2, and $V = V_1 + V_2$ are portfolios with normal returns, show that

$$\text{VaR}_V(N, C) \leq \text{VaR}_{V_1}(N, C) + \text{VaR}_{V_2}(N, C),$$

where $\text{VaR}_V(N, C)$, $\text{VaR}_{V_1}(N, C)$, and $\text{VaR}_{V_2}(N, C)$ are the N day C% VaRs of V, V_1, and V_2, respectively.

Solution: Recall that, if a portfolio W has normal return, then

$$\text{VaR}_W(N, C) = \sqrt{\frac{N}{252}}z_C\sigma_{R_W}W(0) - \frac{N}{252}\mu_{R_W}W(0), \tag{9.27}$$

where μ_{R_W} and σ_{R_W} are the expected value and the standard deviation of the portfolio return R_V, and z_C is the z-score of the standard normal distribution corresponding to C. Since the portfolios V_1, V_2, and V are assumed to have normal returns, we obtain from (9.27) that

$$\mathrm{VaR}_V(N,C) \;\leq\; \mathrm{VaR}_{V_1}(N,C) + \mathrm{VaR}_{V_2}(N,C) \tag{9.28}$$

$$\Longleftrightarrow \quad \sqrt{\frac{N}{252}}\, z_C \sigma_{R_V} V(0) - \frac{N}{252}\mu_{R_V} V(0)$$

$$\leq \sqrt{\frac{N}{252}}\, z_C \left(\sigma_{R_{V_1}} V_1(0) + \sigma_{R_{V_2}} V_2(0)\right) - \frac{N}{252}\left(\mu_{R_{V_1}} V_1(0) + \mu_{R_{V_2}} V_2(0)\right)$$

$$\Longleftrightarrow \quad \sqrt{\frac{N}{252}}\, z_C \sigma_{R_V} V(0) \;\leq\; \sqrt{\frac{N}{252}}\, z_C \left(\sigma_{R_{V_1}} V_1(0) + \sigma_{R_{V_2}} V_2(0)\right)$$

$$-\frac{N}{252}\left(\mu_{R_{V_1}} V_1(0) + \mu_{R_{V_2}} V_2(0) - \mu_{R_V} V(0)\right)$$

$$\Longleftrightarrow \quad \sigma_{R_V} V(0) \;\leq\; \sigma_{R_{V_1}} V_1(0) + \sigma_{R_{V_2}} V_2(0)$$

$$-\frac{1}{z_C}\sqrt{\frac{N}{252}}\left(\mu_{R_{V_1}} V_1(0) + \mu_{R_{V_2}} V_2(0) - \mu_{R_V} V(0)\right)$$

$$\Longleftrightarrow \quad \sigma_{R_V} \;\leq\; \frac{V_1(0)}{V(0)}\sigma_{R_{V_1}} + \frac{V_2(0)}{V(0)}\sigma_{R_{V_2}}$$

$$-\frac{1}{z_C}\sqrt{\frac{N}{252}}\left(\frac{V_1(0)}{V(0)}\mu_{R_{V_1}} + \frac{V_2(0)}{V(0)}\mu_{R_{V_2}} - \mu_{R_V}\right)$$

$$\Longleftrightarrow \quad \sigma_{R_V} \;\leq\; w_1\sigma_{R_{V_1}} + w_2\sigma_{R_{V_2}} - \frac{1}{z_C}\sqrt{\frac{N}{252}}\left(w_1\mu_{R_{V_1}} + w_2\mu_{R_{V_2}} - \mu_{R_V}\right) \tag{9.29}$$

where $w_1 = \frac{V_1(0)}{V(0)}$ and $w_2 = \frac{V_2(0)}{V(0)}$ denote the weights of the first and second portfolio in the combined portfolio, respectively.

Recall that the return R_V of the combined portfolio is the weighted average of the returns R_{V_1} and R_{V_2} of the two portfolios, i.e.,

$$R_V \;=\; w_1 R_{V_1} + w_2 R_{V_2},$$

and therefore

$$\mu_{R_V} \;=\; w_1\mu_{R_{V_1}} + w_2\mu_{R_{V_2}}; \tag{9.30}$$

$$\sigma_{R_V}^2 \;=\; w_1^2\sigma_{R_{V_1}}^2 + 2w_1 w_2 \mathrm{corr}(R_{V_1}, R_{V_2})\sigma_{R_{V_1}}\sigma_{R_{V_2}} + w_2^2\sigma_{R_{V_2}}^2. \tag{9.31}$$

Using (9.30) and (9.31), we find from (9.28) and (9.29) that

$$\mathrm{VaR}_V(N,C) \;\leq\; \mathrm{VaR}_{V_1}(N,C) + \mathrm{VaR}_{V_2}(N,C)$$

$$\Longleftrightarrow \quad \sigma_{R_V} \;\leq\; w_1\sigma_{R_{V_1}} + w_2\sigma_{R_{V_2}}$$

$$\Longleftrightarrow \quad \sigma_{R_V}^2 \;\leq\; (w_1\sigma_{R_{V_1}} + w_2\sigma_{R_{V_2}})^2$$

$$\Longleftrightarrow \quad w_1^2\sigma_{R_{V_1}}^2 + 2w_1 w_2 \mathrm{corr}(R_{V_1}, R_{V_2})\sigma_{R_{V_1}}\sigma_{R_{V_2}} + w_2^2\sigma_{R_{V_2}}^2$$

$$\leq\; w_1^2\sigma_{R_{V_1}}^2 + 2w_1 w_2 \sigma_{R_{V_1}}\sigma_{R_{V_2}} + w_2^2\sigma_{R_{V_2}}^2$$

$$\Longleftrightarrow \quad w_1 w_2 \mathrm{corr}(R_{V_1}, R_{V_2}) \;\leq\; w_1 w_2,$$

which holds true, since we made the implicit assumption that the two portfolios have positive value, and therefore $w_1, w_2 > 0$, and $|\mathrm{corr}(R_{V_1}, R_{V_2})| \leq 1$.

We conclude that the VaR of the combined portfolio is smaller than the sum of the VaRs of the two portfolios, under the assumption that all the portfolios have normally distributed returns. $\square$

Problem 15: Consider three assets with multivariate normal distribution of their returns. The expected values, standard deviations, and correlations of their returns are

$$\begin{aligned}
\mu_1 &= 0.08; & \sigma_1 &= 0.25; & \rho_{1,2} &= -0.25; \\
\mu_2 &= 0.12; & \sigma_2 &= 0.25; & \rho_{2,3} &= -0.25; \\
\mu_3 &= 0.16; & \sigma_3 &= 0.30; & \rho_{1,3} &= 0.25.
\end{aligned}$$

(i) What is the 5–day 95% VaR of $100 million portfolios fully invested in the first asset, fully invested in the second asset, and fully invested in the third asset, respectively?

(ii) What is the minimal 5–day 95% VaR of a $100 million portfolio fully invested in the first and second asset, fully invested in the second and third asset, and fully invested in the first and third asset, respectively?

(iii) What is the minimal 5–day 95% VaR of a $100 million portfolio fully invested in all three assets?

Solution: Recall that the N day $C\%$ VaR of a portfolio can be approximated as

$$\text{VaR}(N, C) \approx \sqrt{\frac{N}{252}} \sigma_{RV} z_C V(0), \tag{9.32}$$

where σ_{RV} is the annualized standard deviation of the return of the portfolio, z_C is the z–score of the standard normal distribution corresponding to C, and $V(0)$ is the current value of the portfolio.

(i) Using the approximate VaR formula (9.32) with $V(0) = \$100\text{mil}$, $N = 5$, and $z_{95} = 1.6449$, we obtain that the 5–day 95% VaR of $100 million portfolios fully invested in the first asset, fully invested in the second asset, and fully invested in the third asset, are, respectively,

$$\text{VaR}_1(5 \text{ days}, 95\%) \approx \sqrt{\frac{5}{252}} \sigma_1 z_{95} \cdot \$100\text{mil} = \$5.792\text{mil};$$

$$\text{VaR}_2(5 \text{ days}, 95\%) \approx \sqrt{\frac{5}{252}} \sigma_2 z_{95} \cdot \$100\text{mil} = \$5.792\text{mil};$$

$$\text{VaR}_3(5 \text{ days}, 95\%) \approx \sqrt{\frac{5}{252}} \sigma_3 z_{95} \cdot \$100\text{mil} = \$6.951\text{mil}.$$

(ii) The covariance matrix of the returns of the first and second asset is

$$\Sigma_{R,1,2} = \begin{pmatrix} \sigma_1^2 & \sigma_1 \sigma_2 \rho_{1,2} \\ \sigma_1 \sigma_2 \rho_{1,2} & \sigma_2^2 \end{pmatrix} = \begin{pmatrix} 0.0625 & -0.0156 \\ -0.0156 & 0.0625 \end{pmatrix}.$$

The asset weights vector of the minimum variance portfolio with no cash position fully invested in the first and second asset is

$$w_{m,1,2} = \frac{1}{1^t \Sigma_{R,1,2}^{-1} 1} \Sigma_{R,1,2}^{-1} 1 = \begin{pmatrix} 0.5 \\ 0.5 \end{pmatrix},$$

where the vector $\Sigma_{R,1,2}^{-1} 1 = \text{linear\_solve\_cholesky} \left(\Sigma_{R,1,2}, \begin{pmatrix} 1 \\ 1 \end{pmatrix} \right)$ was computed using a Cholesky decomposition linear solver. The standard deviation of the return of this minimum variance portfolio is

$$\sigma_{m,1,2} = \sqrt{w_{m,1,2}^t \Sigma_{R,1,2} w_{m,1,2}} = 0.1531 = 15.31\%,$$

and, from (9.32), we find that the minimal 5–day 95% VaR of a $100 million portfolio fully invested in the first and second asset is

$$\text{VaR}_{m,1,2}(5 \text{ days}, 95\%) \approx \sqrt{\frac{5}{252}} \sigma_{m,1,2} z_{95} \cdot \$100\text{mil} = \$3.547\text{mil}.$$

The covariance matrix of the returns of the second and third asset is

$$\Sigma_{R,2,3} = \begin{pmatrix} \sigma_2^2 & \sigma_2 \sigma_3 \rho_{2,3} \\ \sigma_2 \sigma_3 \rho_{2,3} & \sigma_3^2 \end{pmatrix} = \begin{pmatrix} 0.0625 & -0.0187 \\ -0.0187 & 0.0900 \end{pmatrix}.$$

The asset weights vector of the minimum variance portfolio with no cash position fully invested in the second and third asset is

$$w_{m,2,3} = \frac{1}{1^t \Sigma_{R,2,3}^{-1} 1} \Sigma_{R,2,3}^{-1} 1 = \begin{pmatrix} 0.5724 \\ 0.4276 \end{pmatrix},$$

where the vector $\Sigma_{R,2,3}^{-1} 1 = \text{linear\_solve\_cholesky}\left(\Sigma_{R,2,3}, \begin{pmatrix} 1 \\ 1 \end{pmatrix}\right)$ was computed using a Cholesky decomposition linear solver. The standard deviation of the return of this minimum variance portfolio is

$$\sigma_{m,2,3} = \sqrt{w_{m,2,3}^t \Sigma_{R,2,3} w_{m,2,3}} = 0.1666 = 16.66\%,$$

and, from (9.32), we find that the minimal 5–day 95% VaR of a $100 million portfolio fully invested in the second and third asset is

$$\text{VaR}_{m,2,3}(5 \text{ days}, 95\%) \approx \sqrt{\frac{5}{252}} \sigma_{m,2,3} z_{95} \cdot \$100\text{mil} = \$3.860\text{mil}.$$

The covariance matrix of the returns of the first and third asset is

$$\Sigma_{R,1,3} = \begin{pmatrix} \sigma_1^2 & \sigma_1 \sigma_3 \rho_{1,3} \\ \sigma_1 \sigma_3 \rho_{1,3} & \sigma_3^2 \end{pmatrix} = \begin{pmatrix} 0.0625 & 0.0187 \\ 0.0187 & 0.0900 \end{pmatrix}.$$

The asset weights vector of the minimum variance portfolio with no cash position fully invested in the first and third asset is

$$w_{m,1,3} = \frac{1}{1^t \Sigma_{R,1,3}^{-1} 1} \Sigma_{R,1,3}^{-1} 1 = \begin{pmatrix} 0.6196 \\ 0.3804 \end{pmatrix},$$

where the vector $\Sigma_{R,1,3}^{-1} 1 = \text{linear\_solve\_cholesky}\left(\Sigma_{R,1,3}, \begin{pmatrix} 1 \\ 1 \end{pmatrix}\right)$ was computed using a Cholesky decomposition linear solver. The standard deviation of the return of this minimum variance portfolio is

$$\sigma_{m,1,3} = \sqrt{w_{m,1,3}^t \Sigma_{R,1,3} w_{m,1,3}} = 0.2141 = 21.41\%,$$

and, from (9.32), we find that the minimal 5–day 95% VaR of a $100 million portfolio fully invested in the first and third asset is

$$\text{VaR}_{m,1,3}(5 \text{ days}, 95\%) \approx \sqrt{\frac{5}{252}}\sigma_{m,1,3}\, z_{95} \cdot \$100\text{mil} = \$4.961\text{mil}.$$

(iii) The covariance matrix of the returns of the three assets is

$$\Sigma_R = \begin{pmatrix} \sigma_1^2 & \sigma_1\sigma_2\rho_{1,2} & \sigma_1\sigma_3\rho_{1,3} \\ \sigma_1\sigma_2\rho_{1,2} & \sigma_2^2 & \sigma_2\sigma_3\rho_{2,3} \\ \sigma_1\sigma_3\rho_{1,3} & \sigma_2\sigma_3\rho_{2,3} & \sigma_3^2 \end{pmatrix} = \begin{pmatrix} 0.0625 & -0.0156 & 0.0187 \\ -0.0156 & 0.0625 & -0.0187 \\ 0.0187 & -0.0187 & 0.0900 \end{pmatrix}.$$

The asset weights vector of the minimum variance portfolio with no cash position invested in all three assets is

$$w_m = \frac{1}{\mathbf{1}^t\Sigma_R^{-1}\mathbf{1}}\Sigma_R^{-1}\mathbf{1} = \begin{pmatrix} 0.3339 \\ 0.4417 \\ 0.2244 \end{pmatrix},$$

where the vector $\Sigma_R^{-1}\mathbf{1} = \text{linear\_solve\_cholesky}\left(\Sigma_R, \begin{pmatrix} 1 \\ 1 \\ 1 \end{pmatrix}\right)$ was computed using a Cholesky decomposition linear solver. The standard deviation of the return of this minimum variance portfolio is

$$\sigma_m = \sqrt{w_m^t\Sigma_R w_m} = 0.1348 = 13.48\%,$$

and, from (9.32), we find that the minimal 5–day 95% VaR of a $100 million portfolio fully invested in all three assets is

$$\text{VaR}_{m,1,2,3}(5 \text{ days}, 95\%) \approx \sqrt{\frac{5}{252}}\sigma_m\, z_{95} \cdot \$100\text{mil} = \$3.124\text{mil}.$$

Note that, as expected, the VaR of the minimum variance portfolio invested in all three assets, i.e.,

$$\text{VaR}_{m,1,2,3} = \$3.124\text{mil},$$

is significantly smaller than the VaRs of minimum variance portfolios invested in either of the two assets, i.e.,

$$\begin{aligned} \text{VaR}_{m,1,2}(5 \text{ days}, 95\%) &= \$3.547\text{mil}; \\ \text{VaR}_{m,2,3}(5 \text{ days}, 95\%) &= \$3.860\text{mil}; \\ \text{VaR}_{m,1,3}(5 \text{ days}, 95\%) &= \$4.961\text{mil}, \end{aligned}$$

which, in turn, are smaller that the VaRs of any portfolios invested in the only one of the assets, i.e.,

$$\begin{aligned} \text{VaR}_1(5 \text{ days}, 95\%) &= \$5.792\text{mil}; \\ \text{VaR}_2(5 \text{ days}, 95\%) &= \$5.792\text{mil}; \\ \text{VaR}_3(5 \text{ days}, 95\%) &= \$6.951\text{mil}. \end{aligned}$$

This shows how portfolio diversification reduces the portfolio risk exposure. $\square$

Chapter 10

Mathematical appendix and technical results.

10.1   Exercises

1. Show that
$$\det \begin{pmatrix} 1 & a & a^2 \\ 1 & b & b^2 \\ 1 & c & c^2 \end{pmatrix} = (c-a)(c-b)(b-a),$$
where $a, b, c \in \mathbb{R}$.

Note: A matrix of the form
$$\begin{pmatrix} 1 & c_1 & c_1^2 & \cdots & c_1^{n-1} \\ 1 & c_2 & c_2^2 & \cdots & c_2^{n-1} \\ \vdots & \vdots & \vdots & \cdots & \vdots \\ 1 & c_n & c_n^2 & \cdots & c_n^{n-1} \end{pmatrix},$$
where $c_1, c_2, \ldots, c_n$ are constants, is called a Vandermonde matrix. The determinant of the Vandermonde matrix is equal to
$$\prod_{1 \le k < j \le n} (c_j - c_k).$$

2. Show that any orthogonal matrix has determinant 1 or -1. In other words, show that, for any orthogonal matrix Q, either $\det(Q) = 1$, or $\det(Q) = -1$.

3. Let M_1 and M_2 be $n \times n$ symmetric matrices. Show that, if
$$x^t M_1 x = x^t M_2 x, \quad \forall \, x \in \mathbb{R}^n,$$
then $M_1 = M_2$.

4. Show that the quadratic form of a matrix is 0 if and only if the matrix is skew–symmetric, i.e., show that
$$q_A(x) = 0 \quad \text{if and only if} \quad A^t = -A.$$

10.2   Solutions to Chapter 10 Exercises

Problem 1: Show that

$$\det \begin{pmatrix} 1 & a & a^2 \\ 1 & b & b^2 \\ 1 & c & c^2 \end{pmatrix} = (c-a)(c-b)(b-a),$$

where $a, b, c \in \mathbb{R}$.

Solution: Recall that the determinant of the 3×3 matrix can be computed as follows:

$$\det \begin{pmatrix} a_{1,1} & a_{1,2} & a_{1,3} \\ a_{2,1} & a_{2,2} & a_{2,3} \\ a_{3,1} & a_{3,2} & a_{3,3} \end{pmatrix} = a_{1,1}\, a_{2,2}\, a_{3,3} + a_{1,2}\, a_{2,3}\, a_{3,1} + a_{1,3}\, a_{2,1}\, a_{3,2}$$

$$- a_{1,3}\, a_{2,2}\, a_{3,1} - a_{1,1}\, a_{2,3}\, a_{3,2} - a_{1,2}\, a_{2,1}\, a_{3,3}.$$

Then,

$$\det \begin{pmatrix} 1 & a & a^2 \\ 1 & b & b^2 \\ 1 & c & c^2 \end{pmatrix} = bc^2 + ab^2 + a^2c - a^2b - b^2c - ac^2$$

$$= (bc^2 - ac^2) + (a^2c - b^2c) + (ab^2 - a^2b)$$
$$= c^2(b-a) - c(b-a)(a+b) + ab(b-a)$$
$$= (b-a)(c^2 - ac - bc + ab)$$
$$= (b-a)(c-a)(c-b). \quad \square$$

Problem 2: Show that any orthogonal matrix has determinant 1 or -1. In other words, show that, for any orthogonal matrix Q, either $\det(Q) = 1$, or $\det(Q) = -1$.

Solution: Recall that the determinant of the transpose of a matrix is the same as the determinant of the matrix. Thus,

$$\det(Q^t) = \det(Q). \tag{10.1}$$

Since Q is an orthogonal matrix, it follows that

$$Q^t Q = I. \tag{10.2}$$

From (10.1) and (10.2), we obtain that

$$1 = \det(I) = \det(Q^t Q) = \det(Q^t)\det(Q) = (\det(Q))^2.$$

Thus, $(\det(Q))^2 = 1$, and therefore $\det(Q) = 1$ or $\det(Q) = -1$. $\quad \square$

Problem 3: Let M_1 and M_2 be $n \times n$ symmetric matrices. Show that, if

$$x^t M_1 x = x^t M_2 x, \quad \forall\, x \in \mathbb{R}^n,$$

then $M_1 = M_2$.

Solution: Recall from Theorem 10.3 (ii) on Page 295 of Stefanica [3] that, if A and B are square matrices of the same size, then

$$x^t A x = x^t B x, \quad \forall\, x \in \mathbb{R}^n \quad \Longleftrightarrow \quad A + A^t = B + B^t. \tag{10.3}$$

Using (10.3), we obtain that

$$
\begin{aligned}
x^t M_1 x = x^t M_2 x, \quad \forall\, x \in \mathbb{R}^n \quad &\Longleftrightarrow \quad M_1 + M_1^t = M_2 + M_2^t \\
&\Longleftrightarrow \quad 2M_1 = 2M_2 \\
&\Longleftrightarrow \quad M_1 = M_2,
\end{aligned}
$$

since M_1 and M_2 are symmetric matrices and therefore $M_1^t = M_1$ and $M_2^t = M_2$.
□

Problem 4: Show that the quadratic form of a matrix is 0 if and only if the matrix is skew–symmetric, i.e., show that

$$q_A(x) = 0 \quad \text{if and only if} \quad A^t = -A.$$

Solution: Recall from Theorem 10.3 (i) on Page 295 of Stefanica [3] that

$$q_A(x) = \frac{1}{2} x^t \left(A + A^t \right) x. \tag{10.4}$$

Using (10.4), it follows that

$$
\begin{aligned}
q_A(x) = 0, \quad \forall\, x \in \mathbb{R}^n \quad &\Longleftrightarrow \quad x^t \left(A + A^t \right) x = 0, \quad \forall\, x \in \mathbb{R}^n \\
&\Longleftrightarrow \quad A + A^t = 0 \\
&\Longleftrightarrow \quad A^t = -A. \quad \square
\end{aligned}
$$

Bibliography

[1] Gene H. Golub and Charles F. Van Loan. *Numerical Linear Algebra*. The Johns Hopkins University Press, Baltimore, Maryland, 3rd edition, 1996.

[2] Dan Stefanica. *A Mathematical Primer with Numerical Methods for Financial Engineering*. Financial Engineering Advanced Background Series. FE Press, New York, 2nd edition, 2011.

[3] Dan Stefanica. *Numerical Linear Algebra Methods for Financial Engineering Applications*. Financial Engineering Advanced Background Series. FE Press, New York, 2014.

[4] Gilbert Strang. *Introduction to Linear Algebra*. Wellesley-Cambridge Press, Wellesley, Massachusetts, 4th edition, 2009.

[5] Lloyd N. Trefethen and David Bau III. *Numerical Linear Algebra*. SIAM, Philadelphia, Pennsylvania, 1997.

Made in the USA
Middletown, DE
20 February 2022

61587149R00157